Totally Nonnegative Matrices

Princeton Series in Applied Mathematics

The Princeton Series in Applied Mathematics publishes high quality advanced texts and monographs in all areas of applied mathematics. Books include those of a theoretical and general nature as well as those dealing with the mathematics of specific applications areas and real-world situations.

Series Editors: Ingrid Daubechies (Princeton University); Weinan E (Princeton University); Jan Karel Lenstra (Eindhoven University); Endre Sli (University of Oxford)

Books in the series:

Chaotic Transitions in Deterministic and Stochastic Dynamical Systems: Applications of Melnikov Processes in Engineering, Physics, and Neuroscience, by Emil Simiu

Selfsimilar Processes, by Paul Embrechts and Makoto Maejima

Self-Regularity: A New Paradigm for Primal-Dual Interior-Point Algorithms, by Jiming Peng, Cornelis Roos, and Tams Terlaky

Analytic Theory of Global Bifurcation: An Introduction, by Boris Buffoni and John Toland

Entropy, by Andreas Greven, Gerhard Keller, and Gerald Warnecke

Auxiliary Signal Design for Failure Detection, by Stephen L. Campbell and Ramine Nikoukhah

Thermodynamics: A Dynamical Systems Approach, by Wassim M. Haddad, Vijay Sekhar Chellaboina, and Sergey G. Nersesov

Optimization: Insights and Applications, by Jan Brinkhuis and Vladimir Tikhomirov

Max Plus at Work, by Bernd Heidergott, by Geert Jan Olsder, and Jacob van der Woude Wave

Scattering by Time-Dependent Perturbations: An Introduction, by Gary C. Roach

The Traveling Salesman Problem: A Computational Study, by David L. Applegate, Robert E. Bixby, Vasek Chvatal, and William J. Cook

Positive Definite Matrices, by Rajendra Bhatia

Genomic Signal Processing, by Edward Dougherty and Ilya Shmulevich

Algebraic Curves over a Finite Field, by J.W.P. Hirschfeld, G. Korchmros, and F. Torres

Matrix Completions, Moments, and Sums of Hermitian Squares, by Mihly Bakonyi and Hugo J. Woerdeman

Totally Nonnegative Matrices, by Shaun Fallat and Charles Johnson

Totally Nonnegative Matrices

Shaun M. Fallat
Charles R. Johnson

PRINCETON UNIVERSITY PRESS

PRINCETON AND OXFORD

Requests for permission to reproduce material
from this work should be sent to
Permissions, Princeton University Press

Published by Princeton University Press,
41 William Street, Princeton, New Jersey 08540

In the United Kingdom: Princeton University Press,
6 Oxford Street, Woodstock, Oxfordshire OX20 1TW

press.princeton.edu

Library of Congress Control Number: 2010052042
ISBN 978-0-691-12157-4

British Library Cataloging-in-Publication Data is available
The publisher would like to acknowledge the authors of this
volume for providing the camera-ready copy from which
this book was printed

10 9 8 7 6 5 4 3 2 1

Totally Nonnegative Matrices

Shaun M. Fallat
Charles R. Johnson

PRINCETON UNIVERSITY PRESS
PRINCETON AND OXFORD

Requests for permission to reproduce material
from this work should be sent to
Permissions, Princeton University Press

Published by Princeton University Press,
41 William Street, Princeton, New Jersey 08540

In the United Kingdom: Princeton University Press,
6 Oxford Street, Woodstock, Oxfordshire OX20 1TW

press.princeton.edu

Library of Congress Control Number: 2010052042
ISBN 978-0-691-12157-4

British Library Cataloging-in-Publication Data is available
The publisher would like to acknowledge the authors of this
volume for providing the camera-ready copy from which
this book was printed

10 9 8 7 6 5 4 3 2 1

Fallat dedicates this book to the loves of his life: to his wife and best friend, Darcie; and to his children, Noah and Emma.

Contents

List of Figures

Preface

A matrix is called totally nonnegative (resp. totally positive) if the determinant of *every* square submatrix is a nonnegative (resp. positive) number. This seemingly unlikely occurrence arises in a remarkable variety of ways (see Section 0.2 for a sample) and carries with it an array of very attractive mathematical structure. Indeed, it is the most useful and aesthetic matricial topic not covered in the broad references [HJ85] and [HJ91]. It is our purpose here to give a largely self-contained development of the most fundamental parts of that structure from a modern theoretical perspective. Applications of totally nonnegative matrices and related numerical issues are not the main focus of this work, but are recognized as being integral to this subject. We also mention a number of more specialized facts with references and give a substantial collection of references covering the subject and some related ideas.

Historically, the books [GK60, GK02, Kar68] and [GM96] and the survey paper [And87] have been the most common references in this area. However, each has a somewhat special perspective and, by now, each is missing some modern material. The most recent of these, [GM96], is more than fifteen years old and is a collection of useful, but noncomprehensive papers broadly in the area. In [Kar68] and [And87] the perspective taken leads to a notation and organization that can present difficulty to the reader needing particular facts. Perhaps the most useful and fundamental reference over a long period of time [GK60], especially now that it is readily available in English, is from the perspective of one motivating and important application and is now about sixty years old.

The present book takes a core, matrix theoretic perspective common to all sources of interest in the subject and emphasizes the utility of the elementary bidiagonal factorizations (see Chapter 2), which, though its roots are rather old, has only recently emerged as the primary tool (among very few available at present) to understand the beautiful and elaborate structure of totally nonnegative matrices. This tool is largely unused in the prior references. Along with the elementary bidiagonal factorization, planar diagrams, a recently appearing concept, are introduced as a conceptual combinatorial tool for analyzing totally nonnegative matrices. In addition to the seemingly unlimited use of these combinatorial objects, they lend themselves to obtaining prior results in an elegant manner, whereas in the past many results in this area were hard fought. As we completed this volume, the book [Pin10] appeared. It takes a different and more classical view of the subject, with many fewer references. Of course, its appearance is testimony to interest in and importance of this subject.

Our primary intent is that the present volume be a useful reference for all those who encounter totally nonnegative matrices. It could also be readily used to guide a seminar on the subject for either those needing to learn the ideas fundamental to the subject or those interested in an attractive segment of matrix analysis. For brevity, we have decided not to include textbook-type problems to solidify the ideas for students, but for purposes for which these might be useful, a knowledgeable instructor could add appropriate ones.

Our organization is as follows. After an initial introduction that sets major notation, gives examples, and describes several motivating areas, Chapter 1 provides much of the needed special background and preliminary results used throughout. (Any other needed background can be found in [HJ85] or a standard elementary linear algebra text.) Chapter 2 develops the ubiquitously important elementary bidiagonal factorization and related issues, such as LU factorization and planar diagrams that allow a combinatorial analysis of much about totally nonnegative matrices. Though an m-by-n matrix has many minors, total nonnegativity/positivity may be checked remarkably efficiently. This and related ideas, such as sufficient collections of minors, are the subject of Chapter 3.

Viewed as a linear transformation, it has long been known that a totally positive matrix cannot increase the sequential variation in the signs of a vector (an important property in several applications). The ideas surrounding this fact are developed in Chapter 4 and several converses are given. The eigenvalues of a totally positive matrix are positive and distinct and the eigenvectors are highly structural as well. A broad development of spectral structure and principal submatrices etc. are given in Chapter 5. Like positive semidefinite and M-matrices, totally nonnegative matrices enjoy a number of determinantal inequalities involving submatrices. These are explored in Chapter 6.

The remaining four chapters contain introductions to a variety of more specialized and emerging topics, including the distribution of rank in a TN matrix (Chapter 7); Hadamard (or entry-wise) products of totally nonnegative matrices (Chapter 8); various aspects of matrix completion problems associated with totally nonnegative matrices (Chapter 9); and a smorgesboard of further interesting topics are surveyed in the final chapter (Chapter 10).

An index and directory of terminology and notation is given at the end, as well as a lengthy list of relevant references, both ones referred to in the text and others of potential interest to readers. It is inevitable, unfortunately, that we may have missed some relevant work in the literature.

The reader should be alerted to an unfortunate schism in the terminology used among authors on this subject. We consistently use "totally nonnegative" and "totally positive" (among a hierarchy of refinements). Some authors use, instead, terms such as "totally positive" (for totally nonnegative) and "strictly totally positive" (for totally positive). This is at odds with conventional terminology for real numbers. We hope that there can be a convergence to the more natural and unambiguous terms over time.

We would like to thank B. Turner for a useful reading of a late draft of our manuscript, and several colleagues and friends for helpful suggestions. Finally, we note that the so-called sign-regular matrices (all minors of a given size share a common sign that may vary from size to size) share many properties with totally nonnegative and totally positive matrices. It could be argued that they present the proper level of generality in which to study certain parts of the structure. We have chosen not to focus upon them (they are mentioned occasionally) in order not to deflect attention from the most important sign-regular classes: namely the totally nonnegative and totally positive matrices.

SMF
CRJ

Chapter

Introduction

The central theme in this book is to investigate and explore various properties of the class of *totally nonnegative matrices*.

At first it may appear that the notion of total positivity is artificial; however, this class of matrices arises in a variety of important applications. For example, totally positive (or nonnegative) matrices arise in statistics (see [GM96, Goo86, Hei94]), mathematical biology (see [GM96]), combinatorics (see [BFZ96, Bre95, Gar96, Peñ98]), dynamics (see [GK02]), approximation theory (see [GM96, Pri68, CP94b]), operator theory (see [Sob75]), and geometry (see [Stu88]). Historically, the theory of totally positive matrices originated from the pioneering work of Gantmacher and Krein ([GK60]) and was elegantly brought together in a revised version of their original monograph [GK02]). Also, under the influence of I. Schoenberg (see, for example, [Sch30]), Karlin published an influential treatise, *Total Positivity* ([Kar68]), which mostly concerns totally positive kernels but also deals with the discrete version, totally positive matrices. Since then there has been a considerable amount of work accomplished on total positivity, some of which is contained in the exceptional survey paper by T. Ando ([And87]). Two more recent survey papers have appeared ([Fal01] [Hog07, Chap. 21]) and both take the point of view of bidiagonal factorizations of totally nonnegative matrices.

Before we continue, as mentioned in the Preface, we issue here a word of warning. In the existing literature, the terminology used is not always consistent with what is presented above. Elsewhere in the literature the term totally positive (see below) corresponds to "strictly totally positive" (see, for example, the text [Kar68]) and the term totally nonnegative defined here sometimes corresponds to "totally positive" (as in some of the papers in [GM96]).

0.0 DEFINITIONS AND NOTATION

The set of all m-by-n matrices with entries from a field $\mathbb{F}$ will be denoted by $M_{m,n}(\mathbb{F})$. If $\mathbb{F} = \mathbb{R}$, the set of all real numbers, then we may shorten this notation to $M_{m,n}$. Our matrices will, for the most part, be real entried. In the case $m = n$, the matrices are square and $M_{n,n}(\mathbb{F})$ or $M_{m,n}$ will be shortened to $M_n(\mathbb{F})$ or M_n. For $A \in M_{m,n}(\mathbb{F})$, the notation $A = [a_{ij}]$ will indicate that the entries of A are $a_{ij} \in \mathbb{F}$, for $i = 1, 2, \ldots, m$ and $j = 1, 2, \ldots, n$. The transpose of a matrix A will be denoted by A^T.

For $A \in M_{m,n}(\mathbb{F})$, $\alpha \subseteq \{1, 2, \ldots, m\}$, and $\beta \subseteq \{1, 2, \ldots, n\}$, the submatrix of A lying in rows indexed by α and columns indexed by β will be denoted by $A[\alpha, \beta]$. Similarly, $A(\alpha, \beta)$ is the submatrix obtained from A by deleting the rows indexed by α and columns indexed by β. Throughout the book, we let α^c denote the *complement* of the index set α. So, in particular, $A(\alpha, \beta) = A[\alpha^c, \beta^c]$. If $A \in M_n(\mathbb{F})$ and $\alpha = \beta$, then the principal submatrix $A[\alpha, \alpha]$ is abbreviated to $A[\alpha]$, and the complementary principal submatrix is $A(\alpha)$. On the other hand, if $\alpha \neq \beta$, then the submatrix $A[\alpha, \beta]$ is referred to as a *nonprincipal submatrix of A*. In the special case when $\alpha = \{1, 2, \ldots, k\}$ with $1 \leq k \leq n$, we refer to the principal submatrix $A[\alpha]$ as a *leading principal submatrix*. In the same manner, for an n-vector $x \in \mathbb{F}^n$, $x[\alpha]$ denotes the entries of x in the positions indexed by α and $x(\alpha)$ denotes the complementary vector. If $x = [x_i] \in \mathbb{F}^n$, then we let $\mathrm{diag}(x_i)$ denote the n-by-n diagonal matrix with main diagonal entries x_i. For brevity, we often denote the sets $\{1, 2, \ldots, m\}$ and $\{1, 2, \ldots, n\}$ by M and N, respectively.

A *minor* in a given matrix A is by definition the determinant of a (square) submatrix of $A \in M_{m,n}(\mathbb{F})$. Here and throughout, $\det(\cdot)$ denotes the determinant of a square matrix. For example, if $|\alpha| = |\beta|$, the minor $\det A[\alpha, \beta]$ may also be denoted by $A_{\alpha,\beta}$, and the principal minor (if A is square) $\det A[\alpha]$ may be abbreviated to A_α.

We are interested here in those matrices $A \in M_{m,n}$ with *all* minors positive (nonnegative) and, occasionally, with all minors $A_{\alpha,\beta}$, with $|\alpha| = |\beta| \leq k$ positive (nonnegative). The former are called the *totally positive (totally nonnegative)* matrices and are denoted by TP (TN); we denote the latter as TP_k (TN_k). TP (TN) is used both as an abbreviation and as a name of a set. The reader should consult [HJ85, HJ91] or any other standard reference on matrix theory for standard notation and terminology not defined herein. In terms of compounds, TP_k, for example, simply means that the first through the kth compounds are entry-wise positive.

Several related classes of matrices are both of independent interest and useful in developing the theory of TP (TN) matrices. We give our notation for those classes here; we continue to use the subscript k to denote a requirement made on all minors of no more than k rows. No subscript indicates that the requirement is made on all minors. Each class is defined irrespective of the m and n in $M_{m,n}$, except that a subscript k makes sense only for $k \leq m, n$.

By InTN (InTN_k), we denote the square and *invertible* TN matrices (TN_k matrices), and by IrTN (IrTN_k), we mean the square and *irreducible* TN matrices (TN_k matrices). The intersection of these two classes is denoted by IITN (IITN_k); this class is also referred to as the *oscillatory* matrices [GK60], which are defined as those TN matrices with a TP integral power. The equivalence will be demonstrated in Chapter 2. We may also use the symbol OSC to denote oscillatory matrices.

A TN (TN_k) matrix may have a zero line (row or column), which often leads to an exception to convenient statements. So we denote TN (TN_k) matrices with no zero lines as TN' (TN'_k).

A square, triangular matrix cannot be TP, but one that is TN *and* has all its minors positive unless they are identically 0 because only of the triangular structure is called *triangular TP* ; these are denoted by ΔTP. Similarly, ΔTN

denotes the triangular, TN matrices. An adjective "upper" or "lower" denotes the form of triangularity. Finally, the positive diagonal matrices ($\mathcal{D}_n$) are the intersections of upper ΔTP and lower ΔTP matrices, and thus are InTN.

For convenience of reference we present these classes in tabular form in Table 1. Note: Omission of subscript k means *all* minors.

TP_k :	all minors on no more than k rows are positive
TN_k :	all minors on no more than k rows are nonnegative
TN'_k :	TN_k with no all zero lines
InTN_k :	TN_k and invertible
IrTN_k :	TN_k and irreducible
IITN_k :	TN_k, irreducible and invertible
ΔTP:	triangular TP; all minors that may be positive are positive
ΔTN :	triangular TN
$\mathcal{D}_n$:	positive diagonal matrices in M_n

Table 1: Various Classes of Matrices of Interest

The ith standard basis vector is the n-vector whose only nonzero entry occurs in the ith component and that entry is a one, and is denoted by e_i; the (i,j)th standard basis matrix, the m-by-n matrix whose only nonzero entry is in the (i,j)th position and this entry is a one, will be denoted by E_{ij}. Observe that if $m = n$, then $E_{ij} = e_i e_j^T$. We also let e denote the n-vector consisting of all ones (the size of e will be determined from the context). Finally, we let $J_{m,n}$ ($J_{n,n} \equiv J_n$) and I_n denote the m-by-n matrix of all ones and the n-by-n identity matrix, respectively. The subscript is dropped when the sizes of these matrices is clear from the context.

0.1 JACOBI MATRICES AND OTHER EXAMPLES OF TN MATRICES

An n-by-n matrix $A = [a_{ij}]$ is called a *diagonal* matrix if $a_{ij} = 0$ whenever $i \neq j$. For example,

$$A = \begin{bmatrix} 1 & 0 & 0 & 0 \\ 0 & 4 & 0 & 0 \\ 0 & 0 & 3 & 0 \\ 0 & 0 & 0 & 3 \end{bmatrix}$$

is a 4-by-4 diagonal matrix. We may also write A as $A = \text{diag}\,(1, 4, 3, 3)$.

An n-by-n matrix $A = [a_{ij}]$ is referred to as a *tridiagonal* matrix if $a_{ij} = 0$ whenever $|i - j| > 1$. For example,

$$A = \begin{bmatrix} 1 & 4 & 0 & 0 \\ 3 & 4 & 1 & 0 \\ 0 & 0 & 3 & 4 \\ 0 & 0 & 1 & 3 \end{bmatrix}$$

is a 4-by-4 tridiagonal matrix. The entries of $A = [a_{ij}]$ that lie in positions $(i, i+1)$ $i = 1, \ldots n-1$ are referred to as the *superdiagonal*, and the entries of A in positions $(i, i-1)$ $i = 2, \ldots, n$ are called *subdiagonal*.

Hence a matrix is tridiagonal if the only nonzero entries of A are contained on its sub-, main-, and superdiagonal.

Tridiagonal matrices are perhaps one of the most studied classes of matrices. Much of the reason for this is that many algorithms in linear algebra require significantly less computational labor when they are applied to tridiagonal matrices. Some elementary examples include

(1) eigenvalues,

(2) solving linear systems,

(3) LU factorization,

(4) evaluating determinants.

For instance, the determinant of a tridiagonal matrix $A = [a_{ij}]$ can be evaluated by solving the recursive equation

$$\det A = a_{11} \det A[\{2, \ldots, n\}] - a_{12}a_{21} \det A[\{3, \ldots, n\}]. \qquad (1)$$

Equation (1) is a simple consequence of the Laplace expansion of $\det A$ along row (or column) 1. Furthermore, both $A[\{2, \ldots, n\}]$ and $A[\{3, \ldots, n\}]$ are (smaller sized) tridiagonal matrices.

The recursive relation (1) also ties the eigenvalues of a (symmetric) tridiagonal matrix to certain orthogonal polynomials (Chebyshev), as the characteristic polynomials of A and its principal submatrices also satisfy a relation like (1). In addition, (1) arises in the Runge-Kutta methods for solving partial differential equations.

Suppose the tridiagonal matrix A has the form

$$A = \begin{bmatrix} a_1 & b_1 & 0 & \cdots & & 0 \\ c_1 & a_2 & \ddots & \ddots & & \vdots \\ 0 & \ddots & \ddots & \ddots & & \vdots \\ \vdots & \ddots & \ddots & \ddots & & b_{n-1} \\ 0 & \cdots & \cdots & c_{n-1} & & a_n \end{bmatrix}. \qquad (2)$$

Observe that a tridiagonal matrix A in the form (2) is called symmetric if $b_i = c_i$, $i = 1, \ldots, n-1$ and is called irreducible if both $b_i \neq 0$ and $c_i \neq 0$, for $i = 1, \ldots, n-1$.

An n-by-n matrix is called a P_0-matrix if all its principal minors are non-negative. We let P_0 denote the class of all P_0-matrices. Furthermore, an n-by-n matrix is called a P-matrix if all its principal minors are positive. We let P denote the class of all P-matrices.

Suppose A in (2) is an irreducible nonnegative $(a_i, b_i, c_i \geq 0)$ tridiagonal matrix. If $a_i \geq b_i + c_{i-1}$ $i = 1, \ldots, n$, where $c_0 \equiv 0, b_n \equiv 0$, then A is a P_0-matrix. To verify this claim we need the following fact. If A is tridiagonal, then the principal submatrix $A[\{1, 2, 4, 6, 7, 8\}]$ can be written as $A[\{1, 2, 4, 6, 7\}] = A[\{1, 2\}] \oplus A[4] \oplus A[\{6, 7, 8\}]$, where $\oplus$ denotes direct sum. In particular,

$$\det A[\{1, 2, 4, 6, 7\}] = \det A[\{1, 2\}] \cdot \det A[\{4\}] \cdot \det A[\{6, 7, 8\}].$$

Hence to calculate any principal minor of a tridiagonal matrix, it is enough to compute principal minors based on associated contiguous index sets. Moreover, any principal submatrix of a tridiagonal matrix based on contiguous index sets is again a tridiagonal matrix. To verify that any tridiagonal of the form (2) that is nonnegative, irreducible and satisfies $a_i \geq b_i + c_{i-1}$ (row diagonal dominance) is a P_0-matrix, it is sufficient, by induction, to verify that $\det A \geq 0$.

Consider the case $n = 3$. Then by (1),

$$\begin{aligned}
\det A &= a_1 \det A[\{2, 3\}] - b_1 c_1 \det A[\{3\}] \\
&= a_1(a_2 a_3 - b_2 c_2) - b_1 c_1 a_3 \\
&= a_1 a_2 a_3 - a_1 b_2 c_2 - b_1 c_1 a_3 \\
&\geq a_1(c_1 + b_2)a_3 - a_1 b_2 c_2 - b_1 c_1 a_3 \\
&= a_1 b_2(a_3 - c_2) + a_3 c_1(a_1 - b_1) \\
&> 0.
\end{aligned}$$

In addition, since $\det A[\{3\}] > 0$ it follows from (1) that

$$\det A \leq a_1 \det A[\{2, 3\}]. \tag{3}$$

Consider the general case and let A' be the n-by-n matrix obtained from A by using an elementary row operation to eliminate c_1. Then, as $a_1 > b_1 > 0$,

$$\det A = a_1 \det \begin{bmatrix} a_2 - \frac{b_1 c_1}{a_1} & b_2 & 0 & \cdots & 0 \\ c_2 & a_3 & b_3 & \cdots & 0 \\ \vdots & \ddots & \ddots & \ddots & \vdots \\ 0 & \vdots & \ddots & \ddots & b_{n-1} \\ 0 & \cdots & \cdots & c_{n-1} & a_n \end{bmatrix}.$$

Observe that this matrix is an irreducible, nonnegative tridiagonal matrix that still satisfies the row dominance relation and hence by induction has a positive determinant. Thus $\det A > 0$.

Furthermore, we note that by (1) we have

$$\det A \leq a_1 \det A[\{2, \ldots, n\}].$$

So, if a nonnegative, irreducible tridiagonal matrix satisfies the simple dominance relation it is a P_0-matrix. In fact, even more is true.

Since tridiagonal matrices satisfy $a_{ij} = 0$ whenever $|i-j| > 1$ it is a simple consequence that

$$\det A[\{i_1, \ldots, i_k\}, \{j_1, \ldots, j_k\}] = 0$$

whenever there exists an $s \in \{1, \ldots, k\}$ such that $|i_s - j_s| > 1$.

Furthermore, if $A[\{i_1, \ldots, i_k\}, \{j_1, \ldots, j_k\}]$ is any submatrix of A that satisfies $|i_s - j_s| \leq 1, s = 1, \ldots, k$, then $\det A[\{i_1, \ldots, i_k\}, \{j_1, \ldots, j_k\}]$ is a product of principal minors and nonprincipal minors, where the principal minors are based on contiguous index sets and the nonprincipal minors are simply entries from the super- and/or subdiagonal.

For example, suppose A is a 10-by-10 tridiagonal matrix of the form in (2).

Then

$$\det A[\{1, 2, 3, 5, 6, 8, 10\}, \{2, 3, 4, 5, 6, 8, 9\}] = b_1 b_2 b_3 \det A[\{5, 6\}] a_8 c_9.$$

Consequently, any irreducible, nonnegative tridiagonal matrix that satisfies the condition $a_i \geq b_i + c_{i-1}$ has all its minors nonnegative.

We remark here that the irreducibility assumption was needed for convenience only. If the tridiagonal matrix is reducible, then we simply concentrate on the direct summands and apply the same arguments as above.

The upshot of the previous discussion is that if A is a nonnegative tridiagonal matrix that also satisfies $a_i \geq b_i + c_{i-1}$, then A is TN.

In the pivotal monograph [GK60], the chapter on oscillatory matrices began by introducing a class of tridiagonal matrices the authors called *Jacobi matrices*. An n-by-n matrix $\mathcal{J} = [a_{ij}]$ is a Jacobi matrix if $\mathcal{J}$ is a tridiagonal matrix of the form

$$\begin{bmatrix} a_1 & -b_1 & \cdots & & 0 \\ -c_1 & \ddots & & \ddots & \vdots \\ 0 & \ddots & & \ddots & -b_{n-1} \\ 0 & \cdots & & -c_{n-1} & a_n \end{bmatrix}.$$

Further, $\mathcal{J}$ is called a *normal Jacobi matrix* if, in addition, $b_i, c_i \geq 0$. Jacobi matrices were of interest to Gantmacher and Krein for at least two reasons. First, many of the fundamental properties of general oscillatory matrices can be obtained by an examination of analogous properties of Jacobi matrices. Second Jacobi matrices are among the basic types of matrices that arose from studying the oscillatory properties of an elastic segmental continuum (no supports between the endpoints) under small transverse oscillations.

One of the first things that we observe about Jacobi matrices is that they are not TN or even entrywise nonnegative. However, they are not that far off, as we shall see in the next result.

The following result on the eigenvalues of a tridiagonal matrix is well known, although we present a proof here for completeness.

Lemma 0.1.1 *Let T be an n-by-n tridiagonal matrix in the form*

$$T = \begin{bmatrix} a_1 & b_1 & 0 & 0 & 0 & \cdots & 0 \\ c_1 & a_2 & b_2 & 0 & 0 & \cdots & 0 \\ 0 & c_2 & a_3 & b_3 & 0 & \cdots & 0 \\ \vdots & & \ddots & \ddots & \ddots & & \vdots \\ 0 & \cdots & 0 & c_{n-3} & a_{n-2} & b_{n-2} & 0 \\ 0 & \cdots & 0 & 0 & c_{n-2} & a_{n-1} & b_{n-1} \\ 0 & \cdots & 0 & 0 & 0 & c_{n-1} & a_n \end{bmatrix}. \tag{4}$$

If $b_i c_i > 0$ for $i = 1, 2, \ldots, n-1$, then the eigenvalues of T are real and have algebraic multiplicity one (i.e., are simple). Moreover, T is similar (via a positive diagonal matrix) to a symmetric nonnegative tridiagonal matrix.

Proof. Let $D = [d_{ij}]$ be the n-by-n diagonal matrix where $d_{11} = 1$, and for $k > 1$, $d_{kk} = \sqrt{\frac{b_1 b_2 \cdots b_{k-1}}{c_1 c_2 \cdots c_{k-1}}}$. Then it is readily verified that

$$DTD^{-1} = \begin{bmatrix} a_1 & \sqrt{b_1 c_1} & 0 & \cdots & & 0 \\ \sqrt{b_1 c_1} & a_2 & \sqrt{b_2 c_2} & \cdots & & 0 \\ 0 & \ddots & \ddots & \ddots & & \vdots \\ \vdots & & \sqrt{b_{n-2} c_{n-2}} & a_{n-1} & \sqrt{b_{n-1} c_{n-1}} \\ 0 & \cdots & & 0 & \sqrt{b_{n-1} c_{n-1}} & a_n \end{bmatrix}.$$

Since DTD^{-1} is symmetric and T and DTD^{-1} have the same eigenvalues, this implies that the eigenvalues of T are real. Suppose λ is an eigenvalue of T. Then $\lambda I - T$ is also of the form (4). If the first row and last column of $\lambda I - T$ are deleted, then the resulting matrix is an $(n-1)$-by-$(n-1)$ upper triangular matrix with no zero entries on its main diagonal, since $b_i c_i > 0$, for $i = 1, 2, \ldots, n-1$. Hence this submatrix has rank $n-1$. It follows that $\lambda I - T$ has rank at least $n-1$. However, $\lambda I - T$ has rank at most $n-1$ since λ is an eigenvalue of T. So by definition λ has geometric multiplicity one. This completes the proof, as DTD^{-1} is also a nonnegative matrix. $\square$

A diagonal matrix is called a *signature matrix* if all its diagonal entries are ± 1. Furthermore, two n-by-n matrices A and B are said to be *signature similar* if $A = SBS^{-1}$, where S is a signature matrix. It is not difficult to verify that a normal Jacobi matrix is signature similar to a nonnegative tridiagonal matrix. Furthermore, the eigenvalues of a normal Jacobi matrix are real and distinct.

Observe that if an irreducible normal Jacobi matrix satisfies the dominance condition ($a_i \geq |b_i| + |c_i|$), then it is similar to a symmetric TN tridiagonal matrix. In particular (since a symmetric TN matrix is an example of a positive semidefinite matrix [HJ85]), the eigenvalues are nonnegative and distinct by Lemma 0.1.1.

Moreover, if the above normal Jacobi matrix is also invertible, then the eigenvalues are positive and distinct.

Since the eigenvalues are distinct, each eigenvalue has exactly one eigenvector (up to scalar multiple), and these (unique) eigenvectors satisfy many more interesting "sign" properties.

Let

$$\mathcal{J} = \begin{bmatrix} a_1 & b_1 & 0 & \cdots & & 0 \\ c_1 & a_2 & \ddots & \ddots & & \vdots \\ 0 & \ddots & \ddots & \ddots & & \vdots \\ \vdots & \ddots & \ddots & \ddots & & b_{n-1} \\ 0 & \cdots & \cdots & c_{n-1} & a_{n-1} \end{bmatrix}, \quad \text{with } b_i c_i > 0.$$

Given $\mathcal{J}$ above, define

$$D_0(\lambda) \equiv 1,$$
$$D_k(\lambda) = \det((\mathcal{J} - \lambda I)[1, \dots, k]),$$
$$D_n(\lambda) = \text{char. poly. of } \mathcal{J}.$$

Recall that the functions D_k satisfy the recurrence relation

$$D_{k+1}(\lambda) = (a_{k+1} - \lambda)D_k(\lambda) - b_k c_k D_{k-1}(\lambda) \qquad (5)$$

for $k = 1, 2, \dots, n-1$. Evidently,

$$D_0(\lambda) \equiv 1,$$
$$D_1(\lambda) = a_1 - \lambda,$$
$$D_2(\lambda) = (a_2 - \lambda)D_1(\lambda) - b_1 c_1 D_0(\lambda)$$
$$= (a_2 - \lambda)(a_1 - \lambda) - b_1 c_1,$$

and so on.

A key observation that needs to be made is the following:

"When $D_k(\lambda)$ vanishes $(1 < k < n)$, the polynomials $D_{k-1}(\lambda)$ and $D_{k+1}(\lambda)$ differ from zero and have opposite sign."

Suppose that, for some k, $1 < k < n$, this property fails. That is, for fixed λ_0,

$$D_k(\lambda_0) = D_{k-1}(\lambda_0) = D_{k+1}(\lambda_0) = 0.$$

Go back to the recurrence relation (5), and notice that if the above inequalities hold, then

$$D_{k-2}(\lambda_0) = 0, \text{ and } D_{k+2}(\lambda_0) = 0.$$

Consequently, $D_k(\lambda_0) = 0$ for all $k = 2, 3, \dots, n$. But then observe

$$D_3(\lambda_0) = (a_k - \lambda)D_2(\lambda_0) - b_2 c_2 D_1(\lambda_0),$$

which implies $D_1(\lambda_0) = 0$. Now we arrive at a contradiction since

$$D_2(\lambda_0) = (a_2 - \lambda)D_1(\lambda_0) - b_1 c_1 D_0(\lambda_0) = 0 = 0 - b_1 c_1,$$

and $b_1 c_1 > 0$.

Since we know that the roots are real and distinct, when λ passes through a root of $D_k(\lambda)$, the product $D_k D_{k-1}$ must reverse sign from $+$ to $-$. A consequence of this is that the roots of successive D_k's strictly interlace. Using this interlacing property we can establish the following simple but very important property of the functions $D_k(\lambda)$. Between each two adjacent roots of the polynomial $D_k(\lambda)$ lies exactly one root of the polynomial $D_{k-1}(\lambda)$ (that is, the roots *interlace*).

Observation: The sequence

$$D_{n-1}(\lambda), D_{n-2}(\lambda), \ldots, D_0(\lambda)$$

has $j - 1$ sign changes when $\lambda = \lambda_j$ where $\lambda_1 < \lambda_2 < \cdots < \lambda_n$ are the distinct eigenvalues of $\mathcal{J}$.

Further note that $D_k(\lambda) = (-\lambda)^k + \cdots$, and so it follows that $\lim_{\lambda \to -\infty} D_k(\lambda) = \infty$.

Finally, we come to the coordinates of the eigenvectors of $\mathcal{J}$. Assume for the moment that for each i, $b_i = 1$ and $c_i = 1$ (for simplicity's sake).

Suppose

$$x = \begin{bmatrix} x_1 \\ x_2 \\ \vdots \\ x_n \end{bmatrix}$$

is an eigenvector of $\mathcal{J}$ corresponding to the eigenvalue λ. Then examining the vector equation $\mathcal{J}x = \lambda x$ yields the system of equations

$$
\begin{aligned}
(a_1 - \lambda)x_1 - x_2 &= 0, \\
-x_1 + (a_2 - \lambda)x_2 - x_3 &= 0, \\
-x_{n-2} + (a_{n-1} - \lambda)x_{n-1} - x_n &= 0, \\
-x_{n-1} + (a_n - \lambda)x_n &= 0.
\end{aligned}
$$

Since $D_n(\lambda) = 0$, it follows that $D_{n-1}(\lambda) \neq 0$; hence the first $n-1$ equations are linearly independent, and the nth equation is a linear combination of the previous $n - 1$.

Looking more closely at the first $n - 1$ equations we arrive at the relation

$$x_k = (a_{k-1} - \lambda)x_{k-1} - x_{k-2}.$$

Hence the coordinates of x satisfy the same relation as $D_{k-1}(\lambda)$.

Hence $x_k = C D_{k-1}(\lambda)$, $C \neq 0$ constant. Thus, since at $\lambda = \lambda_j$ the sequence $D_{n-1}, \ldots, D_0$ has exactly $j - 1$ sign changes, it follows that the eigenvector corresponding to λ_j has exactly $j - 1$ sign changes. The interlacing of sign changes in the eigenvectors follows as well.

We close with a remark that offers a nice connection between tridiagonal TN matrices and a certain type of additive closure (a property not enjoyed by general TN matrices). Suppose A is a TN matrix. Then a simple examination of the 2-by-2 minors of A involving exactly one main diagonal entry will be enough to verify the following statement. The matrix A is tridiagonal if and only if $A + D$ is TN for all positive diagonal matrices D. This result also brings out the fact that tridiagonal P_0 matrices are, in fact, TN.

0.1.1 Other Examples of TN Matrices

In the previous subsection we demonstrated that any entrywise nonnegative tridiagonal P_0-matrix was in fact TN. Other consequences include (1) an entry-wise nonnegative invertible P_0 tridiagonal matrix is both InTN and a P-matrix, and (2) an entrywise nonnegative invertible P_0 tridiagonal matrix is IITN and hence oscillatory.

From tridiagonal matrices, it is natural to consider their inverses as another example class of InTN matrices.

Example 0.1.2 (Inverses of tridiagonal matrices) Let T be an InTN tridiagonal matrix. Then $ST^{-1}S$ is another InTN matrix. For example, if

$$T = \begin{bmatrix} 2 & 1 & 0 \\ 1 & 2 & 1 \\ 0 & 2 & 1 \end{bmatrix}, \text{ then}$$

$$ST^{-1}S = \begin{bmatrix} 1 & 1 & 1 \\ 1 & 2 & 2 \\ 1 & 2 & 3 \end{bmatrix}.$$

In fact, note that the matrix

$$[\min\{i,j\}] = \begin{bmatrix} 1 & 1 & 1 & \cdot & 1 \\ 1 & 2 & 2 & \cdot & 2 \\ 1 & 2 & 3 & \cdot & 3 \\ \vdots & \vdots & \vdots & \ddots & \vdots \\ 1 & 2 & 3 & \cdots & n \end{bmatrix}$$

is the inverse of a tridiagonal matrix and is also InTN.

We also acknowledge that these matrices are referred to as "single-pair matrices" in [GK60, pp. 79–80], are very much related to "Green's matrices" in [Kar68, pp. 110–112], and are connected with "matrices of type D" found in [Mar70a].

Tridiagonal and inverse tridiagonal TN matrices have appeared in numerous places throughout mathematics. One instance is the case of the Cayley transform. Let $A \in M_n(\mathbb{C})$ such that $I + A$ is invertible. The *Cayley transform* of A, denoted by $\mathcal{C}(A)$, is defined to be

$$\mathcal{C}(A) = (I + A)^{-1}(I - A).$$

The Cayley transform was defined in 1846 by A. Cayley. He proved that if A is skew-Hermitian, then $\mathcal{C}(A)$ is unitary, and conversely, provided, of course, that $\mathcal{C}(A)$ exists. This feature is useful, for example, in solving matrix equations subject to the solution being unitary by transforming them into equations for skew-Hermitian matrices. One other important property of the Cayley transform is that it can be viewed as an extension to matrices of the conformal mapping

$$T(z) = \frac{1 - z}{1 + z}$$

from the complex plane into itself. In [FT02] the following result was proved connecting essentially nonnegative tridiagonal matrices and TN matrices. Recall that a matrix is called *essentially nonnegative* if all its off-diagonal entries are nonnegative.

Theorem 0.1.3 *Let $A \in M_n(\mathbb{R})$ be an irreducible matrix. Then A is an essentially nonnegative tridiagonal matrix with $|\lambda| < 1$ for all eigenvalues λ of A if and only if $I + A$ and $(I - A)^{-1}$ are TN matrices. In particular, $\mathcal{C}(-A) = (I - A)^{-1}(I + A)$ is a factorization into TN matrices.*

Proof. To verify necessity, observe that if $I + A$ is TN, then A is certainly essentially nonnegative. Also, since $(I - A)^{-1}$ is TN, it follows that $I - A$ is invertible and has the checkerboard sign pattern (i.e., the sign of its (i,j)th entry is $(-1)^{i+j}$). Hence $a_{i,i+2} = 0$ and $a_{i+2,i} = 0$ for all $i \in \{1, 2, \ldots, n-2\}$, and since A is irreducible and $I + A$ is TN, $a_{i,j} = 0$ for $|i - j| > 1$. That is, A is tridiagonal. The remaining conclusion now readily follows.

For the converse, if A is an essentially nonnegative irreducible tridiagonal matrix with $|\lambda| < 1$ for all eigenvalues λ of A, then $I + A$ is a nonnegative tridiagonal P-matrix and thus totally nonnegative (see previous section). Similarly, $I - A$ is a tridiagonal M-matrix since $|\lambda| < 1$ for all eigenvalues λ of A (recall that an n-by-n P-matrix is called an M-matrix if all off-diagonal entries are nonpositive), and hence $(I - A)^{-1}$ is TN (this follows from the remarks in the previous section). Since TN matrices are closed under multiplication, $\mathcal{C}(-A) = (I - A)^{-1}(I + A)$ is TN. $\square$

A natural question arising from Theorem 0.1.3 is whether in every factorization $\hat{F} = (I - A)^{-1}(I + A)$ of a totally nonnegative matrix $\hat{F}$ the factors $(I - A)^{-1}$ and $(I + A)$ are TN. We conclude with an example showing that neither of these factors need be TN.

Consider the TN matrix

$$\hat{F} = \begin{bmatrix} 1 & .9 & .8 \\ .9 & 1 & .9 \\ 0 & .9 & 1 \end{bmatrix}$$

and consider $A = -\mathcal{C}(\hat{F})$. Then $\hat{F} = \mathcal{C}(-A) = (I - A)^{-1}(I + A)$, where neither

$$(I - A)^{-1} = \begin{bmatrix} 1 & .45 & .4 \\ .45 & 1 & .45 \\ 0 & .45 & 1 \end{bmatrix} \quad \text{nor} \quad I + A = \begin{bmatrix} .8203 & .3994 & .2922 \\ .6657 & .5207 & .3994 \\ -.2996 & .6657 & .8203 \end{bmatrix}$$

is TN.

Example 0.1.4 (Vandermonde) Vandermonde matrices have been an example class of matrices that have garnered much interest for some time. In many respects it is not surprising that certain Vandermonde matrices are TP (see Chapter 1 for more discussion). For n real numbers $0 < x_1 < x_2 <$

$\ldots x_n$, the Vandermonde matrix $V(x_1, x_2, \cdots < x_n)$ is defined to be

$$V(x_1, x_2, \ldots, x_n) = \begin{bmatrix} 1 & x_1 & x_1^2 & \cdots & x_1^{n-1} \\ 1 & x_2 & x_2^2 & \cdots & x_2^{n-1} \\ \vdots & \vdots & & \vdots & \vdots \\ 1 & x_n & x_n^2 & \cdots & x_n^{n-1} \end{bmatrix},$$

and is a TP matrix (see [GK60, p. 111]). Recall the classical determinantal formula,

$$\det V(x_1, x_2, \ldots, x_n) = \prod_{i>j} (x_i - x_j),$$

which is positive whenever $0 < x_1 < x_2 < \cdots < x_n$.

The reader is also encouraged to consult [GK60] for similar notions involving generalized Vandermonde matrices. Totally positive Vandermonde matrices were also treated in the survey paper on bidiagonal factorizations of TN matrices [Fal01], which we discuss further below.

It is really quite remarkable that the inequalities $0 < x_1 < x_2 < x_3$ are actually sufficient to guarantee that *all* minors of V above are positive. Indeed, any 2-by-2 submatrix of V has the form

$$A = \begin{bmatrix} x_i^{\alpha_1} & x_i^{\alpha_2} \\ x_j^{\alpha_1} & x_j^{\alpha_2} \end{bmatrix},$$

where $i, j \in \{1, 2, 3\}$, $i < j$, $\alpha_1, \alpha_2 \in \{0, 1, 2\}$, and $\alpha_1 < \alpha_2$. Since $0 < x_i < x_j$, it follows that $\det A > 0$. Hence, in this case, A is TP. For an arbitrary k-by-k submatrix of V, consider the following argument, which is similar to that given in [GK60] and in [Fal01]. Any k-by-k submatrix of $V(x_1, \ldots, x_n)$ is of the form $A = [x_{l_i}^{\alpha_j}]$, where $l_1, \ldots, l_k \in \{1, \ldots, n\}$, $l_1 < l_2 < \cdots < l_k$, $\alpha_1, \ldots, \alpha_k \in \{0, 1, \ldots, n-1\}$, and $\alpha_1 < \alpha_2 < \cdots < \alpha_k$. Let $f(x) = c_1 x^{\alpha_1} + c_2 x^{\alpha_2} + \cdots + c_k x^{\alpha_k}$ be a real polynomial. Descartes's rule of signs ensures that the number of positive real roots of $f(x)$ does not exceed the number of sign changes among the coefficients $c_1, \ldots, c_k$, so $f(x)$ has at most $k-1$ positive real roots. Therefore, the system of equations

$$f(x_{l_i}) = \sum_{j=1}^{k} c_j x_{l_i}^{\alpha_j} = 0, \ i = 1, 2, \ldots, k$$

has only the trivial solution $c_1 = c_2 = \cdots = c_k = 0$, so $\det A = \det[x_{l_i}^{\alpha_j}] \neq 0$.

To establish that $\det A > 0$, we use induction on the size of A. We know this to be true when the size of A is 2, so assume $\det A > 0$ whenever the size of A is at most $k-1$. Let $A = [x_{l_i}^{\alpha_j}]$ be k-by-k, in which $l_1, \ldots, l_k \in \{1, \ldots, n\}$, $l_1 < l_2 < \cdots < l_k$, $\alpha_1, \ldots, \alpha_k \in \{0, 1, \ldots, n-1\}$, and $\alpha_1 < \alpha_2 < \cdots < \alpha_k$. Then expanding $\det A$ along the last row gives

$$\det A = g(x_{l_k}) = a_k x_{l_k}^{\alpha_k} - a_{k-1} x_{l_k}^{\alpha_{k-1}} + \cdots + (-1)^{k-1} a_1 x_{l_k}^{\alpha_1}.$$

But $a_k = \det[x_{l_s}^{\alpha_t}]$, where $s, t \in \{1, 2, \ldots, k-1\}$, so the induction hypothesis ensures that $a_k > 0$. Thus $0 \neq g(x_{l_k}) \to \infty$ as $x_{l_k} \to \infty$, so $\det A = g(x_{l_k}) > 0$. The conclusion is that $V(x_1, \ldots, x_n)$ is TP whenever $0 < x_1 < \cdots < x_n$.

Example 0.1.5 (Cauchy matrix) An n-by-n matrix $C = [c_{ij}]$ is called a *Cauchy matrix* if the entries of C are given by

$$c_{ij} = \frac{1}{x_i + y_j},$$

where $x_1, x_2, \ldots, x_n$ and $y_1, y_2, \ldots, y_n$ are two sequences of real numbers (chosen accordingly so that c_{ij} is well defined). A Cauchy matrix is TP whenever $0 < x_1 < x_2 < \cdots < x_n$ and $0 < y_1 < y_2 < \cdots < y_n$ (see [GK02, pp. 77–78]).

We observe that the above claim readily follows from the well-known (Cauchy) identity

$$\det C = \frac{\prod_{i<k}(x_i - x_k) \prod_{i<k}(y_i - y_k)}{\prod_{i,k}(x_i + y_k)}.$$

As a particular instance, consider the matrix $A_a = [1/(a^{i-j} + 1)]$, where $a \neq 1$. Then $A_a = B_a D$, where $D = \mathrm{diag}(a, a^2, \cdots a^n)$, and $B_a = [1/(a^i + a^j)]$. It is now easy to see that both B_a and A_a are TP for all $a > 0$ and $a \neq 1$.

Example 0.1.6 (Pascal matrix) Consider the n-by-n matrix $P_n = [p_{ij}]$ whose first row and column entries are all equal to 1, and, for $2 \leq i, j \leq n$, define $p_{ij} = p_{i-1,j} + p_{i,j-1}$ (Pascal's identity). Then a 4-by-4 symmetric Pascal matrix is given by

$$P_4 = \begin{bmatrix} 1 & 1 & 1 & 1 \\ 1 & 2 & 3 & 4 \\ 1 & 3 & 6 & 10 \\ 1 & 4 & 10 & 20 \end{bmatrix}.$$

The fact that P_n is TP will follow from the existence of a bidiagonal factorization, which is explained in much more detail in Chapter 2.

Example 0.1.7 (Routh-Hurwitz matrix) Let $f(x) = \sum_{i=0}^{n} a_i x^i$ be an nth degree polynomial in x. The n-by-n *Routh-Hurwitz matrix* is defined to be

$$RH = \begin{bmatrix} a_1 & a_3 & a_5 & a_7 & \cdots & 0 & 0 \\ a_0 & a_2 & a_4 & a_6 & \cdots & 0 & 0 \\ 0 & a_1 & a_3 & a_5 & \cdots & 0 & 0 \\ 0 & a_0 & a_2 & a_4 & \cdots & 0 & 0 \\ \vdots & \vdots & \vdots & \vdots & \cdots & \vdots & \vdots \\ 0 & 0 & 0 & 0 & \cdots & a_{n-1} & 0 \\ 0 & 0 & 0 & 0 & \cdots & a_{n-2} & a_n \end{bmatrix}.$$

A specific example of a Routh-Hurwitz matrix for an arbitrary polynomial of degree six, $f(x) = \sum_{i=0}^{6} a_i x^i$, is given by

$$RH = \begin{bmatrix} a_1 & a_3 & a_5 & 0 & 0 & 0 \\ a_0 & a_2 & a_4 & a_6 & 0 & 0 \\ 0 & a_1 & a_3 & a_5 & 0 & 0 \\ 0 & a_0 & a_2 & a_4 & a_6 & 0 \\ 0 & 0 & a_1 & a_3 & a_5 & 0 \\ 0 & 0 & a_0 & a_2 & a_4 & a_6 \end{bmatrix}.$$

A polynomial $f(x)$ is *stable* if all the zeros of $f(x)$ have negative real parts. It is proved in [Asn70] that $f(x)$ is stable if and only if the Routh-Hurwitz matrix formed from f is TN.

Example 0.1.8 (More examples)

The matrix $A = [e^{-\sigma(\alpha_i - \beta_j)^2}]$ is TP whenever $\sigma > 0$, $0 < \alpha_1 < \alpha_2 < \cdots < \alpha_n$, and $0 < \beta_1 < \beta_2 < \cdots < \beta_n$. It is worth noting that A may also be viewed as a positive diagonal scaling of a generalized Vandermonde matrix given by $V = [e^{2\sigma\alpha_i\beta_j}]$, which is also TP.

If the main diagonal entries of a TP matrix A are all equal to one, then it is not difficult to observe a "drop-off" effect in the entries of A as you move away from the main diagonal. For example, if $i < j$, then

$$a_{ij} \le a_{i,i+1}a_{i+1,i+2} \cdots a_{j-1,j}.$$

A similar inequality holds for $i > j$.

Thus a natural question to ask is how much of a drop-off is necessary in a positive matrix to ensure that the matrix is TP. An investigation along these lines was carried in [CC98]. They actually proved the following interesting fact. If $A = [a_{ij}]$ is an n-by-n matrix with positive entries that also satisfy

$$a_{ij}a_{i+1,j+1} \ge c_0 a_{i,j+1}a_{i+1,j},$$

where $c_0 \approx 4.07959562349$, then A is TP. The number c_0 is actually the unique real root of $x^3 - 5x^2 + 4x - 1$.

More recently, this result was refined by [KV06] where they prove that, if $A = [a_{ij}]$ is an n-by-n matrix with positive entries and satisfies $a_{ij}a_{i+1,j+1} \ge ca_{i,j+1}a_{i+1,j}$, with $c \ge 4\cos^2(\frac{\pi}{n+1})$, then A is TP.

These conditions are particularly appealing for both Hankel and Toeplitz matrices. Recall that a *Hankel matrix* is an $(n+1)$-by-$(n+1)$ matrix of the form

$$\begin{bmatrix} a_0 & a_1 & \cdots & a_n \\ a_1 & a_2 & \cdots & a_{n+1} \\ \vdots & \vdots & \cdots & \vdots \\ a_n & a_{n+1} & \cdots & a_{2n} \end{bmatrix}.$$

So if the positive sequence $\{a_i\}$ satisfies $a_{k-1}a_{k+1} \ge ca_k^2$, then the corresponding Hankel matrix is TP. An $(n+1)$-by-$(n+1)$ *Toeplitz matrix* is of the form

$$\begin{bmatrix} a_0 & a_1 & a_2 & \cdots & a_n \\ a_{-1} & a_0 & a_1 & \cdots & a_{n-1} \\ a_{-2} & a_{-1} & a_0 & \cdots & a_{n-2} \\ \vdots & \vdots & \vdots & \cdots & \vdots \\ a_{-n} & a_{-(n-1)} & a_{-(n-2)} & \cdots & a_0 \end{bmatrix}.$$

Hence if the positive sequence $\{a_i\}$ satisfies $a_k^2 \ge ca_{k-1}a_{k+1}$, then the corresponding Toeplitz matrix is TP.

A *sequence* $a_0, a_1, \ldots$ *of real numbers is called totally positive* if the two-way infinite matrix given by

$$\begin{bmatrix} a_0 & 0 & 0 & \cdots \\ a_1 & a_0 & 0 & \cdots \\ a_2 & a_1 & a_0 & \cdots \\ \vdots & \vdots & \vdots & \end{bmatrix}$$

is TP. An infinite matrix is TP if all its minors are positive. Notice that the above matrix is a Toeplitz matrix. Studying the functions that generate totally positive sequences was a difficult and important step in the area of TP matrices; $f(x)$ generates the sequence $a_0, a_1, \ldots$ if $f(x) = a_0 + a_1 x + a_2 x^2 + \cdots$. In [ASW52], and see also [Edr52], it was shown that the above two-way infinite Toeplitz matrix is TP (i.e., the corresponding sequence is totally positive) if and only if the generating function $f(x)$ for the sequence $a_0, a_1, \ldots$ has the form

$$f(x) = e^{\gamma x} \frac{\prod_{\nu=1}^{\infty}(1 + \alpha_v x)}{\prod_{\nu=1}^{\infty}(1 - \beta_v x)},$$

where $\gamma, \alpha_v, \beta_v \geq 0$, and $\sum \alpha_v$, and $\sum \beta_v$ are convergent.

0.2 APPLICATIONS AND MOTIVATION

Positivity has roots in every aspect of pure, applied, and numerical mathematics. The subdiscipline, total positivity, also is entrenched in nearly all facets of mathematics. Evidence of this claim can be found in the insightful and comprehensive proceedings [GM96]. This collection of papers was inspired by Karlin's contributions to total positivity and its applications. This compilation contains 23 papers based on a five-day meeting held in September 1994. The papers are organized by area of application. In particular, the specialties listed are (in order of appearance)

(1) Spline functions

(2) Matrix theory

(3) Geometric modeling

(4) Probability and mathematical biology

(5) Approximation theory

(6) Complex analysis

(7) Statistics

(8) Real analysis

(9) Combinatorics

(10) Integral equations

The above list is by no means comprehensive, but it is certainly extensive and interesting (other related areas include differential equations, geometry, and function theory). A recent application includes new advances in accurate eigenvalue computation and connections with singular value decompositions on TN matrices (see, for example, [KD02, Koe05, Koe07]).

Historically speaking, total positivity came about in essentially three forms:

(1) Gantmacher/Krein—oscillations of vibrating systems

(2) Schoenberg—real zeros of a polynomials, spline function with applications to mathematical analysis, and approximation

(3) Karlin—integral equations, kernels, and statistics.

All these have led to the development of the theory of TP matrices by way of rigorous mathematical analysis.

We will elaborate on the related applications by choosing two specific examples to highlight.

(1) Statistics/probability

(2) CAGD/spline functions

On our way to discussing these specific applications, which in many ways represent a motivation for exploring TP matrices, we begin with a classic definition, which may have been the impetus for both Karlin and Schoenberg's interest in TP matrices.

A real function (or kernel) $k(x, y)$ in two variables, along with two linearly ordered sets X, Y is said to be *totally positive of order n* if for every

$$x_1 < x_2 < \cdots < x_n; \; y_1 < y_2 < \cdots < y_n, \; x_i \in X, \; y_j \in Y,$$

we have

$$\det \begin{bmatrix} k(x_1, y_1) & \cdots & k(x_1, y_n) \\ \vdots & \ddots & \vdots \\ k(x_n, y_1) & \cdots & k(x_n, y_n) \end{bmatrix} > 0.$$

We use the term *TN for a function $k(x, y)$* if the inequalities above are not necessarily strict.

There are many examples of totally positive functions. For instance,

(1) $k(x, y) = e^{xy}$ (see [Kar68, p. 15]), $x, y \in (-\infty, \infty)$,

(2) $k(x, y) = x^y$, $x \in (0, \infty)$, $y \in (-\infty, \infty)$

are both examples (over their respective domains) of totally positive functions.

Upon closer examination of the function $k(x, y)$ in (2), we observe that for a fixed n and given $0 < x_1 < x_2 < \cdots < x_n$ and $y_1 < y_2 < \cdots < y_n$, the n-by-n matrix $(k(x_i, y_j))_{i,j=1}$ is a generalized Vandermonde matrix. For instance, if $n = 4$, then the matrix in question is given by

$$\begin{bmatrix} x_1^{y_1} & x_1^{y_2} & x_1^{y_3} & x_1^{y_4} \\ x_2^{y_1} & x_2^{y_2} & x_2^{y_3} & x_2^{y_4} \\ x_3^{y_1} & x_3^{y_2} & x_3^{y_3} & x_3^{y_4} \\ x_4^{y_1} & x_4^{y_2} & x_4^{y_3} & x_4^{y_4} \end{bmatrix}.$$

An important instance is when a totally positive function of order n, say $k(x, y)$, can be written as $k(x, y) = f(x - y)$ $(x, y \in \mathbb{R})$. In this case, the function $f(u)$ is called a *Pólya frequency function of order* n, and is often abbreviated as PF_n (see [Kar68], for example). Further along these lines, if $k(x, y) = f(x - y)$ but $x, y \in \mathbb{Z}$, then $f(u)$ is said to be a *Pólya frequency sequence of order* n.

As an example, we mention a specific application of totally positive functions of order 2 (and PF_2) to statistics.

Observe that a function $h(u)$, $u \in \mathbb{R}$ is PF_2 if:

(1) $h(u) > 0$ for $-\infty < u < \infty$, and

(2) $\det \begin{pmatrix} h(x_1 - y_1) & h(x_1 - y_2) \\ h(x_2 - y_1) & h(x_2 - y_2) \end{pmatrix} > 0$

for all $-\infty < x_1 < x_2 < \infty$, $-\infty < y_1 < y_2 < \infty$.

In the book [BP75, p. 76], it is noted that conditions (1), (2) on $h(u)$ above are equivalent to either

(3) $\log h(u)$ is concave on $\mathbb{R}$, or

(4) for fixed $\triangle > 0$, $\dfrac{h(u + \triangle)}{h(u)}$ is decreasing in u for $a \le u \le b$ when

$$a = \inf_{h(u)>0} y, \qquad b = \sup_{h(u)>0} u.$$

Recall that for a continuous random variable u, with probability density function $f(u)$, its cumulative distribution function is defined to be $F(u) = \int_{-\infty}^{u} f(t)dt$. Then the reliability function, $\overline{F}$, is given by $\overline{F}(u) = 1 - F(u)$. A distribution function $F(u)$ is called an *increasing failure rate distribution* if the function

$$\overline{F}(x|t) = \frac{\overline{F}(t + x)}{\overline{F}(t)}$$

is decreasing in $t \in \mathbb{R}$ for each $x \ge 0$. It turns out that the condition $\overline{F}(x|t)$ decreasing is equivalent to the failure rate function

$$r(t) = \frac{f(t)}{\overline{F}(t)}$$

being an increasing function.

Hence we have that a distribution function $F(u)$ is an increasing failure rate distribution if and only if $\overline{F} = 1 - F$ is PF_2.

Return to totally positive functions. Suppose k is a function of $X \times Y$; then we will consider the derived "determinant function." Define the set (following [Kar68])

$$\triangle_p(X) = \{\overline{x} = (x_1, x_2, \dots, x_p) | \ x_1 < \dots < x_p, \ x_i \in X\}$$

(ordered p-tuples).

The determinant function

$$K_{[p]}(\overline{x}, \overline{y}) = \det \begin{bmatrix} k(x_1, y_1) & \cdots & k(x_1, y_n) \\ \vdots & \ddots & \vdots \\ k(x_n, y_1) & \cdots & k(x_n, y_n) \end{bmatrix}$$

defined on $\triangle_p(X) \times \triangle(Y)$ is called the *compound kernel of order p* induced by $k(x, y)$. If $k(x, y)$ is a totally positive function of order n, then $K_{[p]}(\overline{x}, \overline{y})$ is a nonnegative function on $\triangle_p(X) \times \triangle_p(Y)$ for each $p = 1, \dots, n$.

As an example, consider the (indicator) function $k(x, y)$ defined

$$k(x, y) = \begin{cases} 1, & 1 \leq x \leq y \leq b, \\ 0, & a \leq y < x \leq b. \end{cases}$$

Then k is a totally nonnegative function, and for arbitrary

$$1 \leq x_1 < x_2 < \dots < x_n \leq b,$$
$$a \leq y_1 < y_2 < \dots < y_n \leq b$$

we have

$$K_p[\overline{x}, \overline{y}] = \begin{cases} 1 & \text{if} \quad x_1 \leq y_1 \leq x_2 \leq y_2 \leq \dots \leq x_n \leq y_n, \\ 0 & \text{otherwise.} \end{cases}$$

The proof is straightforward and is omitted.

Another example class of totally positive functions is the exponential family of (statistical) distributions. Consider the function

$$k(x, y) = \beta(y)e^{xy}$$

with respect to sigma finite measure $d\mu(x)$. Examples of such distributions include the normal distribution with variance 1, exponential distributions, and the gamma distribution.

There is another connection between totally positive matrices and correlation matrices.

Following the paper [SS06], and the references therein, it is known by empirical analysis that correlation matrices of forward interest rates (which have applications in risk analysis) seem to exhibit spectral properties that are similar to TP matrices. In particular, it is shown that the correlation matrix R of yields can be approximately described by the correlation function

$$\rho_{ij} = \exp(-\beta|t_j - t_i|), \beta > 0.$$

If indices with maturities in the above model are identified, then $\rho_{ij} = \rho^{|i-j|}$, where $\rho = \exp(-\beta)$.

Then

$$R = \begin{bmatrix} 1 & \rho & \rho^2 & \cdots & \rho^{n-1} \\ \rho & \ddots & \ddots & \ddots & \vdots \\ \rho^2 & \ddots & \ddots & \ddots & \rho^2 \\ \vdots & \ddots & \ddots & \ddots & \rho \\ \rho^{n-1} & \cdots & \rho^2 & \rho & 1 \end{bmatrix} \quad \text{(special Toeplitz matrix)}.$$

It is well known that R above is oscillatory. For example, appealing to an old result in [Edr52] and referring to [Kar68] (making use of Polyá frequency sequences) to demonstrate that R is TN, before establishing that R is OSC.

Without going into details here, we note that R is essentially (up to re-signing) the inverse of the IITN tridiagonal matrix

$$T = \frac{1}{(1-\rho^2)} \begin{bmatrix} 1 & \rho & 0 & \cdots & 0 \\ \rho & 1+\rho^2 & \rho & \ddots & 0 \\ 0 & \rho & \ddots & \ddots & 0 \\ \vdots & \ddots & \ddots & 1+\rho^2 & \rho \\ 0 & \cdots & 0 & \rho & 1 \end{bmatrix}.$$

Observe that T satisfies the row dominance conditions mentioned in the previous section of this chapter. Consequently T is OSC, and hence R is OSC.

Further applications to risk analysis overlap with the theory of TP matrices.

In [Jew89], where choices between risky prospects are investigated, it is cited under what conditions it is true that

$$E(u(Y+X)) = u(Y+C) \Rightarrow E(v(Y+X)) \le E(v(Y+C)),$$

where

X, Y—risky prospects,

u, v—increasing utilities with v more risk averse than u?

Clearly the answer will depend on what is assumed about the joint distribution of X and Y.

Jewitt proves (along with a slight generalization) that the above condition holds whenever Y is independent of X and has a log-concave density. Recall from the previous discussion that log-concavity is intimately related to the density being a totally positive function of order 2 ($k(x, y) = h(x - y)$ a totally positive function of order 2 is equivalent to being log-concave).

The theory of Pólya frequency functions (PF) is quite immeasurable and its history spans 80 years. Further, their connection to TP matrices is very strong and has produced numerous important results and associated applications.

A *Pólya frequency function* of order k (PF_k) is a function f in a one real variable u, $u \in \mathbb{R}$ for which $k(x, y) = f(x - y)$ is a totally nonnegative function of order k for $x, y \in \mathbb{R}$. Schoenberg was one of the pioneers in the study of Pólya frequency functions.

A basic example of a PF function is

$$f(u) = e^{-\gamma u^2}, \quad \gamma > 0.$$

To verify, observe that

$$f(x - y) = e^{-\gamma(x-y)^2} = e^{-\gamma x^2} e^{-\gamma y^2} e^{2\gamma xy}.$$

Since e^{xy} is a totally nonnegative function, and the other factors are totally positive functions, their product above is a totally positive function.

If the argument of a PF_k function is restricted to the integers, we call the resulting sequence a *Pólya frequency sequence*.

For brevity, we discuss the case of "one-sided" Pólya frequency sequences: $k(m, n) = a_{n-m}$, generates a sequence $\{a_\ell\}$, with the extra stipulation that $a_\ell = 0$ for $\ell < 0$. A (one-sided) sequence is said to be a PF_∞ sequence if the corresponding Kernel written as an infinite matrix

$$A = \begin{bmatrix} a_0 & a_1 & \cdots & \cdots \\ 0 & a_0 & a_1 & \cdots \\ \vdots & 0 & a_0 & a_1 \\ \vdots & \ddots & \ddots & \ddots \end{bmatrix}$$

is TN, and similarly $\{a_\ell\}$ is PF_k if A is TN_k. Observe that if $a_0 > 0$ and A is TP_2, then the sequence $\{a_\ell\}$ is either finite or consists entirely of positive numbers.

To a given sequence $\{a_\ell\}$, we associate the usual generating function $f(z) = \sum_{\ell=0}^{\infty} a_\ell z^\ell$. Further, if $\{a_n\}$ is a PF_2 sequence, then $f(z)$ has a nonzero radius of convergence. Some examples of PF_∞ sequences are

(1) $f(z) = 1 + \alpha z, \quad \alpha > 0,$

(2) $f(z) = \dfrac{1}{1 - \beta z}, \quad \beta > 0.$

Observe that the (discrete) kernel corresponding to (2) has the form

$$A = \begin{pmatrix} 1 & \beta & \beta^2 & \cdots & \cdots \\ 0 & 1 & \beta & \beta^2 & \cdots \\ \vdots & 0 & 1 & \beta & \ddots \\ \vdots & \vdots & \ddots & \ddots & \ddots \\ 0 & 0 & \cdots & 0 & 1 \end{pmatrix}.$$

It can be easily verified that A is TN and that the radius of convergence of $f(z) = (1 - \beta z)^{-1}$ is given by $|z| < 1/\beta$.

There is an "algebra" associated with PF_r sequences. If $f(z)$ and $g(z)$ generate one-sided PF_r sequences, then so do

$$f(z) \cdot g(z) \quad \text{and} \quad \frac{1}{f(-z)}.$$

Hence combining examples (1) and (2) above, we arrive at the function

$$\frac{\prod_{i=1}^{n}(1+\alpha_i z)}{\prod_{j=1}^{n}(1-\beta_j z)},$$

with $\alpha_i > 0$ and $\beta_j > 0$, which generates a one-sided PF sequence. In fact, it can be shown that $e^{\gamma z}$, $\gamma > 0$ generates a PF sequence. Hence the function

$$f(z) = e^{\gamma z} \frac{\prod_{i=1}^{\infty}(1+\alpha_i z)}{\prod_{j=1}^{\infty}(1-\beta_j z)},$$

generates a one-sided PF sequence provided $\alpha_i \geq 0$, $\beta_j \geq 0$ and

$$\sum \alpha_i < \infty, \quad \sum \beta_j < \infty.$$

A crowning achievement along these lines is a representation of one-sided PF sequences in 1952 [Edr52]. A function $f(z) = \sum a_n z^n$ with $f(0) = 1$ generates a one-sided PF sequence if and only if it has the form

$$f(z) = e^{\gamma z} \frac{\prod_{i=0}^{\infty}(1+\alpha_1 z)}{\prod_{j=0}^{\infty}(1-\beta_j z)},$$

where $\gamma \geq 0$, $\alpha_i \geq 0$, $\beta_j \geq 0$, and $\sum(\alpha_i + \beta_i) < \infty$.

It is useful to note that Whitney's reduction theorem for totally positive matrices [Whi52] was a key component to proving one direction of the aforementioned result that appeared in [ASW52].

Edrei later considered doubly infinite sequences and their associated generating functions; see [Edr53a] and [Edr53b]. A doubly infinite sequence of real numbers $\{a_n\}_{-\infty}^{\infty}$ is said to be *totally positive* if for every k and sets of integers $s_1 < s_2 < \cdots < s_k$ and $t_1 < t_2 < \cdots < t_k$, the determinant of $C = [a_{s_i-t_j}]$ for $i, j = 1, 2, \ldots, k$ is nonnegative. We also note that some of Edrei's work on scalar Toeplitz matrices has been extended to block Toeplitz matrices (see, for example, [DMS86]), including a connection via matrix factorization.

We close this section on applications with a discussion on totally positive systems of functions. We incorporate two example applications within such systems of functions:

(1) spline functions,

(2) TP bases.

Given a system of real-valued functions $u = (u_0, u_1, \ldots, u_n)$ defined on $I \subseteq \mathbb{R}$, the *collocation matrix of u at* $t_0 < t_1, < \cdots < t_m$ in I is defined as the $(m+1)$-by-$(n+1)$ matrix

$$M = \begin{pmatrix} u_0(t_0) & u_1(t_0) & \cdots & u_n(t_0) \\ u_0(t_1) & u_1(t_1) & \cdots & u_n(t_1) \\ \vdots & \vdots & \ddots & \vdots \\ u_0(t_m) & u_1(t_m) & \cdots & u_n(t_m) \end{pmatrix}.$$

We say that a *system of functions u is totally positive* if all its collocation matrices are TP. The most basic system of functions that are totally positive is of course $1, x, x^2, \ldots, x^n$ when $I = (0, \infty)$. In this case the associated collocation matrices are just Vandermonde matrices.

We say that $(u_0, u_1, \ldots, u_n)$ is a normalized totally positive system if

$$\sum_{i=0}^{n} u_i(x) = 1 \text{ for } x \in I.$$

If, in addition, the system $(u_0, u_1, \ldots, u_n)$ is linearly independent, then we call $(u_0, u_1, \ldots, u_n)$ a *totally positive basis* for the space $\mathrm{Span}\{u_0, \ldots, u_n\}$. For example, $\{1, x, x^2, \ldots, x^n\}$ is a totally positive basis for $\mathbb{P}_n$, the space of polynomials of degree at most n.

Let $u = (u_0, \ldots, u_n)$ be a TP basis defined on I, and let $U = \mathrm{span}(u_0, \ldots, u_n)$. It follows that for any TP matrix A the basis v defined by

$$v^T = u^T A$$

is again a TP basis. The converse, which is an extremely important property of TP bases, is known to hold, namely, that all TP bases are generated in this manner. The key of course is choosing the appropriate starting basis. A TP basis u defined on I is called a *B-basis* if all TP bases v of U satisfy $v^T = A^T u$ for some invertible TN matrix A.

Some examples of *B*-bases include

(1) the Bernstein basis defined on $[a, b]$

$$b_i(t) = \frac{\dbinom{n}{i}}{(b-a)^n}(t-a)^i(b-t)^{n-i};$$

(2) the *B*-spline basis of the space of polynomial splines on a given interval with a prescribed sequence of knots.

TP bases and normalized TP bases have a natural geometric meaning which leads to representations with optimal shape-preserving properties.

Given a sequence of positive functions $(u_0, \ldots, u_n)$ on $[a, b]$ with $\sum u = 1$ and a sequence $(A_0, A_1, \ldots, A_n)$ of points in $\mathbb{R}^k$, we define the curve

$$\gamma(t) = \sum_{i=0}^{n} c_0 \cdot u_i(t), \quad t \in [a, b].$$

The functions $u_0, \ldots, u_n$ are typically called the *blending functions* and the points on $A_0, A_1, \ldots, A_n$ the *control points*. Let $P(A_0, \ldots, A_n)$ denote the polygonal arc with vertices $A_o, \ldots, A_n$. Often $P(A_0, \ldots, A_n)$ is called the *control polygon of the curve* γ.

If $u_0, \ldots, u_n$ is a normalized TP basis, then the curve γ preserves many of the shape properties of the control polygon. For example, the variation-diminution properties of TP matrices implies that the monotonicity and convexity of the control polygon are inherited by the curve γ. It is for these reasons that TP and B-bases are important in computer-aided geometric design.

As mentioned in example (2), the so-called B-spline's are deeply connected with certain aspects of total positivity. We describe briefly the concepts of spline functions and B-splines and their relationships to TP matrices.

Let $\triangle = \{x_i, i = 1, \ldots, r$ with $x_1 < x_2 < x_3 < \cdots < x_r\}$, $k > 0$, and let $m = (m_1, m_2, \ldots, m_r)$ be a sequence of integers with $1 \leq m_i \leq k + 1$ for each i.

Suppose $[a, b]$ is a given interval with $x_i \in [a, b]$, and define $x_0 = a$, $x_{r+1} = b$. A piecewise polynomial function f of degree k on each of the intervals $[x_i, x_u)$ for $i \leq r - 1$ and on $[x_r, x_{r+1}]$, and continuous of order $k - m_i$ (the function and the associated derivatives) at the knot x_i is called a *spline of degree k with knots x_i of multiplicity m_i*.

Spline functions have a rich and deep history and have applications in many areas of mathematics, including in approximation theory.

One important aspect to the theory for splines is determining an appropriate basis for the space spanned by the splines described above. B-splines turn out to be a natural such basis. To describe B-splines we need to introduce a finer partition based on the knots $x_1 < \cdots < x_r$ above and on $[a, b]$ (here we assume $[a, b]$ is a finite interval for simplicity). Define the new knots

$$y_1 \leq y_2 \leq \cdots < y_{k+1} < a,$$
$$y_{k+1} = y_{k+3} = \cdots = y_{k+m_1+1} = x_1,$$
$$y_{k+m_1+2} = \cdots = y_{k+m_1+m_2+1} = x_2,$$
$$\vdots$$
$$y_{k+n-m_r+2} = y_{k+n-m_r+3} = \cdots = y_{k+n+1} = x_r,$$
$$(n = \textstyle\sum m_i)$$
$$b \leq y_{k+n+2} \leq \cdots \leq y_{n+2k+2}.$$

Observe that $y_{i+k+1} > y_i$ for all i. Given this finer partition we define the functions

$$B_i(x) = (y_{i+k+1} - y_i)[y_i \cdots y_{i+k+1}](y - x)_t^k,$$

$$a \leq x \leq b, \qquad 1 \leq i \leq n + k + 1,$$

when $u_t = \max\{0, u\}$ and $[y_i, \ldots, y_{i+k+1}]f(x, y)$ denotes the $(k+1)$st divided difference of f with respect to the variable y with arguments $y_i, \ldots, y_{i+k+1}$.

The functions $B_i(x)$ are called B-splines and form a basis for the space spanned (with compact support) all spline of degree k with knots x_i of multiplicity m_i.

Consider strictly increasing real numbers $y_1 < \cdots < y_{n+k+1}$, and let $B_i(x)$ be as defined above, with $[y_i, y_{i+k+1}]$ as the support of B_i.

For distinct real numbers $t_1 < t_2 < \cdots < t_n$, we let

$$
M = \begin{pmatrix}
B_1(t_1) & \cdots & B_n(t_1) \\
B_1(t_2) & \cdots & B_n(t_2) \\
\vdots & \vdots & \vdots \\
B_1(t_n) & \cdots & B_n(t_n)
\end{pmatrix}
$$

be the corresponding collocation matrix.

Schoenberg/Whitney [SW 53] proved that M is invertible if and only if $y_i < t_i < y_{i+k+1}$ for $i = 1, \ldots, n$.

In fact, much more is true about the matrix M when $y_2 < t_i < y_{i+k+1} \forall i$. M is a TN matrix, as demonstrated by Karlin in [Kar68]. In [dB76] it was verified that all minors of M are nonnegative and a particular minor is positive (when $y_i < t_i < y_{i+k+1}$) if and only if all main diagonal entries of that minor are positive. Such matrices are also known as *almost totally positive matrices*.

0.3 ORGANIZATION AND PARTICULARITIES

This book is divided into 11 chapters including this introductory chapter on TN matrices. The next chapter, Chapter 1, carefully spells out a number of basic and fundamental properties of TN matrices along with a compilation of facts and results from core matrix theory that are useful for further development of this topic. For our purposes, as stated earlier, a matrix will be considered over the real field, unless otherwise stated. Further, it may be the case that some of the results will continue to hold over more general fields, although they will only be stated for matrices over the reals.

Chapter 2 introduces and methodically develops the important and useful topic of bidiagonal factorization. In addition to a detailed description regarding the existence of such factorizations for TN matrices, we also discuss a number of natural consequences and explain the rather important associated combinatorial objects known as planar diagrams.

The next four chapters, Chapters 3–6, highlight the fundamental topics: recognition of TN matrices (Chapter 3); sign variation of vectors (Chapter 4); spectral properties (Chapter 5); and determinantal inequalities (Chapter 6).

The remaining four chapters cover a wide range of topics associated with TN matrices that are of both recent and historical interest. Chapter 7 contains a detailed account of the distribution of rank deficient submatrices within a TN matrix, including a discussion of the important notion of row and column inclusion. Chapter 8 introduces the Hadamard (or entrywise) product and offers a survey of the known results pertaining to Hadamard products of TN matrices, including a note on Hadamard powers. In Chapter 9, we explore the relatively modern idea of matrix completion problems for TN matrices. This chapter also includes a section on single entry perturbations, known as retractions, which turn out to be a useful tool for other problems on TN matrices. In Chapter 10 we conclude with a brief review of a number of subtopics connected with TN matrices, including powers and roots of TN matrices, TP/TN polynomial matrices, subdirect sums of TN matrices, and Perron complements of TN matrices.

For the reader we have included a brief introductory section with each chapter, along with a detailed account of the necessary terms and notation used within that chapter. We have aimed at developing, in detail, the theory of totally nonnegative matrices, and as such this book is sequential in nature. Hence some results rely on notions in earlier chapters, and are indicated accordingly by references or citations as necessary. On a few occasions some forward references have also been used (e.g., in section 2.6 a reference is made to a result to appear in Chapter 3). We have tried to keep such instances of forward references to a minimum for the benefit of the reader.

This work is largely self-contained with some minor exceptions (e.g., characterization of sign-regular transformations in Chapter 4). When a relevant proof has been omitted, a corresponding citation has been provided. In addition, some of the proofs have been extracted in some way from existing proofs in the literature, and in this case the corresponding citation accompanies the proof.

We have also included, at the end, a reasonably complete list of references on all facets of total nonnegativity. Finally, for the benefit of the reader, an index and glossary of symbols (nomenclature) have also been assembled.

Chapter One

Preliminary Results and Discussion

1.0 INTRODUCTION

Along with the elementary bidiagonal factorization, to be developed in the next chapter, rules for manipulating determinants and special determinantal identities constitute the most useful tools for understanding TN matrices. Some of this technology is simply from elementary linear algebra, but the less well-known identities are given here for reference. In addition, other useful background facts are entered into the record, and a few elementary and frequently used facts about TN matrices are presented. Most are stated without proof, but are accompanied by numerous references where proofs and so on, may be found. The second component features more modern results, some of which are novel to this book. Accompanying proofs and references are supplied for the results in this portion. Since many of the results in later chapters rely on the results contained within this important ground-laying chapter, we trust that all readers will benefit from this preparatory discussion.

1.1 THE CAUCHY-BINET DETERMINANTAL FORMULA

According to the standard row/column formula, an entry of the product of two matrices is a sum of pairwise products of entries, one from each of the factors. The Cauchy-Binet formula (or identity) simply says that much the same is true for minors of a product. A minor of AB is a sum of pair-wise products of minors, one from each factor.

Theorem 1.1.1 (Cauchy-Binet) *Let $A \in M_{m,n}(\mathbb{F})$ and $B \in M_{n,p}(\mathbb{F})$. Then for each pair of index sets $\alpha \subseteq \{1, 2, \ldots, m\}$ and $\beta \subseteq \{1, 2, \ldots, p\}$ of cardinality k, where $1 \leq k \leq \min(m, n, p)$, we have*

$$\det AB[\alpha, \beta] = \sum_{\gamma, |\gamma|=k} \det A[\alpha, \gamma] \det B[\gamma, \beta]. \tag{1.1}$$

Proof. The claim is clear for $k = 1$. In general, it suffices to prove (1.1) for $m = p = k$ by replacing A by $A[\alpha, N]$ and B by $B[N, \beta]$, in which $N = \{1, 2, \ldots, n\}$; then it suffices to show

$$\det AB = \sum_{\gamma \subset N, |\gamma|=k} \det A[K, \gamma] \det B[\gamma, K],$$

in which $K = \{1, 2, \ldots, k\}$. If $n = k$, this is just the multiplicativity of the determinant, and if $n < k$, there is nothing to show. If $k < n$, we need only note that the equality follows since the sum of the k-by-k principal minors of BA (the expression on the right above) equals the sum of the k-by-k principal minors of AB (the expression on the left), as AB and BA have the same nonzero eigenvalues ([HJ85]).   □

This useful identity bears a resemblance to the formula for matrix multiplication (and in fact can be thought of as a generalization of matrix multiplication), and a special case of the above identity is the classical fact that the determinant is multiplicative, that is, if $A, B \in M_n(\mathbb{F})$, then $\det AB = \det A \det B$.

Another very useful consequence of the Cauchy-Binet formula is the multiplicativity of compound matrices. If $A \in M_n(\mathbb{F})$ and $k \leq m, n$, the $\binom{m}{k}$-by-$\binom{n}{k}$ matrix of k-by-k minors (with index sets ordered lexicographically) of A is called the kth compound of A and is denoted by $C_k(A)$. The Cauchy-Binet formula is simply equivalent to the statement that each kth compound is multiplicative,

$$C_k(AB) = C_k(A)C_k(B).$$

This means that TP (TN) matrices are closed under the usual product. In fact,

Theorem 1.1.2 *For $C \in \{\mathrm{TP}_k, \mathrm{TN}_k, \mathrm{TN}'_k, \Delta\mathrm{TP}, \Delta\mathrm{TN}, \mathrm{InTN}_k, \mathrm{IrTN}_k, \mathrm{IITN}_k\}$, if $A \in C \cap M_{m,n}$, $B \in C \cap M_{n,p}$, and $k \leq m, n$, then $AB \in C$.*

We note, though, that neither TN nor TP matrices in $M_{m,n}$ are closed under addition.

Example 1.1.3 Observe that both $\begin{bmatrix} 4 & 15 \\ 1 & 4 \end{bmatrix}$ and $\begin{bmatrix} 4 & 1 \\ 15 & 4 \end{bmatrix}$ are TP but their sum $\begin{bmatrix} 8 & 16 \\ 16 & 8 \end{bmatrix}$ has a negative determinant.

It has been shown, however, that any positive matrix is the sum of (possibly several) TP matrices [JO04].

1.2 OTHER IMPORTANT DETERMINANTAL IDENTITIES

In this section we list and briefly describe various classical determinantal identities that will be used throughout this text. We begin with the well-known identity attributed to Jacobi (see, for example, [HJ85]).

Jacobi's identity: If $A \in M_n(\mathbb{F})$ is nonsingular, then the minors of A^{-1} are related to those of A by Jacobi's identity. Jacobi's identity states that for $\alpha, \beta \subseteq N$, both nonempty, in which $|\alpha| = |\beta|$,

$$\det A^{-1}[\alpha, \beta] = (-1)^s \frac{\det A[\beta^c, \alpha^c]}{\det A}, \qquad (1.2)$$

where $s = \sum_{i \in \alpha} i + \sum_{j \in \beta} j$. Observe that if α and β are both singletons, that is, $\alpha = \{i\}$, $\beta = \{j\}$ $(1 \leq i, j \leq n)$, then (1.2) becomes

$$a_{ij}^{-1} = (-1)^{i+j} \frac{\det A(\{j\}, \{i\})}{\det A},$$

in which a_{ij}^{-1} denotes the (i, j) entry of A^{-1}. This expression is the classical adjoint formula for the inverse of a matrix, so that Jacobi's identity may be viewed as a generalization of the adjoint formula. When $\alpha = \beta$, (1.2) takes the form

$$\det A^{-1}[\alpha] = \frac{\det A[\alpha^c]}{\det A}. \qquad (1.3)$$

We now present some identities discovered by Sylvester (see [HJ85]).

Sylvester's identities: Let $A \in M_n(\mathbb{F})$, $\alpha \subseteq N$, and suppose $|\alpha| = k$. Define the $(n - k)$-by-$(n - k)$ matrix $B = [b_{ij}]$, with $i, j \in \alpha^c$, by setting $b_{ij} = \det A[\alpha \cup \{i\}, \alpha \cup \{j\}]$, for every $i, j \in \alpha^c$. Then Sylvester's identity states that for each $\delta, \gamma \subset \alpha^c$, with $|\delta| = |\gamma| = l$,

$$\det B[\delta, \gamma] = (\det A[\alpha])^{l-1} \det A[\alpha \cup \delta, \alpha \cup \gamma]. \qquad (1.4)$$

Observe that a special case of (1.4) is that $\det B = (\det A[\alpha])^{n-k-1} \det A$. A very useful consequence of (1.4) is that if $A \in M_n \cap \mathrm{TP}$ (or is TN and $\det A[\alpha] > 0$), then B is also a TP (TN) matrix. This fact is actually quite useful for inductive purposes, for example, when considering distributions of ranks and shadows (see [dBP82] and Chapter 7).

Another very useful special case is the following. Let $A \in M_n(\mathbb{F})$ be the partitioned matrix

$$A = \begin{bmatrix} a_{11} & a_{12}^T & a_{13} \\ a_{21} & A_{22} & a_{23} \\ a_{31} & a_{32}^T & a_{33} \end{bmatrix},$$

where $A_{22} \in M_{n-2}(\mathbb{F})$ and a_{11}, a_{33} are scalars. Define the matrices

$$B = \begin{bmatrix} a_{11} & a_{12}^T \\ a_{21} & A_{22} \end{bmatrix}, \quad C = \begin{bmatrix} a_{12}^T & a_{13} \\ A_{22} & a_{23} \end{bmatrix},$$

$$D = \begin{bmatrix} a_{21} & A_{22} \\ a_{31} & a_{32}^T \end{bmatrix}, \quad E = \begin{bmatrix} A_{22} & a_{23} \\ a_{32}^T & a_{33} \end{bmatrix}.$$

If we let $\tilde{b} = \det B$, $\tilde{c} = \det C$, $\tilde{d} = \det D$, and $\tilde{e} = \det E$, then by (1.4) it follows that

$$\det \begin{bmatrix} \tilde{b} & \tilde{c} \\ \tilde{d} & \tilde{e} \end{bmatrix} = \det A_{22} \det A.$$

Hence, provided $\det A_{22} \neq 0$, we have

$$\det A = \frac{\det B \det E - \det C \det D}{\det A_{22}}. \tag{1.5}$$

Karlin's identities: The next identity that we record for minors has appeared in a number of places regarding TP matrices; see, for example, [Kar68, Cry76, FZ99]. In [And87] this identity also appears and is used (as it was in [Kar68]) to prove that membership in the TP can be checked by verifying that all minors based on contiguous rows and columns are positive. We confess that referring to these identities as "Karlin's identities" may give the impression that they are due to Karlin. This is most certainly not the case. This reference is more to indicate that, within the umbrella of total positivity, Karlin was one of the first to utilize this identity and apply it to one of the most fundamental facts about this class of matrices (see [Kar68, pp. 8–10, 58–60]).

This useful identity can be stated in the following manner. Suppose A is an n-by-n matrix, and let α, ω be two index sets of $\{1, 2, \ldots, n\}$ with $|\omega| = n - 2$, $|\alpha| = n - 1$ and $\omega \subset \alpha$. Then the complement of ω, ω^c, is given by $\omega^c = \{a, b\}$, and the complement of α may be assumed to be $\alpha^c = \{a\}$. Then the identity referred to above can be written as

$$\det A(\{a, b\}, \{1, n\}) \det A(\{a\}, \{q\}) \tag{1.6}$$
$$= \det A(\{a, b\}, \{1, q\}) \det A(\{a\}, \{n\}) \det A(\{a, b\}, \{q, n\}) \det A(\{a\}, \{1\}),$$

where q satisfies $1 < q < n$.

We will demonstrate the validity of this identity in the case that A is invertible. If A is singular, then one may apply a continuity argument making use of Schur complements and Sylvester's identity (both of which are mentioned above).

Given that A is invertible, we may invoke Jacobi's identity (1.2) to each of the six minors above and realize the equivalent equation

$$\det B[\{1, n\}, \{a, b\}] \det B[\{q\}, \{a\}]$$
$$= \det B[\{1, q\}, \{a, b\}] \det B[\{n\}, \{a\}] + \det B[\{q, n\}, \{a, b\}] \det B[\{1\}, \{a\}],$$

where $B = A^{-1}$. However, the relation is a simple identity to verify on an arbitrary 3-by-2 matrix (namely, the submatrix of B given by $B[\{1, q, n\}, \{a, b\}]$).

Viewing this 3-term identity as a special case of a determinantal identity on a 3-by-2 matrix gives insight regarding the potential origin of this equation as a *short-term Plücker identity*, which is a special case of the more general Plücker relations or quadratic identities for minors.

To motivate the connection with certain types of Plücker coordinates, we consider the following situation. Suppose α and β are given index sets of N with $|\alpha| = |\beta|$. Define a third index set γ

$$\gamma = \alpha \cup \{2n + 1 - i \mid i \in \beta^c\},$$

where β^c is the complement of β relative to N. Observe that γ is a subset of $\{1, 2, \ldots, 2n\}$. For example, suppose that $n = 8$, $\alpha = \{1, 2, 4, 6\}$, and $\beta = \{2, 3, 6, 7\}$. Then $\gamma = \alpha \cup \{16, 13, 12, 9\}$. As we will see, the set γ above has a very natural association to a totally positive element in the Grassmannian $\mathrm{Gr}(n, 2n)$.

Recall that the Grassmannian $\mathrm{Gr}(n, 2n)$ can be interpreted as the collection of all $2n$-by-n matrices of rank n, modulo multiplication on the right by an invertible n-by-n matrix. Thus if we are given any n-by-n matrix A, we can embed A in an $2n$-by-n matrix, $B(A)$, as follows:

$$B(A) = \left[\frac{A}{ST} \right],$$

where S is the alternating sign signature matrix and T is the backward identity matrix, both of which are of order n.

It is clear that $B(A)$ is a matrix representation of an element of $\mathrm{Gr}(n, 2n)$, as clearly the rank of $B(A)$ is n. Recall that the *Plücker coordinate*

$$[p_1, p_2, \ldots, p_n](B(A)),$$

where $1 \le p_1 < p_2 < \cdots < p_n \le 2n$, is defined to be the minor of $B(A)$ lying in rows $\{p_1, p_2, \ldots, p_n\}$ and columns $\{1, 2, \ldots, n\}$. That is,

$$[p_1, p_2, \ldots, p_n](B(A)) = \det B(A)[\{p_1, p_2, \ldots, p_n\}, N].$$

An element of $\mathrm{Gr}(n, 2n)$ is called a totally positive element if it has a matrix representation in which all its Plücker coordinates are positive. We note that these coordinates are determined up to a multiple (often called homogeneous projective coordinates) and hence depend only on A, given the representation $B(A)$ above.

A key fact is the following determinantal identity, which can be verified by Laplace expansion of the determinant. Suppose α, β are two index sets of N of the same size and γ is defined as above. Then for any n-by-n matrix A, we have

$$[\gamma](B(A)) = \det A[\alpha, \beta].$$

Hence it follows that if A is TP, then $B(A)$ is a matrix representation of a totally positive element in $\mathrm{Gr}(n, 2n)$.

One of the fundamental properties or relations that hold for Plücker coordinates are the so-called Plücker relations or quadratic relations. If $\{i_1, i_2, \ldots, i_n\}$ and $\{j_1, j_2, \ldots, j_n\}$ are two index sets of $\{1, 2, \ldots, 2n\}$, then the following identities hold,

$$[i_1, i_2, \ldots, i_n][j_1, j_2, \ldots, j_n]$$
$$= \sum_{t=1}^{n} [i_1, \ldots, i_{s-1}, j_t, i_{s+1}, \ldots, i_n][j_1, \ldots, j_{s-1}, i_s, j_{s+1}, \ldots, j_n]$$

for each $s = 1, 2, \ldots, n$.

By the nature of their definition, if a Plücker coordinate contains a repeated index, it is zero; and if two indices are switched, the coordinate is negated. Taking into account the above, the *short Plücker relation* often refers to

$$[i, i', \Delta][j, j', \Delta] + [i, j', \Delta][i', j, \Delta] = [i, j, \Delta][i', j', \Delta],$$

where $\Delta \subset \{1, 2, \ldots, 2n\}$ with $|\Delta| = n - 2$, and where $1 \le i < i' < j < j' \le 2n$ with $i, i', j, j' \notin \Delta$.

Consider some illustrative examples. Suppose $n = 6$, and let $\Delta = \{2, 4, 5, 9\}$ and $i = 1$, $i' = 3$, $j = 6$, and $j' = 11$. Then the corresponding (short) Plücker relation is given by

$$[1, 6, 2, 4, 5, 9][3, 11, 2, 4, 5, 9]$$
$$= [1, 3, 2, 4, 5, 9][6, 11, 2, 4, 5, 9] + [1, 11, 2, 4, 5, 9][3, 6, 2, 4, 5, 9].$$

Making use of the representation $B(A)$ and the connection between these coordinates and the minors of A, it follows that the short relation above is equivalent to

$$\det A[\{1, 2, 4, 5, 6\}, \{1, 2, 3, 5, 6\}] \det A[\{2, 3, 4, 5\}, \{1, 3, 5, 6\}]$$
$$= \det A[\{1, 2, 3, 4, 5\}, \{1, 2, 3, 5, 6\}] \det A[\{2, 4, 5, 6\}, \{1, 3, 5, 6\}]$$
$$+ \det A[\{2, 3, 4, 5, 6\}, \{1, 2, 3, 5, 6\}] \det A[\{1, 2, 4, 5\}, \{1, 3, 5, 6\}],$$

which is exactly a three-term identity that we referred to as Karlin's identity above.

On the other hand, if $n = 6$, $\Delta = \{2, 5, 7, 10\}$, $i = 1$, $i' = 4$, $j = 8$, and $j' = 12$, then the corresponding (short) Plücker relation is given by

$$[1, 8, 2, 5, 7, 10][4, 12, 2, 5, 7, 10]$$
$$= [1, 4, 2, 5, 7, 10][8, 12, 2, 5, 7, 10] + [1, 12, 2, 5, 7, 10][4, 8, 2, 5, 7, 10],$$

which is equivalent to

$$\det A[\{1, 2, 5\}, \{1, 2, 4\}] \det A[\{2, 4, 5\}, \{2, 4, 5\}]$$
$$= \det A[\{1, 2, 4, 5\}, \{1, 2, 4, 5\} \det A[\{2, 5\}, \{2, 4\}]$$
$$+ \det A[\{1, 2, 5\}, \{2, 4, 5\} \det A[\{2, 4, 5\}, \{1, 2, 4\}].$$

Evidently, this is not of the form of Karlin's identity above.

Recall specifically that Karlin's identity was written for an n-by-n matrix A as

$$\det A(\{a, b\}, \{1, n\}) \det A(\{a\}, \{q\})$$
$$= \det A(\{a, b\}, \{1, q\}) \det A(\{a\}, \{n\}) \det A(\{a, b\}, \{q, n\}) \det A(\{a\}, \{1\}).$$

Given the above equation, define $\Gamma = N \setminus \{1, q, n\}$ and, assuming that $a < b$, define $x = 2n + 1 - b$ and $y = 2n + 1 - a$, so that $x < y$. Then define $\Delta = \Gamma \cup \{x\}$, $i = 1$, $i' = q$, $j = n$, and $j' = y$, so that $i < i' < j < j'$ and none of them lie in Δ. It then follows that the identity above is equivalent to the short Plücker relation given by

$$[i, i', \Delta][j, j', \Delta] + [i, j', \Delta][i', j, \Delta] = [i, j, \Delta][i', j', \Delta].$$

1.3 SOME BASIC FACTS

A TP_1 (TN_1) matrix is simply an entrywise positive (nonnegative) matrix. The theory of these matrices, which is a precursor to the theory of TP (TN) matrices, goes under the heading Perron-Frobenius theory and is a valuable ingredient in the theory of TP/TN matrices. See [HJ85, Ch. 8] for details. Recall that the *spectral radius* of an n-by-n matrix A is defined to be $\rho(A) = \max\{|\lambda| : \lambda \in \sigma(A)\}$. As usual, we use $\sigma(A)$ to denote the collection of all eigenvalues of A, counting multiplicity. We state here the version of Perron's theorem on positive matrices that is most useful for us.

Theorem 1.3.1 *If $A \in M_n \cap TP_1$, then*

(a) *the spectral radius $\rho(A) \in \sigma(A)$;*

(b) *the algebraic multiplicity of $\rho(A)$ as an eigenvalue of A is one;*

(c) *if $\lambda \in \sigma(A)$, $\lambda \neq \rho(A)$, then $|\lambda| < \rho(A)$; and*

(d) *associated with $\rho(A)$ is an entrywise positive eigenvector x of A.*

If A is primitive (nonnegative and some power is positive), the same conclusions remain valid. If A is only entrywise nonnegative, the spectral radius may be zero, but (a) remains valid, while (b), (c), and (d) are weakened in a manner dictated by continuity: the multiplicity can be higher; there can be ties for spectral radius; and a nonnegative eigenvector may have zero components.

A general fact about left ($y^*A = \lambda y^*$) and right ($Ax = \lambda x$) eigenvectors of a square matrix is the *principle of biorthogonality*.

Proposition 1.3.2 *If A is an n-by-n complex matrix with $\lambda, \mu \in \sigma(A)$, $\lambda \neq \mu$, then any left eigenvector of A corresponding to μ is orthogonal to any right eigenvector of A corresponding to λ.*

One use of the principle of biorthogonality is that only one eigenvalue may have a positive right eigenvector if there is a positive left eigenvector, as in the case of positive matrices, and the proof is a simple consequence of the calculation;

$$\mu y^* x = y^* A x = \lambda y^* x.$$

If $A \in InTN$, A^{-1} cannot be, unless A is diagonal. However, except for zeros, A^{-1} has a *checkerboard* sign pattern because of the cofactor form of the inverse and because A is TN (and has a positive determinant). If S (or S_n) is the n-by-n alternating sign signature matrix ($S = \text{diag}(1, -1, \ldots, \pm 1)$), a notation we reserve throughout, this means that $A^\# := SA^{-1}S$ is an entrywise nonnegative matrix. Note that $(\det A)A^\# = C_{n-1}(A)$, the $(n-1)$st compound of A, if $A \in M_n$, so that the nonnegativity is clear. However, not only is $A^\#$ nonnegative, it is TN, a form of closure under inversion.

Theorem 1.3.3 *If $A \in M_n$ and $A \in TP$ (InTN), then $A^{\#} \in TP$ (InTN).*

Proof. This follows from a careful application of Jacobi's identity (1.2), but it will be even more transparent using the elementary bidiagonal factorization of the next chapter. One need only note that the claim is correct for an elementary bidiagonal matrix.   □

1.4 TN AND TP PRESERVING LINEAR TRANSFORMATIONS

Invertible transformations on matrices that transform one class onto itself can be quite useful as a tool for normalization when analyzing the class.

There are three natural linear transformations on matrices that preserve (each of) the TN classes under study. Since multiplication on the left or right by a matrix from $\mathcal{D}_n$ changes the sign of no minor, the positive diagonal equivalence $A \mapsto DAE$, $A \in M_{m,n}$, invertible $D \in \mathcal{D}_m$, $E \in \mathcal{D}_n$, with D and E fixed, is an invertible linear transformation on $M_{m,n}$ that maps each of our classes onto itself. The same is true of transposition, which maps $M_{m,n}$ onto $M_{n,m}$. The third is more subtle. Most permutation similarities necessarily map TP (TN) outside itself, as some nonprincipal minors are negated by the transformation. Of course, the identity transformation is a permutation similarity on M_n that preserves each class, but there is another fixed permutation similarity that does as well. Let $\tilde{T} = [t_{ij}]$ be the permutation matrix such that $t_{i,n-i+1} = 1$ for $i = 1, 2, \ldots, n$ be the "backward identity" permutation matrix. Of course, $\tilde{T}$ is symmetric, so that the invertible linear transformation given by $A \mapsto \tilde{T} A \tilde{T}$ is a permutation similarity on M_n.

By the Cauchy-Binet identity, any minor (whose location may be moved) that is negated by left multiplication by $\tilde{T}$ is negated again by right multiplication by $\tilde{T}$, thus preserving any TN class. Note that this permutation similarity essentially reads the matrix from the bottom right instead of the top left. If $A \in M_{m,n}$ and the size of $\tilde{T}$ is allowed to vary with context, $A \mapsto \tilde{T} A \tilde{T}$ similarly preserves the classes. As we shall see, TN is the topological closure of TP, so that an invertible linear transformation preserving TP necessarily preserves TN. We formally record these observations, as each will be used from time to time.

Theorem 1.4.1 *Let $\mathcal{C} \in \{$TN, TP, InTN, IrTN, IITN, TN_k TP_k, $InTN_k$, $IrTN_k$, $IITN_k\}$ and $A \in M_{m,n}$, $D \in \mathcal{D}_m$, $E \in \mathcal{D}_n$ and $\tilde{T}$ as defined above. With the understanding that $m = n$ in the case of InTN, IrTN, IITN, etc., then each of the following three linear transformations map $\mathcal{C}$ onto $\mathcal{C}$:*

(i) $A \mapsto DAE$;

(ii) $A \mapsto A^T$; *and*

(iii) $A \mapsto \tilde{T} A \tilde{T}$.

Of course, this means that any composition of this transformations does the same. In [BHJ85] it was shown that these are all such linear transformations that map TN onto itself. Other invertible linear transformations can map TN or TP into itself.

Example 1.4.2 Let $F \in M_{m,n}$ have all its entries equal to one, except the (1,1) entry, which is greater than one. Then the invertible linear transformation $A \mapsto F \circ A$, in which $\circ$ denotes the entrywise, or Hadamard, product, maps TP into TP (and TN into TN). Of course $A \mapsto A^\#$ is a (nonlinear) map that takes TP or IITN onto itself. A basic consequence of this fact is that increasing the (1,1) entry of a TN/TP matrix preserves the property of being TN/TP.

1.5 SCHUR COMPLEMENTS

The next notion we wish to discuss here is the so-called *Schur complement* of a principal submatrix of a given matrix A (see [HJ85]).
Schur complements: For $\phi \neq \alpha \subseteq N$ and $A \in M_n(\mathbb{F})$, if $A[\alpha]$ is non-singular, then the *Schur complement of $A[\alpha]$ in A* is the matrix $A[\alpha^c] - A[\alpha^c, \alpha](A[\alpha])^{-1}A[\alpha, \alpha^c]$. Suppose A is partitioned as

$$A = \begin{bmatrix} A_{11} & A_{12} \\ A_{21} & A_{22} \end{bmatrix},$$

in which A_{11} is k-by-k and nonsingular. Then

$$\begin{bmatrix} I & 0 \\ -A_{21}A_{11}^{-1} & I \end{bmatrix} \begin{bmatrix} A_{11} & A_{12} \\ A_{21} & A_{22} \end{bmatrix} \begin{bmatrix} I & -A_{11}^{-1}A_{12} \\ 0 & I \end{bmatrix} = \begin{bmatrix} A_{11} & 0 \\ 0 & S \end{bmatrix}, \quad (1.7)$$

where S is the Schur complement of A_{11} in A. Hence it follows that $\det A = \det A_{11} \det S$. Finally, we note that if $A[\alpha]$ is a nonsingular principal submatrix of a nonsingular matrix A, then $A^{-1}[\alpha^c] = S^{-1}$, where now S is the Schur complement of $A[\alpha]$ in A.

An immediate consequence of Theorem 1.3.3 is the next fact about the Schur complement of TN matrices (see [And87]). The Schur complement of an invertible submatrix $A[\gamma]$ in A will be denoted by $A/A[\gamma]$, and we know that $A/A[\gamma] = A[\gamma^c] - A[\gamma^c, \gamma](A[\gamma])^{-1}A[\gamma, \gamma^c]$.

For a given index $\alpha = \{i_1, i_2, \ldots, i_k\}$, with $i_j < i_{j+1}$, $j = 1, 2, \ldots, k-1$, the *dispersion of α*, denoted by $d(\alpha)$, is defined to be

$$d(\alpha) = \sum_{j=1}^{k-1}(i_{j+1} - i_j - 1) = i_k - i_1 - (k-1).$$

The convention for singletons is that their dispersion is taken to be zero. The dispersion of a set represents a measure of the gaps in that set. In particular, observe that the dispersion is zero whenever the set is a contiguous set (i.e., a set based on consecutive indices).

Proposition 1.5.1 *If A is a square invertible TN matrix (or TP), then $A/A[\alpha^c]$, the Schur complement of $A[\alpha^c]$ in A, is TN (or TP), for all index sets α with $d(\alpha) = 0$.*

We note here that an auxiliary assumption was made in the above, namely the matrix was assumed to be invertible. However, "Schur complements" can exist when A is singular, as long as the appropriate principal submatrix is nonsingular.

Proposition 1.5.2 *If A is a square TN matrix with $A[\alpha^c]$ invertible, then $A/A[\alpha^c]$, the Schur complement of $A[\alpha^c]$ in A, is TN, for all index sets α with $d(\alpha) = 0$.*

Proof. Firstly, assume that $\alpha = \{k+1, k+2, \ldots, n\}$ and suppose that $A[\alpha^c]$ is invertible. Define the $(n-k)$-by-$(n-k)$ matrix $C = [c_{ij}]$ by $c_{ij} = \det A[\alpha^c \cup \{i\}, \alpha^c \cup \{j\}]$, where $i, j \in \alpha$. Then Sylvester's determinantal identity (1.4) can be written as

$$\det C[\{i_1, \ldots, i_s\}, \{j_1, \ldots, j_s\}]$$
$$= (\det A[\alpha^c])^{s-1} \det A[\alpha^c \cup \{i_1, \ldots, i_s\}, \alpha^c \cup \{j_1, \ldots, j_s\}].$$

From the above identity it follows that $C = (\det A[\alpha^c])(A/A[\alpha^c])$. Moreover, since A is TN, by the identity above, C is TN and hence $A/A[\alpha^c]$ is totally nonnegative. A similar argument holds for the case when $\alpha = \{1, 2, \ldots, k\}$. Thus we have shown that $A/A[\alpha^c]$ is TN whenever $\alpha = \{1, 2, \ldots, k\}$ or $\alpha = \{k, k+1, \ldots, n\}$. The next step is to use the so-called quotient property of Schur complements. Recall that if $\beta \subseteq \alpha$, then $A/A[\alpha] = [(A/A[\beta])/(A[\alpha]/A[\beta])]$. Suppose α is a fixed index set with $d(\alpha) = 0$ and $|\alpha| < n$. Then we may write α as $\alpha = \{i, i+1, \ldots, i+k\}$, where $1 \le i < i+k \le n$. Thus $\alpha^c = \{1, 2, \ldots, i-1, i+k+1, \ldots, n\}$, and define $\beta = \{1, 2, \ldots, i-1\}$ if $i > 1$, otherwise $\beta = \{i+k+1, \ldots, n\}$ if $i+k < n$. Then by the quotient property described above we have that $A/A[\alpha^c] = [(A/A[\beta])/(A[\alpha^c]/A[\beta])]$. By the previous remarks $A/A[\beta]$ is totally nonnegative. Moreover, since $A[\alpha^c]/A[\beta]$ is a submatrix of $A/A[\beta]$ in the bottom right corner, it follows that $(A/A[\beta])/(A[\alpha^c]/A[\beta])$ is TN, and hence that $A/A[\alpha^c]$ is TN. This completes the proof. $\square$

We note here that the requirement of $d(\alpha) = 0$ is necessary, as in general Schur complements of TN matrices need not be TN.

For index sets that are not contiguous, we certainly do not have closure under Schur complementation. For example, if

$$A = \begin{bmatrix} 1 & 1 & 1 \\ 1 & 2 & 3 \\ 1 & 3 & 6 \end{bmatrix},$$

then A is TP, and if $\alpha = \{1, 3\}$, then

$$A/A[\alpha^c] = \begin{bmatrix} 1/2 & -1/2 \\ -1/2 & 3/2 \end{bmatrix}.$$

Thus, in this instance, the Schur complement is not TP. However, such Schur complements are close to being TP.

Proposition 1.5.3 *Let A be an invertible n-by-n TN matrix. Then $A/A[\alpha]$, the Schur complement of $A[\alpha]$ in A, is signature similar to a TN matrix.*

Proof. Recall that $A/A[\alpha] = (A^{-1}[\alpha^c])^{-1}$. Since A is TN, we have that $A^{-1} = SBS$, for $S = \text{diag}(1, -1, \ldots, \pm 1)$ and B totally nonnegative. Then $A/A[\alpha] = (SBS[\alpha^c])^{-1} = (S[\alpha^c]B[\alpha^c]S[\alpha^c])^{-1} = S[\alpha^c](B[\alpha^c])^{-1}S[\alpha^c]$. Since B is TN, $B[\alpha^c]$ is TN, and hence $(B[\alpha^c])^{-1}$ is signature similar to a TN matrix C; that is, $(B[\alpha^c])^{-1} = S'CS'$, where S' is a signature matrix. Therefore

$$A/A[\alpha] = S[\alpha^c]S'CS'S[\alpha^c] = S''CS'',$$

where $S'' = S[\alpha^c]S'$ is a signature matrix. This completes the proof. □

This result can be extended in a natural way to Schur-complements with respect to non-principal submatrices, but such a topic is not considered here.

1.6 ZERO-NONZERO PATTERNS OF TN MATRICES

In this section we are concerned with zero-nonzero patterns of totally nonnegative matrices. As observed in the previous section, inserting a zero row (or column) into a TN matrix preserves the property of being TN. Thus for the remainder of this section we assume that our matrices do not contain any zero rows or columns. To begin our analysis, we give the following definitions.

Definition 1.6.1 An m-by-n *sign pattern* is an m-by-n array of symbols chosen from $\{+, 0, -\}$. A *realization* of a sign pattern, S, is a real m-by-n matrix A such that

$$a_{ij} > 0 \text{ when } s_{ij} = +; \quad a_{ij} < 0 \text{ when } s_{ij} = -; \quad \text{and } a_{ij} = 0 \text{ when } s_{ij} = 0.$$

We also let $*$ denote a nonzero entry of a matrix whenever a sign is not specified. There are two natural mathematical notions associated with various sign-pattern problems. They are the notions of *require* and *allow*. We say an m-by-n sign pattern S *requires property* P if every realization of S has property P. On the other hand, we say a sign pattern S *allows property* P if there exists a realization of S with property P.

Recall that an m-by-n matrix is TN_k (TP_k) for $1 \leq k \leq \min(m, n)$, if all minors of size at most k are nonnegative (positive). We call an m-by-n matrix TN_+ if each square submatrix either is sign-singular (singular by virtue of its sign-pattern) or has positive determinant.

Definition 1.6.2 An m-by-n matrix is said to be *in double echelon form* if

(i) Each row of A has one of the following forms:

(1) $(*, *, \ldots, *)$,

(2) $(*, \ldots, *, 0, \ldots, 0)$,

(3) $(0, \ldots, 0, *, \ldots, *)$, or

(4) $(0, \ldots, 0, *, \ldots, *, 0, \ldots, 0)$.

(ii) The first and last nonzero entries in row $i+1$ are not to the left of the first and last nonzero entries in row i, respectively ($i = 1, 2, \ldots, n-1$).

Note that there is an obvious analog of the above definition for m-by-n sign-patterns. Thus a matrix in double echelon form may appear as

$$
\begin{bmatrix}
* & * & 0 & \cdots & 0 \\
* & \ddots & \ddots & \ddots & \vdots \\
0 & \ddots & \ddots & \ddots & 0 \\
\vdots & \ddots & \ddots & \ddots & * \\
0 & \cdots & 0 & * & *
\end{bmatrix}.
$$

Recall that increasing the $(1,1)$ or (m,n) entry of an m-by-n TN matrix preserves the property of being totally nonnegative (see Example 1.4.2). This observation is needed in the proof of the next result.

Theorem 1.6.3 *Let S be an m-by-n $(0, +)$-sign pattern in double echelon form. Then S allows a TN_+ matrix.*

Proof. Since S has no zero rows or columns, $s_{11} = +$. Assign the value 1 to all the positive positions in row 1. Consider the leftmost $+$ in row 2, say $s_{2j} = +$, but $s_{2k} = 0$ for all $k < j$. Let $B = S[\{1, 2\}, \{1, 2, \ldots, j\}]$. Then s_{2j} enters positively into any minor of B, hence there exists a choice for the value of s_{2j} so that B is TN_+. Continue to choose values s_{2l} ($l > j$) large enough so that there exists a 2-by-l matrix with sign pattern $S[\{1, 2\}, \{1, 2, \ldots, l\}]$ that is TN_+ until a zero element is encountered. Thus we have a 2-by-n TN_+ matrix C that is a realization of $S[\{1, 2\}, \{1, 2, \ldots, n\}]$. Suppose, by induction, there exists a p-by-q matrix B, a realization of $S[\{1, 2, \ldots, p\}, \{1, 2, \ldots, q\}]$, in which the (p, q) has not been specified, but both $B[\{1, 2, \ldots, p\}, \{1, \ldots, q-1\}]$ and $B[\{1, \ldots, p-1\}, \{1, 2, \ldots, q\}]$ are TN_+ matrices. Observe that any submatrix that does not involve b_{pq} is contained in either

$$B[\{1, 2, \ldots, p\}, \{1, \ldots, q-1\}] \text{ or } B[\{1, \ldots, p-1\}, \{1, 2, \ldots, q\}],$$

and hence is TN_+. Thus we are only concerned with submatrices that involve b_{pq}. Then we claim that the (p, q)th entry of B can be chosen so that B is TN_+. If $s_{pq} = 0$, then since S is in double echelon form, either row p or column q of B is zero, and hence B is a TN_+ matrix. Otherwise $s_{pq} = +$, and we may choose a value b_{pq} for s_{pq} large enough so that B is a TN matrix. If B is TN_+, then we are done. Thus assume there exists a k-by-k submatrix of B (which must involve b_{pq}) that is singular but not sign-singular. Denote this submatrix by C and suppose C is in the form $C = \left[\begin{array}{c|c} C_{11} & c_{12} \\ \hline c_{21} & b_{pq} \end{array}\right]$. Since C is not sign-singular, it follows that C_{11} is not sign-singular. To see this suppose that C_{11} is sign-singular. Then there exists a zero block in C_{11} of size at least k. Since C is in double echelon form, it follows that there exists a zero block of size at least $k + 1$, and hence C is sign-singular, which is a contradiction. Since C_{11} is not sign-singular, it must be nonsingular, and in this case we may increase the value b_{pq} so that C is nonsingular. The proof now follows by induction. □

A *line* in a matrix will apply to any given row or column. For example, a matrix in double echelon form has no zero lines. We now come to our main observations for this section.

Theorem 1.6.4 *Let S be an m-by-n $(0, +)$-pattern with no zero rows or columns. Then the following are equivalent:*

(1) S allows a TN_+ matrix;

(2) S allows a TN matrix;

(3) S allows a TN_2 matrix;

(4) S is a double echelon pattern.

Proof. The implications $(1) \Rightarrow (2) \Rightarrow (3)$ are trivial since the containments $TN_+ \subseteq TN \subseteq TN_2$ are obvious. Suppose S allows a TN_2 matrix denoted by A, and assume that some entry, say $s_{pq} = 0$; otherwise S has no zero entries, and we are done. Hence $a_{pq} = 0$. Since S has no zero rows or columns, some entry of A in row p is nonzero. Assume $a_{pt} > 0$. There are two cases to consider: $t > q$ or $t < q$. Suppose $t > q$. Since A is TN_2, it follows that $a_{sq} = 0$ for all $s > p$. Since S has no zero lines, some entry in the qth column must be nonzero. Moreover, this entry must occur with a row index less than p. Again since A is TN_2, it follows that $a_{ps} = 0$ for all $s < q$. Continuing this argument implies that $a_{ij} = 0$ for all $p \le i \le m$ and $1 \le j \le q$. For case (2), a similar argument shows that $a_{ij} = 0$ for all $1 \le i \le p$ and $q \le j \le n$. A routine induction implies that A, and hence S is in double echelon form. The implication $(4) \Rightarrow (1)$ is Theorem 1.6.3. □

Observe that an argument similar to that used in the proof of $(3) \Rightarrow (4)$ may also be used to the prove the following result.

Corollary 1.6.5 *Let A be an m-by-n TN matrix with no zero rows or columns. Then A is in double echelon form.*

In fact, we can say even more regarding the zero-nonzero pattern of a TN matrix in the event of a zero line. For instance, suppose $A = [a_{ij}]$ is TN and m-by-n and satisfies $a_{ij} = 0$, but neither the ith row nor the jth column of A is zero. Then, applying arguments similar to the ones above, we may conclude that $a_{pq} = 0$ for $p = 1, \ldots, i$ and $q = j, \ldots, n$ or for $p = i, \ldots, m$ and $q = 1, \ldots, j$.

Corollary 1.6.6 *Let A be an n-by-n TN matrix with no zero rows or columns. Then A is irreducible if and only if $a_{ij} > 0$ for all $|i - j| \leq 1$.*

A result similar to the above corollary, though stated slightly differently, appeared in [Rad68]. We conclude this section with some results on $(0,1)$-TN matrices.

Lemma 1.6.7 *Let A be an n-by-n $(0,1)$-lower (-upper) triangular matrix with ones on and below (above) the main diagonal. Then A is a TN matrix.*

Proof. Consider the $(0,1)$-matrix, $A = \begin{bmatrix} 1 & 0 \\ 1 & 1 \end{bmatrix}$. It is easy to verify that A is totally nonnegative. Bordering A by a column of zeros on the right and then repeating the bottom row of A preserves total nonnegativity. Next, we may increase the lower right entry to 1, which does not disturb the property of being TN. Thus, our new matrix, $\tilde{A} = \begin{bmatrix} 1 & 0 & 0 \\ 1 & 1 & 0 \\ 1 & 1 & 1 \end{bmatrix}$, is TN. We may continue this procedure to show that an n-by-n full lower triangular $(0,1)$-matrix is TN. The upper triangular version follows by transposition. $\quad\square$

Lemma 1.6.8 *Any $(0,1)$-matrix in the form*

$$\begin{bmatrix} & \begin{smallmatrix} 0 & \cdots & 0 \\ & \ddots & \vdots \\ & & 0 \end{smallmatrix} \\ & 1\text{'s} & \end{bmatrix}$$

is TN.

We can now characterize all the $(0,1)$ matrices that are TN based on a single forbidden submatrix.

Theorem 1.6.9 *Let $A = [a_{ij}]$ be an m-by-n $(0,1)$-matrix with no zero rows or columns. Then A is TN if and only if A is in double echelon form and does not contain the submatrix*

$$B = \begin{bmatrix} 1 & 1 & 0 \\ 1 & 1 & 1 \\ 0 & 1 & 1 \end{bmatrix}.$$

Proof. First, observe that the matrix B is not TN because $\det(B) = -1$. The necessity is trivial since B is not TN and the property of total nonnegativity is inherited. To prove sufficiency we will proceed by induction on $m+n$. Note that this statement is not difficult to verify if $n \leq 3$. Suppose $m + n = k$, and that the result holds for all m'-by-n' such matrices with $m' + n' < k$. Consider the m-by-n matrix A. We may assume that A has no repeated consecutive rows or columns; otherwise the submatrix $\hat{A}$ of A obtained by deleting a repeated row or column satisfies the induction hypothesis and hence is TN. Therefore A is TN. Hence we may assume that $a_{11} = 1$. If $a_{12} = 0$, then the submatrix $A[\{2, 3, \ldots, m\}, \{1, 2, \ldots, n\}]$ satisfies the induction hypothesis and hence is TN. In this case A is again TN by Example 1.4.2. Therefore we assume $a_{12} = 1$ and $a_{21} = 1$ (by transposition). Observe that since A is in double echelon form and $a_{12} = a_{21} = 1$, it follows that $a_{22} = 1$. If $a_{33} = 0$, then either $a_{13} = a_{23} = 0$ or $a_{31} = a_{32} = 0$. In each case, since A is in double echelon form, either row $1 = $ row 2 (when $a_{13} = a_{23} = 0$) or column $1 = $ column 2 (otherwise), and hence we may delete a row (column) and apply the induction hypothesis. Therefore, we may assume $a_{33} = 1$, and similarly, we assume that $a_{23} = a_{32} = 1$, because otherwise either row $1 = $ row 2 or column $1 = $ column 2. There are three cases to consider: $a_{13} = 0$; $a_{13} = 1$ and $a_{31} = 0$; and $a_{13} = a_{31} = 1$.

Case 1: $a_{13} = 0$.

$$\begin{bmatrix} 1 & 1 & 0 & \cdots & \\ 1 & 1 & 1 & \cdots & \\ 1 & 1 & 1 & & \\ \vdots & \vdots & & * & \\ & & & & \ddots \end{bmatrix}.$$

In this case we have $a_{31} = 1$, because if $a_{31} = 0$, then the submatrix B would appear in the upper left corner of A. Now, if $a_{44} = 0$, then either row $2 = $ row 3 (if $a_{34} = a_{24} = 0$), or column $1 = $ column 2 (if $a_{42} = a_{43} = 0$). Thus, we may assume $a_{44} = 1$. Similarly, $a_{42} = a_{43} = 1$; otherwise column $1 = $ column 2. Additionally, $a_{34} = 1$, or else row $2 = $ row 3. Together, this implies that $a_{41} = 1$:

$$\begin{bmatrix} 1 & 1 & 0 & * & \cdots & \\ 1 & 1 & 1 & * & \cdots & \\ 1 & 1 & 1 & 1 & & \\ 1 & 1 & 1 & 1 & & \\ \vdots & \vdots & & & \ddots & \end{bmatrix}.$$

If $a_{55} = 0$, then either column $1 = $ column 2 (if $a_{51} = a_{52} = a_{53} = a_{54} = 0$) or row $3 = $ row 4 (if $a_{35} = a_{45} = 0$), and thus we may assume $a_{55} = 1$. Similarly, $a_{52} = a_{53} = a_{54} = 1$, because otherwise column 1

= column 2. Therefore, $a_{51} = 1$, because if $a_{51} = 0$, then B would appear as a submatrix of A. We may also assume that $a_{45} = 1$, for otherwise, row 3 = row 4. Thus, our matrix has the form

$$\begin{bmatrix} 1 & 1 & 0 & * & * & \cdots \\ 1 & 1 & 1 & * & * & \cdots \\ 1 & 1 & 1 & 1 & * & \\ 1 & 1 & 1 & 1 & 1 & \\ 1 & 1 & 1 & 1 & 1 & \\ \vdots & \vdots & & & & \ddots \end{bmatrix}.$$

Continuing this method of analysis implies that every entry of A on and below the main diagonal must be equal to 1. Thus, A is TN by Lemma 1.6.8.

Case 2: $a_{13} = 1$, $a_{31} = 0$.

Observe that this case follows from Case 1 and transposition. The conclusion in this case, however, will be that every entry of A on and *above* the main diagonal must be equal to 1.

Case 3: $a_{13} = 1$ and $a_{31} = 1$.

Applying reasoning similar to that in the previous two cases we can assume that $a_{44} = a_{24} = a_{34} = a_{43} = a_{42} = 1$. Then the argument proceeds in the same manner as before with three cases to consider: (1) $a_{14} = 0$; (2) $a_{14} = 1$ and $a_{41} = 0$; or (3) $a_{14} = a_{41} = 1$. Each of these cases is handled by arguments similar to the cases (1), (2), and (3) above, and the conclusions are similar in each case. Continuing in this manner will prove that any (0,1)-matrix in double echelon form that does not contain B as a submatrix is TN.

$\square$

It is worth noting that the same 3-by-3 submatrix will arise in the context of closure under Hadamard multiplication (see Chapter 8).

Chapter Two

Bidiagonal Factorization

2.0 INTRODUCTION

Factorization of matrices is one of the most important topics in matrix theory, and plays a central role in many related applied areas such as numerical analysis and statistics. In fact, a typical (and useful) preliminary topic for any course in linear algebra is Gaussian elimination and elementary row operations, which essentially represents the groundwork for many forms of matrix factorization. In particular, triangular factorizations are a byproduct of many sorts of elimination strategies including Gaussian elimination.

Investigating when a class of matrices (such as positive definite or TP) admit a particular type of factorization is an important study, which historically has been fruitful. Often many intrinsic properties of a particular class of matrices can be deduced via certain factorization results. For example, it is a well-known fact that any (invertible) M-matrix can be factored into a product of a lower triangular (invertible) M-matrix and an upper triangular (invertible) M-matrix. This LU factorization result allows us to immediately conclude that the class of M-matrices is closed under Schur complementation, because of the connection between LU factorizations and Schur complements.

Matrix factorizations for the class of TN (TP) matrices have been well studied and continue to be a vital avenue for research pertaining to this class. Specifically, we are most interested in preserving the property of TN (or TP) for various factorization results. This constraint aids the focus for using these factorizations to exhibit properties of TN (or TP) matrices (cf. with the example above for M-matrices).

In this chapter our main focus will be on triangular factorization extended beyond just LU factorization. The most important factorization that will be studied here is the so-called "bidiagonal factorization." A list of references for this topic is lengthy; see, for example, [And87, BFZ96, CP94b, CP97, Cry73, Cry76, FZ99, GP92b, GP93b, GP94b, GP96, Loe55, Peñ97, RH72, Whi52] for further discussion on factorizations of TN matrices.

The basic observation that any TN matrix A can be written as $A = LU$, where L and U are lower and upper triangular TN matrices, respectively, was established in the nonsingular case in 1972 and later generalized to the singular case; see [Cry73] and [Cry76]. To obtain this stronger result, Cryer focused on unusual row and column operations not typically used in classical Gaussian elimination. Great care must be exercised in such an argument, as many row and column operations do not preserve total nonnegativity. Even though Whitney did not prove explicitly that any nonsingular TN matrix can be factored into lower and upper triangular TN matrices, she provided an observation that, in fact, implies more.

Suppose $A = [a_{ij}]$ is an n-by-n matrix with $a_{j1}, a_{j+1,1} > 0$, and $a_{k1} = 0$ for $k > j + 1$. Then, let B be the n-by-n matrix obtained from A by using row j to eliminate $a_{j+1,1}$. Whitney essentially proved that A is TN if and only if B is TN. This observation serves as the foundation for a bidiagonal factorization result for TP matrices.

It is not clear why Whitney did not continue with this elimination scheme, as it would have led to a factorization of a nonsingular TN matrix into a product of TN bidiagonal matrices. In 1955 such a bidiagonal factorization result appeared for nonsingular TN matrices in [Loe55], where it was attributed to Whitney.

As far as we can discern, Theorem 2.2.2 was first stated in [Loe55], where it was used for an application to Lie-group theory. In [Loe55] the reference to a bidiagonal factorization of an invertible TN matrix was attributed to Whitney and her work in [Whi52], although Whitney did not make such a claim in her paper. More accurately, in [Whi52] a lemma was proved that is needed to establish the existence of such a factorization. More recently, such authors as Gasca, Peña, and others considered certain bidiagonal factorizations for which a specified order of row operations is required to ensure a special and appealing form for the factorization. They refer to this particular factorization as a Neville factorization or elimination (see [CP97, GP92b, GP93b, GP94b, GP96]). The work in [Cry76], and previously in [Cry73], seems to be the first time it is shown that a singular TN matrix also admits such a bidiagonal factorization. See [BFZ96] and the sequel [FZ99] for an excellent treatment of the combinatorial and algebraic aspects of bidiagonal factorization. Some of the graphical representations provided in [BFZ96] (and also [Bre95]) are employed throughout this book, and have been very useful for proving many results.

Other factorizations have also been considered for TN matrices; see, for example, [Cry76, GP94b, Peñ97, RH72]. The following factorization seems to be useful for various spectral problems for TN matrices (see also Chapter 5).

Theorem 2.0.1 *If A is an n-by-n TN matrix, then there exist a TN matrix S and a tridiagonal TN matrix T such that*

(i) $TS = SA$, and

(ii) the matrices A and T have the same eigenvalues.

Moreover, if A is nonsingular, then S is nonsingular.

This result was first proved in the nonsingular case in [RH72], and the general case was later proved by in [Cry76]. In Chapter 5 we will make use of a slightly modified version of this factorization to aid in characterizing properties about the positive eigenvalues of an irreducible TN matrix.

2.1 NOTATION AND TERMS

We begin this chapter by defining necessary terms and setting the required notation for understanding the topic of bidiagonal factorization.

Let I denote the n-by-n identity matrix, and for $1 \leq i, j \leq n$, we let E_{ij} denote the n-by-n standard basis matrix whose only nonzero entry is a 1 that occurs in the (i, j) position. An n-by-n matrix $A = [a_{ij}]$ is called:

(1) *diagonal* if $a_{ij} = 0$ whenever $i \neq j$;

(2) *tridiagonal* if $a_{ij} = 0$ whenever $|i - j| > 1$;

(3) *upper (lower) triangular* if $a_{ij} = 0$ whenever $i > j$ ($i < j$).

A tridiagonal matrix that is also upper (lower) triangular is called an *upper (lower) bidiagonal matrix*. Statements referring to just triangular or bidiagonal matrices without the adjectives "upper" or "lower" may be applied to either case.

It is necessary (for clarity of composition) to distinguish between two important classes of bidiagonal matrices. This will aid the development of the bidiagonal factorization over the next few sections.

Definition 2.1.1 (elementary bidiagonal matrices) For any positive integer n and complex numbers s, t, we let

$$L_i(s) = I + sE_{i,i-1} \quad \text{and} \quad U_j(t) = I + tE_{j-1,j}$$

for $2 \leq i, j \leq n$. Matrices of the form $L_i(s)$ or $U_j(t)$ above will be called *elementary bidiagonal matrices*, and the class of elementary bidiagonal matrices will be denoted by EB.

Definition 2.1.2 (generalized elementary bidiagonal matrices) Matrices of the form $D + sE_{i,i-1}$ or $D + tE_{j-1,j}$, where n is a positive integer, s, t are complex numbers, and $2 \leq i, j \leq n$, will be called *generalized elementary bidiagonal matrices*, and this class will be denoted by GEB.

Since EB matrices are contained among the class of the so-called "elementary" matrices, EB matrices correspond to certain elementary row and column operations. For example, if A is an n-by-n matrix, then the product $L_i(s)A$ replaces row i of A by row i plus s times row $i - 1$. Similarly, the product $AU_j(t)$ replaces column j by column j plus t times column $j - 1$.

Also, because $L_i(s)$ corresponds to an elementary row operation, it immediately follows that

$$(L_i(s))^{-1} = L_i(-s).$$

Notice that all row/column operations using EB matrices only involve consecutive rows and columns.

Observe that any GEB matrix is TN whenever D is entrywise nonnegative and $s, t \geq 0$, and any EB matrix is (invertible) TN whenever $s, t \geq 0$.

In an effort to exhibit a factorization of a given matrix A into EB matrices, it is useful to consider the most natural elimination ordering when only considering EB matrices.

Recall that under typical "Gaussian elimination" to lower triangular form the (1,1) entry is used to eliminate all other entries in the first column (so-called *pivoting*), then moving to the 2nd column the (2,2) entry is used to eliminate all entries in column 2 below row 2, and so on. In this case we write the elimination (of entries) ordering as follows:

$$(2,1)(3,1)\cdots(n,1)(3,2)(4,2)\cdots(n,2)\cdots(n,n-1).$$

Using only EB matrices this elimination ordering is not appropriate for the simple reason that once the (2,1) entry has annihilated we cannot eliminate the (3,1) entry with EB matrices. This is different, however, from claiming that a matrix of the form

$$A = \begin{bmatrix} 1 & 0 & 0 \\ 0 & 1 & 0 \\ 1 & 0 & 1 \end{bmatrix}$$

has no (lower) EB factorization. In fact,

$$A = L_2(-1)L_3(1)L_2(1)L_3(-1).$$

The key item here is that we first had to multiply A by $L_2(1)$ (on the left), which essentially filled in the (2,1) entry of A, that is,

$$L_2(1)A = \begin{bmatrix} 1 & 0 & 0 \\ 1 & 1 & 0 \\ 1 & 0 & 1 \end{bmatrix}.$$

Now we use the (2,1) entry of $L_2(1)A$ to eliminate the (3,1) entry, and so-on.

Consider another useful example. Let

$$V = \begin{bmatrix} 1 & 1 & 1 \\ 1 & 2 & 4 \\ 1 & 3 & 9 \end{bmatrix}.$$

We begin by using the second row to eliminate the (3,1) entry of V and continue eliminating up the first column to obtain

$$L_2(-1)L_3(-1)V = \begin{bmatrix} 1 & 1 & 1 \\ 0 & 1 & 3 \\ 0 & 1 & 5 \end{bmatrix}.$$

Then we shift to the second column and again use the second row to eliminate the (3,2) entry, yielding

$$L_3(-1)L_2(-1)L_3(-1)V = \begin{bmatrix} 1 & 1 & 1 \\ 0 & 1 & 3 \\ 0 & 0 & 2 \end{bmatrix} = U,$$

or $V = L_3(1)L_2(1)L_3(1)U$. Applying similar row operations to the transpose U^T, we can write U as $U = DU_3(2)U_2(1)U_3(1)$, where $D = \text{diag}(1,1,2)$.

Hence, we can write the (Vandermonde) matrix V as

$$V = \begin{bmatrix} 1 & 0 & 0 \\ 0 & 1 & 0 \\ 0 & 1 & 1 \end{bmatrix} \begin{bmatrix} 1 & 0 & 0 \\ 1 & 1 & 0 \\ 0 & 0 & 1 \end{bmatrix} \begin{bmatrix} 1 & 0 & 0 \\ 0 & 1 & 0 \\ 0 & 1 & 1 \end{bmatrix} \begin{bmatrix} 1 & 0 & 0 \\ 0 & 1 & 0 \\ 0 & 0 & 2 \end{bmatrix}$$

$$\cdot \begin{bmatrix} 1 & 0 & 0 \\ 0 & 1 & 2 \\ 0 & 0 & 1 \end{bmatrix} \begin{bmatrix} 1 & 1 & 0 \\ 0 & 1 & 0 \\ 0 & 0 & 1 \end{bmatrix} \begin{bmatrix} 1 & 0 & 0 \\ 0 & 1 & 1 \\ 0 & 0 & 1 \end{bmatrix}. \qquad (2.1)$$

In general, a better elimination ordering strategy is to begin at the bottom of column 1 and eliminate upward, then repeat this with column 2, and so on. In this case the elimination (of entries) ordering is given by

$$(n,1)(n-1,1)\cdots(2,1)(n,2)(n-1,2)\cdots(3,2)\cdots(n,n-1),$$

which (in general) will reduce A to upper triangular form, say U. The same elimination ordering can be applied to U^T to reduce A to diagonal form. This elimination scheme has also been called "Neville elimination"; see, for example, [GP96].

For example, when $n = 4$, the complete elimination order is given by

$$(4,1)(3,1)(2,1)(4,2)(3,2)(4,3)(3,4)(2,3)(2,4)(1,2)(1,3)(1,4),$$

which is equivalent to (bidiagonal) factorization of the form:

$$L_4(\cdot)L_3(\cdot)L_3(\cdot)L_4(\cdot)L_3(\cdot)L_4(\cdot)DU_4(\cdot)U_3(\cdot)U_2(\cdot)U_4(\cdot)U_3(\cdot)U_4(\cdot). \qquad (2.2)$$

This particular ordering of the factors has been called a *successive elementary bidiagonal factorization* and a *Neville factorization*.

For brevity, we may choose to shorten the factorization in (2.2) by just referring to the subscripts that appear in each of the Ls and Us in (2.2). For example, to represent (2.2) we use

$$(432)(43)(4)D(4)(34)(234).$$

As in [JOvdD99], we refer to the product of the factors within () (like (234)) as a *stretch*.

Recall that an n-by-n matrix A is said to have an *LU factorization* if A can be written as $A = LU$, where L is an n-by-n lower triangular matrix and U is an n-by-n upper triangular matrix.

2.2 STANDARD ELEMENTARY BIDIAGONAL FACTORIZATION: INVERTIBLE CASE

Consider a 3-by-3 Vandermonde matrix V given by

$$V = \begin{bmatrix} 1 & 1 & 1 \\ 1 & 2 & 4 \\ 1 & 3 & 9 \end{bmatrix}. \qquad (2.3)$$

Expressing a matrix as a product of a lower triangular matrix L and an upper triangular matrix U is called an *LU-factorization*. Such a factorization is typically obtained by reducing a matrix to upper triangular form via row operations, that is, by Gaussian elimination. If all the leading principal minors of a matrix are nonzero, then Gaussian elimination can always proceed without encountering any zero pivots; see [HJ85, p. 160]. For example, we can write V in (2.3) as

$$V = LU = \begin{bmatrix} 1 & 0 & 0 \\ 1 & 1 & 0 \\ 1 & 2 & 1 \end{bmatrix} \begin{bmatrix} 1 & 1 & 1 \\ 0 & 1 & 3 \\ 0 & 0 & 2 \end{bmatrix}. \tag{2.4}$$

Such a factorization is unique if L is lower triangular with unit main diagonal, and U is upper triangular.

We use the elementary matrices $L_k(\mu)$ to reduce L to a diagonal matrix with positive diagonal entries. As was shown in the previous section, $V = L_3(1)L_2(1)L_3(1)U$. Applying similar row operations to the transpose U^T, we can write U as $U = DU_3(2)U_2(1)U_3(1)$, where $D = \text{diag}(1, 1, 2)$.

Hence, we can write the (Vandermonde) matrix V as

$$V = \begin{bmatrix} 1 & 0 & 0 \\ 0 & 1 & 0 \\ 0 & 1 & 1 \end{bmatrix} \begin{bmatrix} 1 & 0 & 0 \\ 1 & 1 & 0 \\ 0 & 0 & 1 \end{bmatrix} \begin{bmatrix} 1 & 0 & 0 \\ 0 & 1 & 0 \\ 0 & 1 & 1 \end{bmatrix} \begin{bmatrix} 1 & 0 & 0 \\ 0 & 1 & 0 \\ 0 & 0 & 2 \end{bmatrix}$$
$$\cdot \begin{bmatrix} 1 & 0 & 0 \\ 0 & 1 & 2 \\ 0 & 0 & 1 \end{bmatrix} \begin{bmatrix} 1 & 1 & 0 \\ 0 & 1 & 0 \\ 0 & 0 & 1 \end{bmatrix} \begin{bmatrix} 1 & 0 & 0 \\ 0 & 1 & 1 \\ 0 & 0 & 1 \end{bmatrix}. \tag{2.5}$$

We are leading to the main result that any InTN matrix has a bidiagonal factorization into InTN bidiagonal factors.

There are two very important impediments to proving that such a factorization exists via row operations using only EB matrices:

(1) to ensure that the property of being TN is preserved, and

(2) to avoid accidental cancellation.

Before we discuss these issues we should keep in mind that if A is invertible, then the reduced matrix obtained after a sequence of row operations is still invertible.

Item (1) was answered in the pioneering paper [Whi52], where she actually worked out the very crux of the existence of a bidiagonal factorization for InTN matrices, which is cited in [Loe55]. We now discuss in complete detail A. Whitney's key result and demonstrate its connections to the bidiagonal factorization of InTN matrices.

Theorem 2.2.1 (Whitney's key reduction result) *Suppose* $A = [a_{ij}]$ *has the property that* $a_{j1} > 0$, *and* $a_{t1} = 0$ *for all* $t > j + 1$. *Then* A *is TN if and only if* $L_{j+1}(-a_{j+1,1}/a_{j1})A$ *is TN.*

Proof. Let $B = L_{j+1}(-a_{j+1,1}/a_{j1})A$. Then any minor of B that does not involve row $j + 1$ is certainly nonnegative, since A is TN. In addition, any minor of B that includes both row j and row $j + 1$ is nonnegative since it is equal to the corresponding minor of A by properties of the determinant. Thus let

$$\mu = \det B[\{r_1, r_2, \ldots, r_k, j + 1, r_{k+2}, \ldots, r_t\}, \beta]$$

be any minor of B with $r_k \neq j$ and $|\beta| = t$. Then by the properties of the determinant it follows that

$$\mu = \det A[\{r_1, r_2, \ldots, r_k, j + 1, r_{k+2}, \ldots, r_t\}, \beta]$$
$$- \frac{a_{1,j+1}}{a_{1j}} \det A[\{r_1, r_2, \ldots, r_k, j, r_{k+2}, \ldots, r_t\}, \beta].$$

To show that $\mu \geq 0$, it is enough to verify the following equivalent determinantal inequality:

$$a_{1,j+1} \det A[\{r_1, \ldots, r_k, j, \ldots, r_t\}, \beta] \leq a_{1j} \det A[\{r_1, \ldots, r_k, j + 1, \ldots, r_t\}, \beta]. \tag{2.6}$$

To prove this inequality, first let us consider the case that $1 \notin \beta$. Then let

$$C = A[\{r_1, r_2, \ldots, r_k, j, j + 1, r_{k+2}, \ldots, r_t\}, 1 \cup \beta]. \tag{2.7}$$

In this case C is an $(t + 1)$-by-$(t + 1)$ TN matrix, and the inequality (2.6) is equivalent to proving that

$$c_{j+1,1} \det C(j + 1, 1) \leq c_{j1} \det C(j, 1). \tag{2.8}$$

To prove (2.8) we will instead prove a more general determinantal inequality for TN matrices, which can also be found in [GK02, p. 270].

Claim: Let $A = [a_{ij}]$ be an n-by-n TN matrix. For each i, j, let $A_{ij} = \det A(i, j)$. Then for $p = 1, 2, \ldots \lfloor n/2 \rfloor$ and $q = 1, 2, \ldots, \lfloor (n + 1)/2 \rfloor$ we have

$$a_{11}A_{11} - a_{21}A_{21} + \cdots - a_{2p,1}A_{2p,1} \leq \det A$$
$$\leq a_{11}A_{11} - a_{21}A_{21} + \cdots + a_{2q-1,1}A_{2q-1,1} \tag{2.9}$$

Proof of claim: Note it is enough to assume that the rank of A is at least $n - 1$; otherwise all terms in (2.9) are zero. Now observe that (2.9) is equivalent to

$$(-1)^{s+1} \sum_{i=s}^{n} (-1)^{i+1} a_{i1}A_{i1} \geq 0 \tag{2.10}$$

for $s = 1, 2, \ldots, n$. Moreover, (2.10) is equivalent to

$$(-1)^{s+1} \det(A_s) \geq 0, \tag{2.11}$$

where A_s is obtained from A by replacing the entries $a_{11}, a_{21}, \ldots, a_{s1}$ by 0. When $n = 2$, inequality (2.11) is obvious. Assume that (2.11) is valid for such matrices of order less than n.

Suppose that for some $\alpha \subset \{1, 2, \ldots, n\}$ with $|\alpha| = n - 2$ and $\beta = \{3, \ldots, n\}$, we have $\det A[\alpha, \beta] > 0$. If no such minor exists, then observe that to compute each A_{i1} we could expand along column 1 of A_{i1} (or column 2 of A), in which case each A_{i1} would be zero, and again all terms in (2.9) are zero.

Hence assume that $\{x, y\} = \{1, 2, \ldots, n\} \setminus \alpha$, where $x < s < y$; otherwise we could compute A_{i1} ($i \geq s$) by expanding along row $x < s$. Now permute the rows of A_s such that row x is first and row y is last, and call the new matrix A'_s. Then by Sylvester's determinantal identity applied to $A'[\alpha, \beta]$ we have

$$\det A'_s \det A'[\alpha, \beta]$$
$$= \det A'_s[x \cup \alpha, 1 \cup \beta] \det A'[x \cup \alpha, 2 \cup \beta]$$
$$\times \det A'_s[y \cup \alpha, 1 \cup \beta] \det A'[x \cup \alpha, 2 \cup \beta].$$

Taking into account the effect to the determinant by column switches, the above equality is the same as

$$\det A_s \det A[\alpha, \beta]$$
$$= \det A_s[x \cup \alpha, 1 \cup \beta] \det A[y \cup \alpha, 2 \cup \beta]$$
$$\times \det A_s[y \cup \alpha, 1 \cup \beta] \det A[x \cup \alpha, 2 \cup \beta].$$

By induction both $(-1)^s \det A_s[x \cup \alpha, 1 \cup \beta]$ ($x < s$) and $(-1)^{s-1} \det A_s[y \cup \alpha, 1 \cup \beta]$ ($s < y$) are nonnegative. Hence $(-1)^s \det A_s \geq 0$, which completes the proof of the claim. $\qquad \square$

Now if $1 \in \beta$, define C to be as above with the first column repeated; then the inequality (2.6) is equivalent to (2.8) for this new matrix C.

Finally, using (2.10) for the matrix C as in (2.7), and for $s = j$, we have

$$0 \leq (-1)^{j+1}((-1)^{j+1}c_{j1}C_{j1} + (-1)^{j+2}c_{j+1,1}C_{j+1,1}),$$

since $c_{k1} = 0$, for $k > j + 1$. Clearly the above inequality is exactly the inequality (2.6). This completes the proof.

Observe that there is a similar Whitney reduction result that can be applied to the first row of A by considing the transpose.

Using Whitney's reduction fact, it follows that an invertible TN matrix can be reduced to upper triangular form by applying consecutive row operations starting from the bottom of the leftmost column and working up that column, then shifting columns. This reduction to upper triangular form depends on the fact that there is no accidental cancellation along the way (see item (2) above). Because A. Whitney was the first (as far as we can tell) to prove that the property of TN is preserved under this row operation scheme, we will refer to this elimination scheme as the *Whitney elimination*, instead of, say, the Neville elimination as used by others (see, for example, [GP96]).

The next important topic to address is (item (2) above) how to avoid accidental cancellation when applying the Whitney elimination to an InTN matrix. The answer follows from the simple observation that if $A = [a_{ij}]$ is InTN, then $a_{ii} > 0$. This fact is easily verified, since if $a_{ii} = 0$ for some i, then by the double echelon form of any TN matrix (notice A cannot have any zero lines), A would have a block of zeros of size i-by-$n - i + 1$ (or $n - i + 1$-by-i), which implies that A is singular, as the dimensions of the zero block add to $n + 1$. Since $L_{j+1}(-a_{j+1,1}/a_{j1})A$ is InTN (Theorem 2.2.1) whenever A is InTN, we could never encounter the situation in which $a_{kj} = 0$ and $a_{k+i,j} = 0$ (assuming we are eliminating in the jth column). The reason is that in this case the kth row of this matrix would have to be zero, otherwise this matrix will have some negative 2-by-2 minor, and, in particular, would not be in double echelon form. Thus any InTN matrix can be reduced to upper triangular from (while preserving the property of InTN) via TN EB matrices; such a factorization using the Whitney elimination scheme will be called an *successive elementary bidiagonal factorization* (denoted by SEB). We now state in full detail the complete SEB factorization result for InTN matrices.

Theorem 2.2.2 *Any n-by-n invertible TN matrix can be written as*

$$(L_n(l_k)L_{n-1}(l_{k-1}) \cdots L_2(l_{k-n+2}))(L_n(l_{k-n+1}) \cdots L_3(l_{k-2n+4})) \cdots (L_n(l_1))$$
$$\cdot D(U_n(u_1))(U_{n-1}(u_2)U_n(u_3)) \cdots (U_2(u_{k-n+2}) \cdots U_{n-1}(u_{k-1})U_n(u_k)), (2.12)$$

where $k = \binom{n}{2}$; $l_i, u_j \geq 0$ for all $i, j \in \{1, 2, \ldots, k\}$; and $D = \mathrm{diag}(d_1, \ldots, d_n)$ is a diagonal matrix with all $d_i > 0$.

Proof. The proof is by induction on n. The cases $n = 1, 2$ are trivial. Assume the result is true for all InTN matrices of order less than n. Suppose A is an n-by-n InTN matrix. Then applying Whitney's reduction lemma to the first column of A and the first row of A, we can rewrite A as

$$A = L_n(l_k)L_{n-1}(l_{k-1}) \cdots L_2(l_{k-n+2}) \left[\begin{array}{c|c} \alpha & 0 \\ \hline 0 & A' \end{array} \right]$$
$$\times U_2(u_{k-n+2}) \cdots U_{n-1}(u_{k-1})U_n(u_k).$$

Here $\alpha > 0$ and A' is InTN. Now apply induction to A'. To exhibit the factorization in (2.12) simply direct sum each TN EB matrix with the matrix $[1]$ and the diagonal factor with $[\alpha]$. This completes the proof. □

For example, a 4-by-4 InTN matrix A can be written as

$$A = (L_4(l_6)L_3(l_5)L_2(l_4))(L_4(l_3)L_3(l_2))(L_4(l_1))D$$
$$\cdot (U_4(u_1))(U_3(u_2)U_4(u_3))(U_2(u_4)U_3(u_5)U_4(u_6)). \quad (2.13)$$

Theorem 2.2.2 grew out of the work in [Whi52], which contains a key preliminary reduction result involving certain row operations on TP matrices. This result appeared explicitly in [GP96] in connection with the so-called *Neville elimination*. It actually appeared first (without proof) in [Loe55], where it was referred to as *Whitney's theorem*. In any event, Theorem 2.2.2 is and continues to be an important and useful characterization of TP matrices; see also [FZ00a].

Part of the attraction of the SEB factorization is the natural connection to the Whitney elimination ordering. The multipliers that arise in each EB matrix can be expressed in terms of certain ratios of products of minors of the original matrix, which lead to important and useful determinantal test criterion (see Chapter 3, Section 1) for verifying if a given matrix is TP or InTN.

Specifically, for TP matrices it follows that no multiplier could ever be zero, since all minors are positive. Moreover, the converse holds, which we state as a corollary to Theorem 2.2.2.

Corollary 2.2.3 *An n-by-n matrix is TP if and only if it can be written as*

$$(L_n(l_k)L_{n-1}(l_{k-1})\cdots L_2(l_{k-n+2}))(L_n(l_{k-n+1})\cdots L_3(l_{k-2n+4}))\cdots(L_n(l_1))\cdot$$
$$D(U_n(u_1))(U_{n-1}(u_2)U_n(u_3))\cdots(U_2(u_{k-n+2})\cdots U_{n-1}(u_{k-1})U_n(u_k)), \quad (2.14)$$

where $k = \binom{n}{2}$; $l_i, u_j > 0$ *for all* $i, j \in \{1, 2, \ldots, k\}$; *and* $D = \mathrm{diag}(d_1, \ldots, d_n)$ *is a diagonal matrix with all* $d_i > 0$.

Example 2.2.4 (symmetric Pascal matrix) Consider the special case of the factorization (2.13) of a 4-by-4 matrix in which all the variables are equal to one. A computation reveals that the product (2.13) is then the matrix

$$A = \begin{bmatrix} 1 & 1 & 1 & 1 \\ 1 & 2 & 3 & 4 \\ 1 & 3 & 6 & 10 \\ 1 & 4 & 10 & 20 \end{bmatrix}, \quad (2.15)$$

which is necessarily TP.

On the other hand, consider the n-by-n matrix $P_n = [p_{ij}]$ whose first row and column entries are all ones, and for $2 \le i, j \le n$ let $p_{ij} = p_{i-1,j} + p_{i,j-1}$. The matrix P_n is called the *symmetric Pascal matrix* because of its connection with Pascal's triangle. When $n = 4$,

$$P_4 = \begin{bmatrix} 1 & 1 & 1 & 1 \\ 1 & 2 & 3 & 4 \\ 1 & 3 & 6 & 10 \\ 1 & 4 & 10 & 20 \end{bmatrix}.$$

Thus, P_4 is the matrix A in (2.15), so P_4 is TP. In fact, the relation $p_{ij} = p_{i-1,j} + p_{i,j-1}$ implies that P_n can be written as

$$P_n = L_n(1)\cdots L_2(1) \begin{bmatrix} 1 & 0 \\ 0 & P_{n-1} \end{bmatrix} U_2(1)\cdots U_n(1).$$

Hence, by induction, P_n has the factorization (2.12) in which the variables involved are all equal to one. Consequently, the symmetric Pascal matrix P_n is TP for all $n \ge 1$.

Note that in the above factorization each elementary bidiagonal matrix is in fact TN. If a matrix admits such a factorization, we say that A *has a successive bidiagonal factorization* and denote this by SEB.

The final important topic to discuss is the issue of uniqueness of the SEB factorization for TP and InTN matrices.

In general, the first part of an SEB factorization (reduction only to upper triangular form) that follows from the Whitney elimination ordering (or scheme) can be abbreviated to

$$(n, n-1, n-2, \ldots, 2)(n, n-1, \ldots, 3) \cdots (n, n-1)(n).$$

To prove the uniqueness of this factorization (assuming this order of the TN EB matrices), assume two such factorizations exist for a fixed TP or InTN matrix A. Starting from the left, suppose the first occurrence of a difference in a multiplier occurs in the jth factor in the product $(n, n-1, \ldots, k)$. Then cancel from the left all common EB matrices until we arrive at the EB matrix corresponding to the jth factor in $(n, n, -1, \ldots, k)$. So then the simplified factorization can be written as

$$(j, j-1, j-2, \ldots, k)(n, n-1, \ldots, k+1) \cdots (n, n-1)(n).$$

Suppose the multipliers in the EB matrices in the leftmost product above are denoted by (in order) $m_{j,k-1}, m_{j-1,k-1}, \ldots, m_{k,k-1}$. By construction we are assuming that the multiplier $m_{j,k-1}$ is different.

First observe that $j > k$, since for this simplified factorization the $(k, k-1)$ entry of A is uniquely determined by the factorization and is equal to $m_{j,k-1}$. Furthermore, the entries in positions $(k+1, k-1), (k+2, k-1), \ldots, (j, k-1)$ of A are easily computed in terms of the multipliers above. In particular, for $t = k, k+1, \ldots, j$ the entry in position $(t, k-1)$ of A is equal to $m_{k,k-1} m_{k+1,k-1} \cdots m_{t,k-1}$. Since the entry in position $(k, k-1)$ is uniquely determined and is equal to $m_{k,k-1}$, it follows, assuming that all multipliers involved are positive, that all subsequent multipliers (or EB factors) $m_{t,k-1}$ are the same in both factorizations by working backward from the entries in positions $(k, k-1)$ to $(j, k-1)$. Hence the multiplier $m_{j,k-1}$ must be the same in both factorizations, which is a contradiction.

In the event that some multiplier, say $m_{t,k-1} = 0$ for some $t \geq k$, then observe that $m_{t,k-1} = 0$ for all $t \geq k$. Otherwise (recalling item (2) above regarding no accidental cancellation) we must have encountered a zero row at some point in the reduction process, which contradicts the fact that A was invertible. Hence the multiplier $m_{j,k-1} = 0$ in both factorizations, again a contradiction.

A nice consequence of the existence of an SEB factorization for any n-by-n InTN matrices is that an InTN matrix can be written as a product of $n-1$ InTN tridiagonal matrices. This fact requires a short argument regarding the relations that exist for the matrix algebra generated by such EB matrices.

2.3 STANDARD ELEMENTARY BIDIAGONAL FACTORIZATION: GENERAL CASE

For singular TN matrices it is natural to ask if there are corresponding bidiagonal factorization results as in the case of InTN or TP. As a matter of fact, such results do exist although depending on the preponderance of singular submatrices or the rank deficiency, obtaining such a factorization can be problematic. However, in 1975 C. Cryer established in complete detail the existence of a bidiagonal factorization for any (singular) TN matrix into generalized TN EB matrices. Of course depending on the nature of the singularity, the bidiagonal factors may themselves be singular, so as a result we should longer expect in general that:

(1) the factorization to be unique,

(2) the factors to be EB matrices, or

(3) all the "singularity" to be absorbed into the middle diagonal factor.

Consider the following illustrative example.
Let

$$A = \begin{bmatrix} 1 & 0 & 0 \\ 0 & 0 & 0 \\ 1 & 0 & 1 \end{bmatrix}.$$

Then observe that A is a singular TN matrix. Moreover, notice that A can be factored into bidiagonal TN factors in the following two ways:

$$A = \begin{bmatrix} 1 & 0 & 0 \\ 0 & 0 & 0 \\ 0 & 1 & 0 \end{bmatrix} \begin{bmatrix} 1 & 0 & 0 \\ 1 & 1 & 0 \\ 0 & 0 & 1 \end{bmatrix} \begin{bmatrix} 1 & 0 & 0 \\ 0 & 0 & 1 \\ 0 & 0 & 0 \end{bmatrix}$$

and

$$A = \begin{bmatrix} 1 & 0 & 0 \\ 0 & 0 & 0 \\ 0 & 0 & 1 \end{bmatrix} \begin{bmatrix} 1 & 0 & 0 \\ 0 & 1 & 0 \\ 0 & 1 & 1 \end{bmatrix} \begin{bmatrix} 1 & 0 & 0 \\ 1 & 1 & 0 \\ 0 & 0 & 1 \end{bmatrix} \begin{bmatrix} 1 & 0 & 0 \\ 0 & 0 & 0 \\ 0 & 0 & 1 \end{bmatrix}.$$

The difficulties that arise when considering (1)–(3) above were all taken into account by Cryer when he proved his factorization result for TN matrices. In fact, Cryer was more interested in exhibiting a bidiagonal factorization than in determining the possible number, or order of the associated TN factors. It should be noted that Cryer did pay attention to the types of bidiagonal factors used to produce such a factorization, and in particular verified that the factors were generalized TN EB matrices. Also he was well aware that such a factorization was not unique.

As pointed out in the 1976 paper [Cry76] there are two major impediments when attempting to produce a bidiagonal factorization of a singular TN matrix:

(1) we may encounter a zero row, or

(2) we may encounter a "degenerate" column (i.e., a zero entry occurs above a nonzero entry).

Both (1) and (2) above could come about because of accidental cancellation. Consider the following examples:

$$
\begin{bmatrix} 1 & 1 & 0 \\ 2 & 2 & 0 \\ 1 & 3 & 1 \end{bmatrix} \rightarrow \begin{bmatrix} 1 & 1 & 0 \\ 2 & 2 & 0 \\ 0 & 2 & 1 \end{bmatrix} \rightarrow \begin{bmatrix} 1 & 1 & 0 \\ 0 & 0 & 0 \\ 0 & 2 & 1 \end{bmatrix},
$$

$$
\begin{bmatrix} 1 & 0 & 0 \\ 1 & 0 & 0 \\ 1 & 1 & 0 \end{bmatrix} \rightarrow \begin{bmatrix} 1 & 0 & 0 \\ 1 & 0 & 0 \\ 0 & 1 & 0 \end{bmatrix} \rightarrow \begin{bmatrix} 1 & 0 & 0 \\ 0 & 0 & 0 \\ 0 & 1 & 0 \end{bmatrix}.
$$

Notice that in the second case we have no way of eliminating the (3,2) entry.

Much of the analysis laid out in [Cry76] was to handle these situations. Before we give arguments to circumvent these difficulties, it is worth taking a moment to inspect the zero/nonzero pattern to determine the exact nature of the degeneracies that may occur. For example, if during the elimination process we observe that in column $i \geq 1$ a nonzero entry ends up immediately below a zero entry, then every entry in the row to the right of the zero must also be zero. Moreover, if there is a nonzero entry in column i above the identified zero entry in row k, then the entire kth row must be zero. Finally, if all entries above the nonzero in column i are zero, then there must be a zero in every position (s, t) with $s \leq k$ and $t \geq i$, namely in the upper right block formed by rows $1, 2, \ldots, k$ and column $i, i+1, \ldots, n$.

Assume when it comes time to eliminate in column $i \geq 1$ that we are in the following scenario:

(1) we have been able to perform the desired elimination scheme up to column i, and

(2) $a_{ki} > 0$, $a_{ti} = 0$, for $t > k$; $a_{li} = 0$ for $k > l > j \geq i$; and $a_{ji} > 0$.

Under these assumptions, we know that rows $j+1, j+2, \ldots, k-1$ must all be zero, in which case we (in a manner of speaking) can ignore these rows when devising a matrix to apply the required row operation. Let $K_m(\lambda)$ denote the m-by-m matrix given by $K_m(\lambda) = E_{11} + E_{mm} + \lambda E_{m1}$. To "eliminate" the (k, i) entry, observe that this matrix, call it A, can be factored as $A = LA'$, where $L = I \oplus K_{k-j+1}(a_{ki}/a_{ji}) \oplus I$, where the matrix $K_{k-j+1}(a_{ki}/a_{ji})$ occupies rows and columns $j, j+1, \ldots, k$. It is easy to check that A' has a zero in the (k, i) position, and still possesses the desired triangular form. Moreover, it is clear that L is TN, as $K_m(\lambda)$ is TN whenever $\lambda \geq 0$.

However, the matrix $I \oplus K_{k-j+1}(a_{ki}/a_{ji}) \oplus I$ is not a bidiagonal matrix! This situation can be easily repaired by noting that, for example,

$$
K_4(\lambda) = \begin{bmatrix} 1 & 0 & 0 & 0 \\ 0 & 1 & 0 & 0 \\ 0 & 0 & 1 & 0 \\ 0 & 0 & 1 & 0 \end{bmatrix} \begin{bmatrix} 1 & 0 & 0 & 0 \\ 0 & 0 & 0 & 0 \\ 0 & 1 & 1 & 0 \\ 0 & 0 & 0 & 1 \end{bmatrix} \begin{bmatrix} 1 & 0 & 0 & 0 \\ \lambda & 0 & 0 & 0 \\ 0 & 0 & 0 & 1 \\ 0 & 0 & 0 & 0 \end{bmatrix}.
$$

It is straightforward to determine that $K_m(\lambda)$ can be factored in a similar manner. Thus $K_m(\lambda)$ can be factored into TN generalized EB matrices.

The final vital observation that must be verified is that the resulting (reduced) matrix is still TN. This follows rather easily since zero rows do not affect membership in TN; so, ignoring rows $j + 1, \ldots, k$, the property of preserving TN is a consequence of Whitney's original reduction result (see Theorem 2.2.1).

Upon a more careful inspection of this case, we do need to consider separately the case $j < i$, since we are only reducing to upper triangular form. In this instance (as before) we note that rows $j + 1, \ldots, k - 1$ are all zero, and the procedure here will be to move row k to row i. In essence we will interchange row k with row i via a bidiagonal TN matrix, which is possible only because all rows in between are zero. For example, rows k and $k - 1$ can be interchanged by writing A as $A = GA'$, where A' is obtained from A by interchanging rows k and $k - 1$, and where G is the (singular) bidiagonal TN matrix given by $I \oplus G_2 \oplus I$, with

$$G_2 = \begin{bmatrix} 0 & 0 \\ 1 & 0 \end{bmatrix}.$$

Note that the matrix G_2 in the matrix G is meant to occupy rows and columns $k - 1, k$. The point here is to avoid non-TN matrices (such as the usual transposition permutation matrix used to switch rows), and to derive explicitly the factorization into a product of TN matrices. This is vital since both A and G are singular.

Finally, if we go back to A and notice that all entries in column i above a_{ki} are all zero, then we can apply similar reasoning as in the previous case and move row k to row i, and then proceed.

Up to this point we have enough background to state, and in fact prove, the result on the bidiagonal factorization of general TN matrices ([Cry76]).

Theorem 2.3.1 [Cry76] *Any n-by-n TN matrix A can be written as*

$$A = \prod_{i=1}^{M} L^{(i)} \prod_{j=1}^{N} U^{(j)}, \qquad (2.16)$$

where the matrices $L^{(i)}$ and $U^{(j)}$ are, respectively, lower and upper bidiagonal TN matrices with at most one nonzero entry off the main diagonal.

The factors $L^{(i)}$ and $U^{(j)}$ in (2.16) are not required to be invertible, nor are they assumed to have constant main diagonal entries, as they are generalized EB matrices.

We intend to take Theorem 2.3.1 one step farther and actually prove that every TN matrix actually has an SEB factorization, where now the bidiagonal factors are allowed to be singular. Taking into account the possible row/column interchanges and the fact that possibly non-bidiagonal matrices are used to perform the elimination (which are later factored themselves as was demonstrated earlier), there is a certain amount of bookkeeping required to guarantee that such a factorization exists. To facilitate the proof in general will use induction on the order of the matrix.

To begin consider (separately) the case $n = 2$. Let

$$A = \begin{bmatrix} a & b \\ c & d \end{bmatrix},$$

be an arbitrary 2-by-2 TN matrix. If $a > 0$, then

$$A = \begin{bmatrix} 1 & 0 \\ c/a & 1 \end{bmatrix} \begin{bmatrix} 1 & 0 \\ 0 & (ad-bc)/a \end{bmatrix} \begin{bmatrix} 1 & b/a \\ 0 & 1 \end{bmatrix},$$

and thus A has an SEB factorization into bidiagonal TN matrices.

On the other hand, if $a = 0$, then at least one of b or c is also zero. Suppose without loss of generality (otherwise consider A^T) that $b = 0$. Then in this case,

$$A = \begin{bmatrix} 0 & 0 \\ c & 1 \end{bmatrix} \begin{bmatrix} 1 & 0 \\ 0 & d \end{bmatrix} \begin{bmatrix} 1 & 0 \\ 0 & 1 \end{bmatrix},$$

and again A has an SEB factorization into bidiagonal TN matrices.

Recall that matrices of the form $D + sE_{i,i-1}$ or $D + tE_{j-1,j}$, where n is a positive integer, s, t are real numbers, and $2 \le i, j \le n$, are called GEB matrices. To avoid any confusion with EB matrices from the previous section we devise new notation for generalized bidiagonal matrices. For any $k \in \{2, \ldots, n\}$, we let $E_k(\mu) = [e_{ij}]$ denote the *elementary lower bidiagonal matrix* whose entries are

$$e_{ij} = \begin{cases} d_i & \text{if } i = j, \\ \mu & \text{if } i = k, \ j = k-1, \\ 0 & \text{otherwise,} \end{cases} \quad \text{that is, } E_k(\mu) = \begin{bmatrix} d_1 & 0 & \cdots & 0 \\ 0 & \ddots & \ddots & \vdots \\ \vdots & \mu & \ddots & 0 \\ 0 & \cdots & 0 & d_n \end{bmatrix}.$$

We are now in a position to state and prove that any TN matrix has an SEB factorization into generalized bidiagonal matrices of the form $E_k(\cdot)$.

Theorem 2.3.2 *Any n-by-n TN matrix can be written as*

$$(E_n(l_k)E_{n-1}(l_{k-1})\cdots E_2(l_{k-n+2}))(E_n(l_{k-n+1})\cdots E_3(l_{k-2n+4}))\cdots(E_n(l_1))$$
$$D(E_n^T(u_1))(E_{n-1}^T(u_2)E_n^T(u_3))\cdots(E_2^T(u_{k-n+2})\cdots E_{n-1}^T(u_{k-1})E_n^T(u_k)),$$

$$(2.17)$$

where $k = \binom{n}{2}$; $l_i, u_j \ge 0$ for all $i, j \in \{1, 2, \ldots, k\}$; and $D = \text{diag}(d_1, \ldots, d_n)$ is a diagonal matrix with all $d_i \ge 0$.

Proof. Suppose first that the $(1,1)$ entry of A is positive. We begin eliminating up the first column as governed by the Whitney ordering until we encounter the first instance of a zero entry (in the first column) immediately above a nonzero entry. If this situation does not arise, then we continue eliminating up the first column.

Suppose $a_{k1} > 0$ and $a_{k-1,1} = 0$. Assume that the first positive entry in column 1 above row k is in row j with $j \leq 1$, such a j exists since $a_{11} > 0$. Then as before all rows between the jth and kth must be zero. To 'eliminate' the $(k,1)$ entry observe that A can be factored as $A = LA'$, where $L = I \oplus K_{k-j+1}(a_{k1}/a_{j1}) \oplus I$, where the matrix $K_{k-j+1}(a_{k1}/a_{j1})$ occupies rows and columns $j, j+1, \ldots, k$.

Moreover, we observed that the matrix L can be factored as

$$L = E_k(1)E_{k-1}(1) \cdots E_2(a_{k1}/a_{j1}),$$

which is the correct successive order of generalized bidiagonal matrices needed to perform the Whitney ordering from rows k to j.

Now assume that the $(1,1)$ entry of A is zero. In this case the only remaining situation that could arise is if all of column 1 above row k is zero and $a_{k1} > 0$. Then rows $1, 2, \ldots, k-1$ are zero. Then we can interchange row k with row 1 by factoring A as

$$A = E_k(1)E_{k-1}(1) \cdots E_2(1)A',$$

by using the matrix G_2 in the appropriate positions. Again this is the correct successive order to go from rows k to 1 in the Whitney ordering.

After having completed the elimination procedure on column 1, we proceed by eliminating row 1 in a similar successive manner to achieve

$$A = E_n(l_k)E_{n-1}(l_{k-1}) \cdots E_2(l_{k-n+2}) \begin{bmatrix} \alpha & 0 \\ 0 & B \end{bmatrix}$$
$$\times (E_2^T(u_{k-n+2}) \cdots E_{n-1}^T(u_{k-1})E_n^T(u_k)),$$

where $\alpha \geq 0$ and B is TN. By induction B has an SEB factorization into generalized bidiagonal matrices.

If $\alpha = 0$, then direct sum each generalized bidiagonal factor in an SEB factorization of B with a 1-by-1 zero matrix to obtain an SEB factorization of A. Notice that in this case both row and column 1 of A must have been zero. If $\alpha > 0$, then direct sum each generalized bidiagonal factor in an SEB factorization of B with a 1-by-1 matrix whose entry is a one except for the diagonal factor, which is to be adjoined (by direct summation) with a 1-by-1 matrix whose entry is α. In each case we then obtain an SEB factorization of A, as desired. □

Thinking more about the EB matrices L_i and U_j there are in fact relations satisfied by these matrices that are important and, for example, have been incorporated into the work in [BFZ96]. Some of these relations are:

(1) $L_i(s)L_j(t) = L_j(t)L_i(s)$, if $|i-j| > 1$;

(2) $L_i(s)L_i(t) = L_i(s+t)$, for all i;

(3) $L_{i+1}(r)L_i(s)L_{i+1}(t) = L_i(r')L_{i+1}(s')L_i(t')$, for each i and where for $r, s, t > 0$ we have that $r', s', t' > 0$; and

(4) $L_i(s)U_j(t) = U_j(t)L_i(s)$, if $i \neq j$.

Naturally, relations (1)–(3) also exist for the upper EB matrices U_j. Relations (1)–(3) are simple computations to exhibit. Relation (4) is also straightforward when $i \neq j$, but if $i = j$, then a rescaling by a positive diagonal factor is required. Hence, in this case we have a relation involving GEB matrices. For example, $L_i(s)U_i(t) = DU_i(t')L_i(s')$, and if $s, t > 0$, then $s', t' > 0$.

As a passing note, we acknowledge that there have been advances on bidiagonal factorizations of rectangular TN matrices, but we do not consider this topic here. For those interested, consult [GT06a, GT06b, GT08].

2.4 *LU* FACTORIZATION: A CONSEQUENCE

It is worth commenting that the SEB factorization results discussed in the previous sections are extensions of *LU* factorization. However, it may be more useful or at least sufficient to simply consider *LU* factorizations rather than SEB factorizations.

The following remarkable result is, in some sense, one of the most important and useful results in the study of TN matrices. This result first appeared in [Cry73] (although the reduction process proved in [Whi52] is a necessary first step to an *LU* factorization result) for nonsingular TN matrices and later was proved in [Cry76] for general TN matrices; see also [And87] for another proof of this fact.

Theorem 2.4.1 *Let A be an n-by-n matrix. Then A is (invertible) TN if and only if A has an LU factorization such that both L and U are n-by-n (invertible) TN matrices.*

Using Theorem 2.4.1 and the Cauchy-Binet identity we have the following consequence for TP matrices.

Corollary 2.4.2 *Let A be an n-by-n matrix. Then A is TP if and only if A has an LU factorization such that both L and U are n-by-n ΔTP matrices.*

Proof. Since all the leading principal minors of A are positive, A has an *LU* factorization; the main diagonal entries of L may be taken to be 1, and the main diagonal entries of U must then be positive. Observe that

$$0 < \det A[\{1, 2, \ldots, k\}, \beta] = \sum_{\gamma} \det L[\{1, 2, \ldots, k\}, \gamma] \det U[\gamma, \beta]$$

$$= \det L[\{1, 2, \ldots, k\}] \det U[\{1, 2, \ldots, k\}, \beta]$$

$$= \det U[\{1, 2, \ldots, k\}, \beta].$$

The fact that $\det L[\alpha, \{1, 2, \ldots, k\}] > 0$ if $|\alpha| = k$ follows by applying similar arguments to the minor $\det A[\alpha, \{1, 2, \ldots, k\}]$. To conclude that both L and U are ΔTP we appeal to Proposition 3.1.2 in Chapter 3. □

That any TN matrix can be factored as a product of TN matrices L and U is, indeed, important, but for the following application (see also the section on subdirect sums in Chapter 10) we need rather more than what is contained in Theorem 2.4.1. When the TN matrix A has the form

$$A = \begin{bmatrix} A_{11} & A_{12} & 0 \\ A_{21} & A_{22} & A_{23} \\ 0 & A_{32} & A_{33} \end{bmatrix},$$

we need that, in addition to being lower and upper triangular, respectively, the (correspondingly partitioned) L has its 3,1 block and U its 1,3 block equal to zero. In case A_{11} is invertible, a simple partitioned calculation reveals that this must occur in any LU factorization of A (since L_{11} and U_{11} will be invertible). However, when A_{11} is singular, there can occur TN matrices A of our form that have TN LU factorizations with positive entries in the L_{31} or U_{13} blocks. Fortunately, though, LU factorizations will be highly nonunique in this case, and there will always exist ones of the desired form. Thus, an auxiliary assumption that A_{11} is invertible would avoid the need for Lemma 2.4.3, but this somewhat specialized lemma (perhaps of independent interest, though it requires an elaborate proof) shows that such an auxiliary assumption is unnecessary.

Let

$$F_2 = \begin{bmatrix} 0 & 0 \\ 1 & 1 \end{bmatrix},$$

and let $K_r(\lambda)$, $c \geq 0$, denote the r-by-r matrix

$$K_r(\lambda) = \begin{bmatrix} 1 & 0 & 0 \\ 0 & 0_{r-2} & 0 \\ \lambda & 0 & 1 \end{bmatrix}.$$

Observe that both of the above matrices are TN.

Lemma 2.4.3 *Let*

$$A = \begin{bmatrix} A_{11} & A_{12} & 0 \\ A_{21} & A_{22} & A_{23} \\ 0 & A_{32} & A_{33} \end{bmatrix},$$

in which A_{11}, A_{22}, and A_{33} are square. Suppose A is TN. Then

$$A = LU = \begin{bmatrix} L_{11} & 0 & 0 \\ L_{21} & L_{22} & 0 \\ 0 & L_{32} & L_{33} \end{bmatrix} \cdot \begin{bmatrix} U_{11} & U_{12} & 0 \\ 0 & U_{22} & U_{23} \\ 0 & 0 & U_{33} \end{bmatrix},$$

in which L and U (partitioned conformally with A) are both TN.

Proof. The proof is by induction on the size of A. The case $n = 2$ is trivial. For completeness, however, we handle the case $n = 3$ separately. Suppose A

is a 3-by-3 matrix of the form

$$A = \begin{bmatrix} a_{11} & a_{12} & 0 \\ a_{21} & a_{22} & a_{23} \\ 0 & a_{32} & a_{33} \end{bmatrix}.$$

There are two cases to consider. Suppose $a_{11} > 0$. Then A can be written as

$$A = \begin{bmatrix} 1 & 0 & 0 \\ a_{21}/a_{11} & 1 & 0 \\ 0 & 0 & 1 \end{bmatrix} \cdot \begin{bmatrix} a_{11} & 0 & 0 \\ 0 & * & * \\ 0 & * & * \end{bmatrix} \cdot \begin{bmatrix} 1 & a_{12}/a_{11} & 0 \\ 0 & 1 & 0 \\ 0 & 0 & 1 \end{bmatrix}.$$

Now we can apply induction to the bottom right block of the matrix in the middle. On the other hand if $a_{11} = 0$, then we can assume that $a_{12} = 0$, in which case A can be written as

$$A = (F_2 \oplus [1]) \begin{bmatrix} a_{21} & 0 & 0 \\ 0 & a_{22} & a_{23} \\ 0 & a_{32} & a_{33} \end{bmatrix}.$$

Again we can apply induction to the bottom right block of the second factor above. The general case is handled similarly.

If the (1,1) entry of A is positive, then no row or column switches are needed to reduce A to

$$A = E_k(l_k)E_{k-1}(l_{k-1})\cdots E_2(l_2) \begin{bmatrix} \alpha & 0 \\ 0 & A' \end{bmatrix} E_2^T(l_2)E_3^T(l_3)\cdots E_k^T(l_k).$$

Here we allow the possibility of some of the generalized EB matrices to be equal to I, and note that A' will have the same form as A. Hence induction applies.

On the other hand if the (1,1) entry of A is zero, then we may assume without loss of generality that the entire first row of A is zero. If column 1 of A is zero, then apply induction. Otherwise suppose that $a_{j1} > 0$ is the first positive entry in column 1. Then A can be reduced without row interchanges to

$$A' = \left[\begin{array}{c|ccc} 0 & 0 & \cdots & 0 \\ \hline \vdots & & & \\ a_{j1} & & A'' & \\ 0 & & & \end{array} \right].$$

Now we can apply the matrix F_2, as used in the case $n = 3$, to reduce A' to

$$\left[\begin{array}{c|ccc} a_{j1} & 0 & \cdots & 0 \\ \hline \vdots & & & \\ 0 & & A'' & \\ 0 & & & \end{array} \right].$$

Now induction can be applied to the matrix A''. In either case we obtain the desired LU factorization. This completes the proof. $\square$

2.5 APPLICATIONS

As indicated by the lengthy discussion, SEB factorizations of TN matrices are not only important in their own right, but they also arise in numerous applications (see [GM96]).

Perhaps the most natural application is simply generating matrices in this class. As noted earlier constructing TP or TN matrices directly from the definition is not only difficult but extremely tedious. The number of conditions implied by ensuring all minors are positive are too numerous to check by hand. However, armed with the SEB factorizations for TN and TP matrices (see Theorem 2.3.2 and Corollary 2.2.3), to generate TN or TP matrices, it suffices to choose the parameters that arise in these factorizations nonnegative or positive, respectively. By this we mean that if all multipliers in an SEB factorization are chosen to be positive along with the diagonal factor, then the resulting product is guaranteed to be TP; while if all multipliers in (2.17) are chosen to be nonnegative, then the resulting product is TN.

Similarly, if the diagonal factor is chosen to be positive in (2.12) and all the multipliers are nonnegative, then the resulting product will be InTN. A subtle issue along these lines is when we can be sure that such an InTN matrix, constructed as above, is actually oscillatory?

In fact the answer is known and will be described completely in the next section. It was shown in [GK02, p. 100] that an InTN matrix A is oscillatory if and only if $a_{i,i+1}, a_{i+1,i} > 0$ for each i.

Another less obvious (but not difficult) application of SEB factorizations is the classical fact (see [GK02, Kar65, Whi52, And87]) that the topological closure of the TP matrices is the TN matrices. We are only concerned with Euclidean topology in this case. Most proofs of this fact require some clever fiddling with the entries of the matrix (often via multiplication with a certain TP matrix– the Polya matrix). Again using only the SEB factorization, the proof of this closure result is immediate. For any TN matrix, write a corresponding SEB factorization into generalized TN EB matrices. Then perturb each such generalized EB matrix by making the multiplier positive (and small), and by making the diagonal entries of each factor positive and small. After this we can pass a positive diagonal factor through the Ls and Us to ensure that each generalized EB matrix is actually an EB TN matrix. By Corollary 2.2.3, the resulting product must be TP. Moreover, this product can be chosen to be as "close" to A as we wish.

A less useful but straightforward implication is that if A is InTN, then $S_n A^{-1} S_n$ is InTN. This result follows trivially from the SEB factorization because $L_k(\alpha)^{-1} = L_k(-\alpha)$, so that $S_n L_k(\alpha)^{-1} S_n = L_k(\alpha)$, which is a TN EB matrix. The same argument applies to the EB matrices U_j, and observe that $S_n D^{-1} S_n$ is a positive diagonal matrix. Hence it follows that $S_n A^{-1} S_n$ is InTN.

Another application we present in this section deals with a specific issue from one-parameter semigroups. In [Loe55] the bidiagonal factorization (2.12) of TN matrices was used to show that for each nonsingular n-by-n TN matrix A there is a piecewise continuous family of matrices $\Omega(t)$ of a special form such that the unique solution of the initial value problem

$$\frac{dA(t)}{dt} = \Omega(t)A(t), \ A(0) = I \tag{2.18}$$

has $A(1) = A$.

Loewner was interested in the theory of one-parameter semigroups. Along these lines, let $A(t) = [a_{ij}(t)]$ be a differentiable matrix-valued function of t such that $A(t)$ is nonsingular and TN for all $t \in [0,1]$, and $A(0) = I$. From this

$$\Omega \equiv \left(\frac{dA(t)}{dt}\right)_{t=0} \tag{2.19}$$

is often referred to as an *infinitesimal element* of the semigroup of nonsingular TN matrices.

Loewner observed that an n-by-n matrix $\Omega = [\omega_{ij}]$ is an infinitesimal element if and only if

$$\omega_{ij} \geq 0 \text{ whenever } |i - j| = 1 \text{ and } \omega_{ij} = 0 \text{ whenever } |i - j| > 1. \tag{2.20}$$

He then showed that the nonsingular TN matrices obtained from a constant $\Omega(t) \equiv \Omega = [\omega_{ij}]$ satisfying (2.20) with only one nonzero entry are either (a) diagonal matrices with positive main diagonal entries or (b) the elementary bidiagonal matrices $L_i(\omega_{ij}t)$ or $U_j(\omega_{ij}t)$ in which $\omega_{ij}t \geq 0$. From here he applied Theorem 2.2.2 to show that every nonsingular TN matrix can be obtained from the solution of the initial value problem (2.18).

Recall the symmetric Pascal matrix P_n in Example 2.2.4. Consider the solution $A(t)$ of (2.18) in which $\Omega(t) \equiv H$, the *creation* or *derivation matrix*, given by the lower bidiagonal TN matrix

$$H = \begin{bmatrix} 0 & & & & \\ 1 & \ddots & & & \\ & 2 & \ddots & & \\ & & \ddots & \ddots & \\ & & & n-1 & 0 \end{bmatrix}.$$

Then $A(t) = e^{Ht} = \sum_{k=0}^{n-1} \frac{t^k}{k!} H^k$ is a polynomial of degree $n-1$ in Ht, and the *Pascal matrix* $P = A(1) = [\pi_{ij}]$, necessarily TN and nonsingular, has entries

$$\pi_{ij} = \begin{cases} \binom{i}{j} & \text{for } i \geq j, \\ 0 & \text{otherwise.} \end{cases}$$

The symmetric Pascal matrix is $P_n = PP^T$, so it is also TN; see [AT01] for more details.

The final application we present here deals with computer-aided geometric design and corner cutting algorithms. Over the past years, there has been a considerable amount of research conducted on developing algorithms in computer-aided geometric design (CAGD) that may be classified as corner cutting algorithms. An example of such algorithms makes use of evaluating a polynomial curve using the Bernstein basis. It is known (see [MP99]) that the Bernstein basis is a basis for the space of all polynomials of degree at most n with optimal shape preserving properties, which is due, in part, to the variation diminishing property of TN matrices (see also Chapter 4). It turns out that bidiagonal matrices and bidiagonal factorization have an important connection to corner cutting algorithms. Following the description in [MP99], an *elementary corner cutting* may be viewed as a mapping of a polygon with vertices $x_0, x_1, \ldots, x_n$ into another polygon with vertices $y_0, y_1, \ldots, y_n$ defined by: for some $0 \leq i \leq n-1$, $y_i = (1-\lambda)x_i + \lambda x_{i+1}$ ($0 \leq \lambda < 1$) and $y_j = x_j$ for $j \neq i$; or for some $1 \leq i \leq n$, $y_i = (1-\lambda)x_i + \lambda x_{i-1}$ ($0 \leq \lambda < 1$) and $y_j = x_j$ for $j \neq i$. Such corner cuttings may be defined in terms of GEB TN matrices, and since a corner cutting algorithm is a composition of elementary corner cuttings, such algorithms may be described by a product of corresponding GEB TN matrices, and hence this product is TN. The reader is urged to consult [MP99] and other related references where corner cutting algorithms are described.

2.6 PLANAR DIAGRAMS AND EB FACTORIZATION

As discussed earlier, TN matrices arise in many areas of mathematics. Perhaps one of the most notable interplays comes from TP matrices and combinatorics. This connection, which will be discussed in more detail shortly, seems to be explicit (in some form) in [Bre95], although some of his analysis builds on existing fundamental results, such as Lindstrom's work on acyclic weighted directed graphs to compute determinants.

We now discuss the connection between SEB factorizations and planar diagrams or networks by introducing the associated diagrams, stating and verifying Lindström's lemma, and then using it to demonstrate how to calculate minors of the corresponding matrix. We also direct the reader to the important work of on transition matrices associated with birth and death processes (see [KM59]).

Roughly speaking we associate a simple weighted digraph to each EB matrix and diagonal factor. Then from this we will obtain a larger weighted directed graph corresponding to an SEB factorization by concatenation of each smaller diagram.

An excellent treatment of the combinatorial and algebraic aspects of bidiagonal factorization of TN matrices along with generalizations for totally positive elements in reductive Lie groups is given in [BFZ96, FZ99, FZ00a]. One of the central tools used in these papers and others is a graphical representation of the bidiagonal factorization in terms of planar diagrams (or networks) that can be described as follows.

An n-by-n diagonal matrix $\operatorname{diag}(d_1, d_2, \ldots, d_n)$ is represented by the elementary diagram on the left in Figure 2.1, while an elementary lower (upper) bidiagonal matrix $L_k(l)$ $(U_j(u))$ is the diagram middle (right) of Figure 2.1.

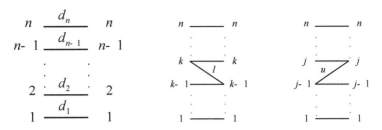

Figure 2.1 Elementary diagrams

Each horizontal edge of the last two diagrams has a weight of 1. It is not difficult to verify that if A is a matrix represented by any one of the diagrams in the figure, then $\det A[\{i_1, i_2, \ldots, i_t\}, \{j_1, j_2, \ldots, j_t\}]$ is nonzero if and only if in the corresponding diagram there is a family of t vertex-disjoint paths joining the vertices $\{i_1, i_2, \ldots, i_t\}$ on the left side of the diagram with the vertices $\{j_1, j_2, \ldots, j_t\}$ on the right side. Moreover, in this case this family of paths is unique, and $\det A[\{i_1, i_2, \ldots, i_t\}, \{j_1, j_2, \ldots, j_t\}]$ is equal to the product of all the weights assigned to the edges that form this family. Given a planar network and a path connecting a source and a sink, the *weight* of this path is defined to be the product of the weights of its edges, and the *weight of a collection* of directed paths is the product of their weights.

To extend this notion further, we consider the case $n = 3$ for clarity of exposition. The general case is a straightforward extension of this case and is omitted here. Interested readers may consult [FZ99, Bre95, BFZ96, FZ00a] and other references.

One of the benefits of considering these weighted planar networks or planar diagrams (or acyclic directed graphs) is the connection to computing minors of the matrices associated with a given planar network. To this end, consider the product of two 3-by-3 EB matrices

$$L_3(a)L_2(b) = \begin{bmatrix} 1 & 0 & 0 \\ b & 1 & 0 \\ ab & a & 1 \end{bmatrix}.$$

Recall the Cauchy-Binet determinantal identity

$$\det L_3(a)L_2(b)[\alpha, \beta] = \sum_\gamma \det L_3(a)[\alpha, \gamma] \det L_2(b)[\gamma, \beta].$$

As observed above, $\det L_3(a)[\alpha, \gamma]$ is obtained by calculating the (sum of the) weights of all collections of vertex disjoint paths from rows (source set) α to columns (sink set) γ. Similarly, $\det L_2(b)[\gamma, \beta]$ is obtained by calculating the (sum of the) weights of all collections of vertex disjoint paths from rows (source set) γ to columns (sink set) β. Thus as γ runs through all k subsets of $\{1, 2, 3\}$ (here $k = |\alpha| = |\beta|$), each such summand (or product) $\det L_3(a)[\alpha, \gamma] \det L_2(b)[\gamma, \beta]$ constitutes the weight of a collection of vertex disjoint paths from source set α to an intermediate set γ, and then to sink set β. Such a collection of paths can be realized as a collection of vertex disjoint paths in the diagram obtained by concatenating the diagrams (in order) associated with $L_3(a)$ and $L_2(b)$ (see Figure 2.2). In other words,

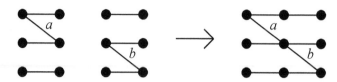

Figure 2.2 Concatenation of elementary diagrams

$\det L_3(a)L_2(b)[\alpha, \beta]$ is equal to the sum of the weights of all families of vertex-disjoint paths joining the vertices α on the left of the concatenated diagram with the vertices β on the right.

This analysis can be extended to an arbitrary finite product of EB matrices in the following manner. Given a product $A = A_1 A_2 \cdots A_l$ in which each matrix A_i is either a diagonal matrix or an (upper or lower) EB matrix, a corresponding diagram is obtained by concatenation left to right of the diagrams associated with the matrices $A_1, A_2, \ldots, A_l$. Then the Cauchy-Binet formula for determinants applied to the matrix A implies the next two results (which are sometimes called Lindstrom's lemma).

Proposition 2.6.1 *Suppose $A = A_1 A_2 \cdots A_l$ in which each matrix A_i is either a diagonal matrix or an (upper or lower) EB matrix. Then:*

(1) *Each (i, j)-entry of A is equal to the sum over all paths joining the vertex i on the left side of the obtained diagram with the vertex j on the right side of products of all the weights along the path.*

(2) *$\det A[\{i_1, i_2, \ldots, i_t\}, \{j_1, j_2, \ldots, j_t\}]$ is equal to the sum over all families of vertex-disjoint paths joining the vertices $\{i_1, i_2, \ldots, i_t\}$ on the left of the diagram with the vertices $\{j_1, j_2, \ldots, j_t\}$ on the right of products of all the weights assigned to edges that form each family.*

For example, in an arbitrary SEB factorization of a 3-by-3 InTN matrix given by

$$A = L_3(l_3)L_2(l_2)L_3(l_1)DU_3(u_1)U_2(u_2)U_3(u_3),$$

the associated planar network is constructed as shown in Figure 2.3.

In particular, the factorization in (2.1) can be written as

$$V = L_3(1)L_2(1)L_3(1)DU_3(2)U_2(1)U_3(1),$$

and the planar network corresponding to this factorization of V is presented in Figure 2.4.

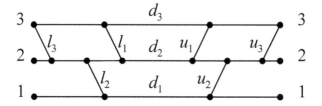

Figure 2.3 General 3-by-3 planar network

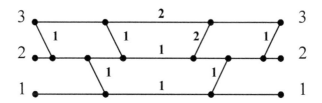

Figure 2.4 Planar network corresponding to the factorization (2.1)

The edges in Figure 2.4 are all oriented left to right. This network contains 3 sources at the left and 3 sinks at the right. Each slanted edge corresponds to the positive entry off the main diagonal of one TN EB matrix in the factorization (2.1). The order of these edges coincides with the order of the factors in (2.1). All horizontal edges have weight one except, possibly, the horizontal edges in the middle of Figure 2.4 that correspond to the main diagonal entries of D.

Theorems 2.2.2 and 2.3.2 imply that every TN matrix can be represented by a weighted planar diagram constructed from building blocks as in Figure 2.1 (in the case of degenerate matrices some of the horizontal edges may have to be erased).

A minor extension along these lines is to associate TN matrices with more general weighted planar digraphs (see [Bre95]). One defines a *planar diagram of order n* as a planar acyclic digraph $\mathcal{D}$ with all edges oriented from left to right and $2n$ distinguished boundary vertices: n sources on the left and n sinks on the right with both sources and sinks labeled $1, \ldots, n$ from bottom to top. To each edge of a planar diagram $\mathcal{D}$ we assign a positive weight, and we denote by W the collection of all weights of the edges. We refer to the pair $(\mathcal{D}, W)$ as a *weighted planar diagram of order n*. Clearly, the diagrams in Figure 2.1 and concatenations thereof are weighted planar diagrams of order n.

Suppose $(\mathcal{D}, W)$ is a given weighted planar diagram of order n. Then we may apply the first statement in Proposition 2.6.1 as a *definition* of an n-by-n matrix $A(\mathcal{D}, W)$ associated with $(\mathcal{D}, W)$. It was first observed in [KM59] that the second statement remains valid in this more general situation and, in particular, $A(\mathcal{D}, W)$ is TN.

A wonderfully useful byproduct to considering planar networks associated to TN, InTN, or TP matrices is that, in this setting, **every** minor of such a matrix is a *subtraction-free* polynomial in the parameters associated with the EB matrices that make up the SEB factorization. For example, if A is represented by the planar network in Figure 2.3, then

$$\det A[23, 23] = d_2 d_3 + d_1 d_3 l_2 u_2 + d_1 d_2 l_1 u_1 l_2 u_2.$$

This consequence is extremely important and particularly useful for studying determinantal inequalities among minors of TN matrices (see also Chapter 6).

For example, consider Fischer's determinantal inequality, which can be stated as follows. Let $A \in M_n$ be TN and let $\alpha \subset \{1, 2, \ldots, n\}$ be any index set. Then

$$\det A \leq \det A[\alpha] \det A[\alpha^c],$$

where α^c denotes the complement of α relative to $\{1, 2, \ldots, n\}$.

The proof of Fischer's inequality given in [GK02, p. 131] relies on Sylvester's determinantal identity and the fact that the TP matrices are dense in the set of TN matrices. Here is a proof that uses the bidiagonal factorization for TP matrices.

Fix an index set α, and factor A as in (2.17). By the Cauchy-Binet identity and the fact that each elementary bidiagonal matrix has ones on the main diagonal we have

$$
\begin{aligned}
&\det A[\alpha] \det A[\alpha^c] \\
&= (\det D[\alpha] + (\text{nonnegative terms}))(\det D[\alpha^c] + (\text{nonnegative terms})) \\
&\geq \det D[\alpha] \det D[\alpha^c] = \det D = \det A,
\end{aligned}
$$

which completes the proof.

Hadamard's inequality for TN can be derived from Fischer's inequality as follows. Fischer's inequality ensures that $\det A \leq \det A[\{1, \ldots, n-1\}] a_{nn}$ for any TN matrix $A = [a_{ij}]$. Now apply induction to the TN submatrix $A[\{1, \ldots, n-1\}]$.

Before we move on to further results, we first present a proof of the fact that if all the multipliers are positive in the SEB factorization (2.14), then the resulting product is totally positive. Notice that it is sufficient to prove that if all the multipliers l_i are positive, then the resulting product is a lower triangular ΔTP matrix. When $n = 2$, this result is immediate. So assume it is true for all such lower triangular matrices of order less than n. Suppose L is a lower triangular matrix obtained from multiplying all the lower TN EB matrices in (2.14) in which all the multipliers are positive. Let Γ be the corresponding planar diagram for L. To prove that L is ΔTP, it is enough to exhibit the existence of a collection of vertex-disjoint paths from source set $\alpha = \{i_1, \ldots, i_k\}$ to sink set $\beta = \{j_1, \ldots, j_k\}$ with the property that $i_t \geq j_t$ for $t = 1, 2, \ldots, k$ in Γ. Since all possible slanted edges are present, it follows from induction (here we only consider the section Γ after the stretch $(n, n-1, \ldots, 2)$ and ignoring the bottom level associated to the vertex 1) that there exists a collection of vertex-disjoint paths from α to β whenever $j_1 > 1$. Thus there are two cases remaining: (i) $1 \in \alpha$ (and hence $1 \in \beta$); or (ii) $1 \notin \alpha$ and $1 \in \beta$. In case (i) the only path from vertex 1 to vertex 1 is along the bottom of Γ, so combining this unique path with the collection of vertex disjoint paths from $\alpha \backslash \{1\}$ to $\beta \backslash \{1\}$, which is guaranteed by induction, we have the existence of a collection of vertex-disjoint paths from α to β. In case (ii) send i_1 to one via the shortest possible path using the first slanted edges possible, and send $\alpha \backslash \{i_1\}$ to $\beta \backslash \{1\}$ by first passing by the stretch corresponding to $(n, n-1, \ldots, 2)$ and then applying induction. As before, this implies existence of a collection of vertex-disjoint paths from α to β. Hence $\det L[\alpha, \beta] > 0$ for all such $\alpha = \{i_1, \ldots, i_k\}$ to sink set $\beta = \{j_1, \ldots, j_k\}$ with the property that $i_t \geq j_t$ for $t = 1, 2, \ldots, k$ in Γ. Therefore L is ΔTP.

A common theme in this book is to interchange the notions of SEB factorizations and planar diagrams when dealing with various properties of TN matrices. In particular, we will often think of a TN matrix by merely representing it as a general planar network (as in Figure 2.3 for $n = 3$). Given such a correspondence, we can recast Theorems 2.2.2 and 2.3.2 and Corollary 2.2.3 in terms of planar networks.

Theorem 2.6.2 *If Γ is a planar network corresponding to (2.12), then:*

(1) The associated matrix is InTN if and only if all the weights of the EB parameters are nonnegative and all $d_i > 0$.

(2) The associated matrix is TP if and only if all the weights of the EB parameters are positive and all $d_i > 0$.

Theorem 2.6.3 *If Γ is a planar network corresponding to (2.17), then the associated matrix is TN if and only if all the weights of the generalized EB parameters are nonnegative (including the main diagonal of each factor) and all $d_i \geq 0$.*

We close this section with a short subsection illustrating the connections between SEB factorizations and oscillatory matrices which were eluded to in Section 2.5.

2.6.1 EB Factorization and Oscillatory Matrices

It is evident from a historical perspective that oscillatory matrices (or IITN matrices) play a key role both in the development of the theory of TN matrices, and also in their related applications (see [GK02], for example). Incorporating the combinatorial viewpoint from the previous section, we recast some of the classical properties of oscillatory matrices. We begin with a characterization (including a proof) of IITN matrices described entirely by the SEB factorization (see [Fal01]).

Theorem 2.6.4 *Suppose A is an InTN matrix with corresponding SEB factorization given by (2.12). Then A is oscillatory if and only if at least one of the multipliers from each of $L_n, L_{n-1}, \ldots, L_2$ and each of $U_n, U_{n-1}, \ldots, U_2$ is positive.*

Proof. Suppose all the multipliers from some L_k are zero. Then there is no path from k to $k-1$ in the planar network associated with (2.12). Hence the entry in position $k, k-1$ of A must be zero, which implies A is not oscillatory.

For the converse, suppose that for every L_i, U_j there is at least one positive multiplier (or parameter). Then since A is invertible all horizontal edges of all factors are present, and hence there exists a path from k to both $k-1$ and $k+1$ for each k (except in extreme cases where we ignore one of the paths). In other words, $a_{ij} > 0$ whenever $|i - j| \leq 1$. Thus it follows that A is oscillatory. $\qquad\square$

In their original contribution to this subject, Gantmacher and Krein did establish a simple yet informative criterion for determining when a TN matrix was oscillatory, which we state now.

Theorem 2.6.5 (GK) (Criterion for being oscillatory) *An n-by-n TN matrix $A = [a_{ij}]$ is oscillatory if and only if*

    (i) A is nonsingular and

    (ii) $a_{i,i+1} > 0$ and $a_{i+1,i} > 0$ for $i = 1, 2, \ldots, n-1$. $\qquad$ (2.21)

As presented in [GK02] their proof makes use of some minor technical results. For example, they needed to define a new type of submatrix referred to as *quasi-principal* to facilitate their analysis. It should be noted that part of the purpose of examining quasi-principal submatrices was to verify the additional claim that if A is an n-by-n oscillatory matrix, then necessarily A^{n-1} is totally positive. Again the fact that the $(n-1)$st power is shown to be a so-called *critical exponent* was also hinted by tridiagonal TN matrices, as the $(n-1)$st power of an n-by-n tridiagonal matrix is the first possible power in which the matrix has no zero entries.

The next observation may be viewed as a combinatorial characterization of IITN matrices, at least in the context of an SEB factorization, may be found in [Fal01].

Theorem 2.6.6 *Suppose A is an n-by-n invertible TN matrix. Then A^{n-1} is TP if and only if at least one of multipliers from each of $L_n, L_{n-1}, \ldots, L_2$ and each of $U_n, U_{n-1}, \ldots, U_2$ is positive.*

We omit the details of the proof (see [Fal01]), but the general idea was to employ the method of induction, incorporate SEB factorizations, and apply the important fact that if A is TN, then A is TP if and only if all corner minors of A are positive (for more details see Chapter 3).

As in the original proof by Gantmacher and Krein we also established that if A is oscillatory, then necessarily A^{n-1} is TP. Furthermore, all the n-by-n oscillatory matrices for which the $(n-1)$st power is TP and no smaller power is TP have been completely characterized by incorporating the powerful combinatorics associated with the planar diagrams (see [FL07a]).

Another nice fact that is a consequence of the results above states when a principal submatrix of an IITN matrix is necessarily IITN. The same fact also appears in [GK60] (see also [GK02, p. 97]).

Corollary 2.6.7 *Suppose A is an n-by-n IITN matrix. Then $A[\alpha]$ is also IITN, whenever α is a contiguous subset of N.*

As an aside, a similar result has appeared in [FFM03] but, given the general noncommutative setting, the proof is vastly different.

The next result connects Theorem 2.6.6 to the classical criterion of Gantmacher and Krein for oscillatory matrices (Theorem 2.6.5); see also [Fal01].

Theorem 2.6.8 *Suppose A is an n-by-n invertible TN matrix. Then $a_{i,i+1} > 0$ and $a_{i+1,i} > 0$, for $i = 1, 2, \ldots, n-1$ if and only if at least one of multipliers from each of $L_n, L_{n-1}, \ldots, L_2$ and each of $U_n, U_{n-1}, \ldots, U_2$ is positive.*

It is apparent that factorization of TN matrices are an extremely important and increasingly useful tool for studying oscillatory matrices. Based on the analysis in [FFM03], another subset of oscillatory matrices was defined and called *basic oscillatory matrices*. As it turns out, basic oscillatory matrices may be viewed as minimal oscillatory matrices in the sense of the fewest number of bidiagonal factors involved.

In [FM04] a factorization of oscillatory matrices into a product of two invertible TN matrices that enclose a basic oscillatory matrix was established. A different proof of this factorization was presented by utilizing bidiagonal factorizations of invertible TN matrices and certain associated weighted planar diagrams in [FL07b], which we state now.

Suppose A is an oscillatory matrix. Then A is called a basic oscillatory matrix if A has *exactly* one parameter from each of the bidiagonal factors $L_2, L_3, \ldots, L_n$ and $U_2, U_3, \ldots, U_n$ is positive. For example, an IITN tridiagonal matrix is a basic oscillatory matrix, and, in some sense, represents the main motivation for considering such a class of oscillatory matrices.

Theorem 2.6.9 *Every oscillatory matrix A can be written in the form $A = SBT$, where B is a basic oscillatory matrix and S, T are invertible TN matrices.*

Chapter Three

Recognition

3.0 INTRODUCTION

When TN matrices are first encountered, it may seem that examples should be rate and that checking for TN or TP is tedious and time consuming.

However, TN matrices enjoy tremendous structure, as a result of requiring all minors to be nonnegative. This intricate structure actually makes it easier to determine when a matrix is TP than to check when it is a P-matrix, which formally involves far fewer minors.

Furthermore, this structure is one of the reasons that TN matrices arise in many applications in both pure and applied mathematics. We touch on one such application here, which also highlights the structure of TN matrices.

Vandermonde matrices arise in the problem of determining a polynomial of degree at most $n - 1$ that interpolates n data points. Suppose that n data points $(x_i, y_i)_{i=1}^n$ are given. The goal is to construct a polynomial $p(x) = a_0 + a_1 x + \cdots + a_{n-1} x^{n-1}$ that satisfies

$$p(x_i) = y_i \text{ for } i = 1, 2, \ldots, n. \tag{3.1}$$

The n equations in (3.1) are expressed by the linear system

$$\begin{bmatrix} 1 & x_1 & x_1^2 & \cdots & x_1^{n-1} \\ 1 & x_2 & x_2^2 & \cdots & x_2^{n-1} \\ \vdots & & \vdots & & \vdots \\ 1 & x_n & x_n^2 & \cdots & x_n^{n-1} \end{bmatrix} \begin{bmatrix} a_0 \\ a_1 \\ \vdots \\ a_{n-1} \end{bmatrix} = \begin{bmatrix} y_1 \\ y_2 \\ \vdots \\ y_n \end{bmatrix}. \tag{3.2}$$

The n-by-n coefficient matrix in (3.2) is called a *Vandermonde matrix*, and we denote it by $V(x_1, \ldots, x_n)$ (see also Chapter 0). The determinant of the n-by-n Vandermonde matrix in (3.2) is given by the formula $\prod_{i>j}(x_i - x_j)$; see [HJ85, p. 29]. Thus, if $0 < x_1 < x_2 < \cdots < x_n$, then $V(x_1, \ldots, x_n)$ has positive entries, positive leading principal minors, and, in particular, a positive determinant.

Consider a 3-by-3 Vandermonde matrix V with $x_1 = 1$, $x_2 = 2$, and $x_3 = 3$:

$$V = \begin{bmatrix} 1 & 1 & 1 \\ 1 & 2 & 4 \\ 1 & 3 & 9 \end{bmatrix}. \tag{3.3}$$

The inequalities $0 < x_1 < x_2 < x_3$ are actually sufficient to guarantee that *all* minors of V in (3.3) are positive. Indeed, any 2-by-2 submatrix of

V has the form

$$A = \begin{bmatrix} x_i^{\alpha_1} & x_i^{\alpha_2} \\ x_j^{\alpha_1} & x_j^{\alpha_2} \end{bmatrix},$$

with $i, j \in \{1, 2, 3\}$, $i < j$, $\alpha_1, \alpha_2 \in \{0, 1, 2\}$, and $\alpha_1 < \alpha_2$. Since $0 < x_i < x_j$, it follows that $\det A > 0$. Hence, in this case, V is TP.

More generally, it is known [GK02, p. 111] that if $0 < x_1 < x_2 < \cdots < x_n$, then $V(x_1, \ldots, x_n)$ is TP. Here is an outline of a proof similar to that given in [GK02]: Consider a k-by-k submatrix of $V(x_1, \ldots, x_n)$, which is of the form $A = [x_{l_i}^{\alpha_j}]$, where $l_1, \ldots, l_k \in \{1, \ldots, n\}$, $l_1 < l_2 < \cdots < l_k$, $\alpha_1, \ldots, \alpha_k \in \{0, 1, \ldots, n-1\}$, and $\alpha_1 < \alpha_2 < \cdots < \alpha_k$. Let $f(x) = c_1 x^{\alpha_1} + c_2 x^{\alpha_2} + \cdots + c_k x^{\alpha_k}$ be a real polynomial. Descartes's rule of signs ensures that the number of positive real roots of $f(x)$ does not exceed the number of sign changes among the coefficients $c_1, \ldots, c_k$; so, $f(x)$ has at most $k-1$ positive real roots. Therefore, the system of equations

$$f(x_{l_i}) = \sum_{j=1}^{k} c_j x_{l_i}^{\alpha_j} = 0, \ i = 1, 2, \ldots, k$$

has only the trivial solution $c_1 = c_2 = \cdots = c_k = 0$, so $\det A = \det[x_{l_i}^{\alpha_j}] \neq 0$.

To establish that $\det A > 0$ we use induction on the size of A. We know this to be true when the size of A is 2, so assume $\det A > 0$ whenever the size of A is at most $k-1$. Let $A = [x_{l_i}^{\alpha_j}]$ be k-by-k, in which $l_1, \ldots, l_k \in \{1, \ldots, n\}$, $l_1 < l_2 < \cdots < l_k$, $\alpha_1, \ldots, \alpha_k \in \{0, 1, \ldots, n-1\}$, and $\alpha_1 < \alpha_2 < \cdots < \alpha_k$. Then expanding $\det A$ along the last row gives

$$\det A = g(x_{l_k}) = a_k x_{l_k}^{\alpha_k} - a_{k-1} x_{l_k}^{\alpha_{k-1}} + \cdots + (-1)^{k-1} a_1 x_{l_k}^{\alpha_1}.$$

But $a_k = \det[x_{l_s}^{\alpha_t}]$, where $s, t \in \{1, 2, \ldots, k-1\}$, so the induction hypothesis ensures that $a_k > 0$. Thus $0 \neq g(x_{l_k}) \to \infty$ as $x_{l_k} \to \infty$, so $\det A = g(x_{l_k}) > 0$. The conclusion is that $V(x_1, \ldots, x_n)$ is TP whenever $0 < x_1 < \cdots < x_n$.

3.1 SETS OF POSITIVE MINORS SUFFICIENT FOR TOTAL POSITIVITY

Given $A \in M_{m,n}$, how may it be determined whether A is TP or TN? Direct verification of the definition (calculation of $\sum_{k=1}^{\min\{m,n\}} \binom{m}{k}\binom{n}{k}$ minors) is dramatically complex (and prohibitive). By comparison, for verification that $A \in M_n$ is a P-matrix, no better than exponential algorithms (that amount to verifying the positivity of each of the $2^n - 1$ principal minors) are known to date. Fortunately, by contrast, not only do fewer than $\sum \binom{m}{k}\binom{n}{k}$ minors suffice for checking for TP, it is actually possible to check in polynomial time in n.

Since permutation of rows or columns generally alters total positivity, the cadence of index sets that determine a minor is especially important.

Definition 3.1.1 For $\alpha \subseteq \{1, 2, \ldots, n\}$, $\alpha = \{\alpha_1, \alpha_2, \ldots, \alpha_k\}$, $0 < \alpha_1 < \alpha_2 < \cdots < \alpha_k \leq n$, the *dispersion* of the index set α is $d(\alpha) = \alpha_k - \alpha_1 - k + 1$.

The intent is to measure roughly how spread out the index set is relative to $\{1, 2, \ldots, n\}$. If $d(\alpha) = 0$, then the index set α is called a *contiguous index set*, which we may shorten to just contiguous. If α and β are two contiguous index sets with $|\alpha| = |\beta| = k$, then the submatrix $A[\alpha, \beta]$ is called a *contiguous submatrix of A* and the minor $\det A[\alpha, \beta]$ is called a *contiguous minor*, which we may also refer to as just contiguous. A contiguous minor in which α or β is $\{1, 2, \ldots, k\}$ is called *initial*.

As an example, consider $A \in M_{m,n}$, with $m = n = 2$; $A = [a_{ij}]$. The initial minors in this case are a_{11}, a_{12}, a_{21}, and $\det A$. But if all the initial minors are positive, then a_{22} must be positive as well. Then, all contiguous minors are as well. In general, the bidiagonal entries of the EB factors (and the diagonal entries of the diagonal factor) in the SEB factorization of a TP matrix may be written as ratios of products of initial minors; for example,

$$\begin{bmatrix} a_{11} & a_{12} \\ a_{21} & a_{22} \end{bmatrix} = \begin{bmatrix} 1 & 0 \\ a_{12}/a_{11} & 1 \end{bmatrix} \begin{bmatrix} a_{11} & 0 \\ 0 & \det A/a_{11} \end{bmatrix} \begin{bmatrix} 1 & a_{12}/a_{11} \\ 0 & 1 \end{bmatrix}.$$

Before we derive formulas for the entries of the desired EB factors in general, we recall the SEB factorization by presenting it in a slightly different context.

Performing the elimination as prescribed by the Whitney ordering produces a sequence of matrices:

$$A = A^{(1)} \to A^{(2)} \to \cdots \to A^{(n)} = U, \tag{3.4}$$

where $A^{(t)}$ has zeros below its main diagonal in the first $t - 1$ columns. If we let $A^{(t)} = [a_{ij}^{(t)}]$, then the entry $p_{ij} = a_{ij}^{(t)}$, $1 \leq j \leq i \leq n$ is called the (i, j)-*pivot* of this elimination, and the number

$$m_{ij} = \begin{cases} a_{ij}^{(t)}/a_{i-1,j}^{(t)} = p_{ij}/p_{i-1,j} & \text{if } a_{i-1,j}^{(t)} \neq 0, \\ 0 & \text{otherwise,} \end{cases}$$

is called the (i, j)-*multiplier* (see [GP92b, GP96]). As described in Chapter 2, the sequence of matrices in (3.4) corresponds to the factorization

$$A = (L_n(m_{n1}) \cdots L_2(m_{21}))(L_n(m_{n2}) \cdots L_3(m_{32})) \cdots (L_n(m_{n,n-1}))DU, \tag{3.5}$$

where D is a diagonal matrix and U is upper triangular. The purpose here in the short term is to derive formulas in terms of minors of A for the multipliers m_{ij}.

Consider separately the case $n = 4$. If A is factored as in (3.5), then using the planar diagram approach derived in Section 2.6 it is not difficult to determine the following formulas for some of the multipliers:

$$m_{41} = \det A[4, 1]/\det A[3, 1]; \quad m_{42} = \frac{\det A[34, 12] \det A[2, 1]}{\det A[23, 12] \det A[3, 1]};$$

$$m_{32} = \frac{\det A[23, 12] \det A[1, 1]}{\det A[12, 12] \det A[2, 1]}; \quad m_{43} = \frac{\det A[234, 123] \det A[12, 12]}{\det A[123, 123] \det A[23, 12]}.$$

In fact, to facilitate the proof of the general case we will make use of the planar networks associated with bidiagonal factorizations. One last observations is that if $D = \text{diag}(d_1, \ldots, d_n)$ in the factorization (3.4), then assuming U is constructed with unit main diagonal, we have

$$d_1 = a_{11}; \quad d_i = \frac{\det A[\{1, 2, \ldots, i\}]}{\det A[\{1, 2, \ldots, i-1\}]}$$

for $i = 2, 3, \ldots, n$.

Proposition 3.1.2 *Suppose A is factored as in (3.4). Then the multipliers m_{ij}, with $i = 2, 3, \ldots, n$ and $j = 1, 2, \ldots, n-1$ and $j < i$, are given by*

$$m_{ij} = \begin{cases} a_{i1}/a_{i-1,1} & \text{if } j = 1, \\ \frac{\det A[\{i-j+1,\ldots,i\},\{1,\ldots,j\}]\det A[\{i-j,\ldots,i-2\},\{1,\ldots,j-1\}]}{\det A[\{i-j+1,\ldots,i-1\},\{1,\ldots,j-1\}]\det A[\{i-j,\ldots,i-1\},\{1,\ldots,j\}]} & \text{if } j \geq 2. \end{cases}$$

Proof. First consider the case $j = 1$. Since for each $i = 2, 3, \ldots, n$ there is a unique path in the associated planar diagram from i to 1, it follows that $m_{i1}a_{i-1,1} = a_{i1}$. Hence the formula for m_{i1} given above follows. Now suppose $j > 1$. In a similar spirit, notice that there is only one collection of vertex-disjoint paths from sink set $\{i-j+1, \ldots, i\}$ to source set $\{1, 2, \ldots, j\}$. We can easily derive a formula for the corresponding minor of A, which in fact is given by

$$\left[\prod_{k=1}^{j} m_{i-(j-k),k} \right] \det A[\{i-j, \ldots, i-1\}, \{1, \ldots, j\}]$$

$$= \det A[\{i-j+1, \ldots, i\}, \{1, \ldots, j\}] \qquad (3.6)$$

Assuming the formulas in the statement of the result are true by induction on i, it follows that the product $\prod_{k=1}^{j-1} m_{i-(j-k),k}$ can be rewritten as

$$\prod_{k=1}^{j-1} m_{i-(j-k),k} = \frac{\det A[\{i-j+1, \ldots, i-1\}, \{1, \ldots, j-1\}]}{\det A[\{i-j, \ldots, i-2\}, \{1, \ldots, j-1\}]}.$$

Hence combining the above formula for $\prod_{k=1}^{j-1} m_{i-(j-k),k}$ with the equation (3.6) we have that m_{ij} is equal to

$$\frac{\det A[\{i-j+1, \ldots, i\}, \{1, \ldots, j\}] \det A[\{i-j, \ldots, i-2\}, \{1, \ldots, j-1\}]}{\det A[\{i-j+1, \ldots, i-1\}, \{1, \ldots, j-1\}] \det A[\{i-j, \ldots, i-1\}, \{1, \ldots, j\}]}.$$

$\square$

Observe that similar formulas exist for the multipliers used to reduce U to diagonal form, by considering A^T in the above proposition. We omit rewriting these formulas, as they simply involve the transposes of the minors above.

This result, then, allows us to write down the SEB factorization of a square TP matrix in terms of its initial minors and to deduce that positivity of initial minors implies that a matrix is TP in the square case.

We now consider nonsquare matrices. Note that an m-by-n matrix has

$$\sum_{k=1}^{\min\{m,n\}} [(1+n-k) + (1+m-k) - 1] = mn$$

initial minors and

$$\sum_{k=1}^{\min\{m,n\}} (1+n-k)(1+m-k)$$

contiguous minors.

Before we come to our next result on the sufficiency of the positivity of initial minors for checking for TP, we need a simple lemma to bridge the gap between the square and rectangular case.

Lemma 3.1.3 *Suppose A is an n-by-$(n+1)$ matrix in which all initial minors are positive. Then A can be extended to an $(n+1)$-by-$(n+1)$ matrix A' all of whose initial minors are positive.*

Proof. Suppose $A = [a_{ij}]$ and that the $(n+1)$st row of A' has entries x_j, for $j = 1, 2, \ldots, n+1$. Begin with x_1. Observe that the only 1-by-1 initial minor involving x_1 is (of course) x_1, so choose x_1 arbitrarily positive. Notice that the only initial 2-by-2 minor that involves x_2 (in A') is $\det A'[\{n, n+1\}, \{1, 2\}]$. Moreover, since all the initial minors of A are positive, x_2 contributes positively into this minor, x_2 can be chosen large enough so that $\det A'[\{n, n+1\}, \{1, 2\}] > 0$. Continuing in this manner, since x_j enters positively into the initial minor $\det A'[\{n+1-j+1, \ldots, n+1\}, \{1, 2, \ldots, j\}]$, it follows that x_j can be chosen large enough to ensure this initial minor is positive. This completes the proof. $\square$

The next result has appeared often in recent times (according to [FZ00a] it first appeared in [GP92a]), and can also be found in [GP96, Thm. 5.2].

Theorem 3.1.4 *If all initial minors of $A \in M_{m,n}$ are positive, then A is $TP_{\min\{m,n\}}$.*

Proof. Use Lemma 3.1.3 to extend A to a square matrix A', with all the initial minors of A' are positive. Then by the remarks preceding Lemma 3.1.3 it follows that A' is TP, and hence A is TP. $\square$

Since the initial minors are contained among the contiguous minors, positivity of the latter implies positivity of the former and, thus, total positivity. This gives the classical result due to Fekete [Fek13] as a corollary.

Corollary 3.1.5 *If all contiguous minors of $A \in M_{m,n}$ are positive, then A is TP.*

We can extend this idea of the sufficiency of checking just the contiguous minors to the class TP_k.

Corollary 3.1.6 *Suppose A is an m-by-n matrix with the property that all its k-by-k contiguous submatrices are TP. Then A is TP_k.*

Proof. Applying Corollary 3.1.5 we have that any submatrix of the form $A[\{i, i+1, \ldots, i+k-1\}, N]$ or $A[M, \{j, j+1, \ldots, j+k-1\}]$ consisting of k consecutive rows or k consecutive columns of A is TP. Now consider any collection of k rows of A, say, $B = A[\{i_1, i_2, \ldots, i_k\}, N]$. Note that any k-by-k contiguous submatrix of B must lie in some collection of k consecutive columns of A. Hence, this submatrix of B must be TP. Thus every k-by-k contiguous submatrix of B is TP, and therefore, by Corollary 3.1.5, B is TP. A similar argument may be applied to any collection of k columns of A. From this we may conclude that every k-by-k submatrix of A is TP, and hence A is TP_k. □

Corollary 3.1.7 *Suppose A is an m-by-n matrix. If all the initial minors of A up to size $k-1$ are positive and all its k-by-k contiguous minors are positive, then A is TP_k.*

We note that the initial minors form a minimal set of minors whose positivity is sufficient for total positivity.

Example 3.1.8 The positivity of no proper subset of the initial minors is sufficient for total positivity. Any $n^2 - 1$ of the minors being positive allows the remaining minor to be chosen negative.

Nonetheless, there are other subsets of minors, noncomparable to the initial minors, whose positivity is sufficient for total positivity. Minimal such sets have been identified in [FZ99, Thm 4.13] (and in [BFZ96] for analogous criteria involving triangular matrices).

In the event that it is somehow known that $A \in M_n$ is TN, positivity of a smaller set of minors is sufficient for total positivity.

Definition 3.1.9 An upper right (lower left) *corner* minor of $A \in M_{m,n}$ is one of the form $\det A[\alpha, \beta]$ in which α consists of the first k (last k) and β consists of the last k (first k) indices, $k = 1, 2, \ldots, \min\{m, n\}$. A *corner minor* is one that is either a lower left or upper right minor.

We note that there are $2n - 1$ distinct corner minors in $A \in M_n$.

Theorem 3.1.10 *Suppose that $A \in M_n$ is TN. Then A is TP if and only if all corner minors of A are positive.*

Proof. Clearly if A is TP, then the corner minors of A are positive. So assume that A is TN and the corner minors of A are positive. If, in addition, the initial minors of A are also positive, then A is TP by Theorem 3.1.4. So assume that some initial minor of A is zero, say $\det A[\{1, 2, \ldots, k\}, \{j, j+1, \ldots, j+k-1\}] = 0$. Then observe that $A[\{1, 2, \ldots, k\}, \{j, j+1, \ldots, j+k-1\}]$ is principal in the (corner) submatrix $A[\{1, 2, \ldots, n-j+1\}, \{j, j+1, \ldots, n\}]$.

So by Fischer's determinantal inequality (see Section 6.0) this corner minor must be zero, which is a contradiction. Hence all initial minors of A must be positive, and thus A is TP. $\qquad\qquad\qquad\qquad\qquad\qquad\qquad\qquad\qquad\qquad\qquad\square$

In case $A \in M_{m,n}$, $m \neq n$, the corner minors need not include all entries of A, and thus Theorem 3.1.10, is not valid in the nonsquare case. What additional minors are needed?

Definition 3.1.11 If $A \in M_{m,n}$, we call any minor that is either a corner minor or an initial minor with a maximal number of rows and columns $(\min\{m,n\})$ a *generalized corner minor*.

One may think of sliding the maximal corner minors to the left/right in case $m < n$.

Theorem 3.1.12 *Suppose* $A \in M_{m,n}$ *is TN. Then* A *is TP if and only if all generalized corner minors of* A *are positive.*

Proof. The proof is by induction on n (assuming $m \leq n$). If $n = m$, then the result holds by Theorem 3.1.10. Suppose A is m-by-n. Observe that since the corner minor $\det A[\{1, 2, \ldots, m\}, \{n - m + 1, \ldots, n\}] > 0$ and A is TN, it follows from Fischer's determinantal inequality (see Section 6.0) that

$$\det A[\{1, 2, \ldots, m - 1\}, \{n - m + 1, \ldots, n - 1\}] > 0,$$

and

$$\det A[\{2, \ldots, m\}, \{n - m, \ldots, n\}] > 0,$$

and $a_{mn} > 0$. Applying the same reasoning to $\det A[\{1, 2, \ldots, m - 1\}, \{n - m, \ldots, n\}] > 0$ implies $\det A[\{1, 2, \ldots, m - 2\}, \{n - m + 1, \ldots, n - 1\}] > 0$. Continuing in this manner proves that all proper corner minors of

$$A[\{1, 2, \ldots, m\}, \{1, 2, \ldots, n - 1\}]$$

are positive. Furthermore, the m-by-m corner minor of

$$A[\{1, 2, \ldots, m\}, \{1, 2, \ldots, n - 1\}]$$

is positive since all maximal contiguous minors are assumed positive. So by induction $A[\{1, 2, \ldots, m\}, \{1, 2, \ldots, n - 1\}]$ is TP. To prove that A is TP we must show that all contiguous minors that include column n are positive. Observe that every such minor is a principal minor of some corner minor, which by assumption is positive. So, by Fischer's determinantal inequality, each of these minors are positive. Hence A is TP. $\qquad\qquad\qquad\square$

For completeness, we offer a different proof of Theorem 3.1.10, one from a combinatorial point of view.

"Suppose that A is TN. Then A is TP if and only if all corner minors of A are positive."

Proof. Since A is nonsingular and TN, A can factored as in (2.12), where the parameters l_i and u_j are nonnegative. Moreover, since A is assumed to

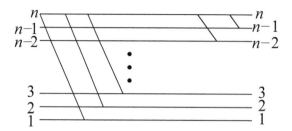

Figure 3.1 Diagram corresponding to A

be invertible, it follows from Fischer's inequality that each $d_i > 0$, as they can be written as ratios of products of principal minors of A. Consider the following figure (see Figure 3.1) which corresponds to the planar diagram corresponding to lower triangular part of (2.12). Since the $(n, 1)$ entry of A is positive, by assumption, there exists a path from n to 1 in Figure 3.1. This implies that all the weights of the edges along the first stretch from n down to 1 must be positive, as this represents the only possible such path. Similarly, since $\det A[\{n-1, n\}, \{1, 2\}] > 0$, it follows that the weights of the edges that make up the second stretch from n down to 2 must all be positive. Continuing inductively using $\det A[\{n-k+1, n-k+2, \ldots, n\}, \{1, 2, \ldots, k\}] > 0$, it follows that each such edge weight in the lower triangular part of (2.12) is positive. Similarly, the weights of the edges corresponding to the upper triangular part of (2.12) may also be shown to be positive. $\qquad\square$

Observe that for any corner minor $\det A[\alpha, \beta]$ of the form $\det A[\{n - k + 1, n-k+2, \ldots, n\}, \{1, 2, \ldots, k\}]$, there is a *unique* collection of vertex-disjoint paths in $P(\alpha, \beta)$. As demonstrated in the proof of Theorem 3.1.10, all edges appear at least once in the union of collections of paths in $P(\alpha, \beta)$ over all corner minors $\det A[\alpha, \beta]$, and, more important, there is at least one edge that appears in $P(\alpha, \beta)$ that does not appear in $P(\gamma, \delta)$ for each other corner minor $\det A[\gamma, \delta]$. Hence if some corner minor is omitted, it follows that the weight of at least one edge is unaccounted for, and therefore it may be zero. However, by Theorem 2.6.2, the corresponding matrix is not TP. Thus no subset of the list of corner minors suffices as a test for total positivity.

3.2 APPLICATION: TP INTERVALS

We turn to an application of our criteria to study "intervals" "between" TP matrices. We use $<, \leq$ to denote the usual entrywise partial order on matrices; that is, for $A = [a_{ij}]$, $B = [b_{ij}] \in M_{m,n}$, $A \leq (<) B$ means $a_{ij} \leq (<) b_{ij}$, for all i, j.

Example 3.2.1 If $0 < A < B$ and $A, B \in M_n$ are TP, then not all matrices in the interval between A and B need be TP; that is, if $A \leq C \leq B$, then C

need not be TP. Let

$$A = \begin{bmatrix} 2 & 1 \\ 1 & 1 \end{bmatrix}, \quad B = \begin{bmatrix} 4 & 5 \\ 5 & 13 \end{bmatrix}.$$

Then A, B are TP and

$$C = \begin{bmatrix} 3 & 4 \\ 4 & 5 \end{bmatrix}$$

satisfies $A < C < B$. But, C is not TP.

There is, however, a partial order on real matrices for which the interval between two matrices that are TP consists only of TP matrices.

Definition 3.2.2 The *"checkerboard" partial order* $\overset{*}{\leq}$ on real matrices is defined as follows. Recall from Section 1.3 that S_k denotes the k-by-k signature matrix whose diagonal entries alternate in sign beginning with $+$: for example,

$$S_5 = \begin{bmatrix} 1 & & & & \\ & -1 & & & \\ & & 1 & & \\ & & & -1 & \\ & & & & 1 \end{bmatrix}.$$

For $A, B \in M_{m,n}$, we write $A \overset{*}{\leq} B$ if and only if $S_m A S_n \leq S_m B S_n$ and $A \overset{*}{<} B$ if and only if $S_m A S_n < S_m B S_n$.

This means that the entries in positions, the sum of whose indices is even, are aligned in the usual way and that order is reversed in the "odd" positions.

Lemma 3.2.3 *If $A, B, C \in M_n$, $A \overset{*}{\leq} C \overset{*}{\leq} B$, and A and B are TP, then $\det C > 0$.*

Proof. If $A \overset{*}{\leq} C \overset{*}{\leq} B$, and A and B are TP, then it follows that $S_n A S_n \leq S_n C S_n \leq S_n B S_n$ and that both $S_n A S_n, S_n B S_n$ are inverse positive. Let $\tilde{A} = S_n A S_n$ with similar meanings applied to $\tilde{B}$, and so on. To complete a proof of this lemma, we need an auxiliary result relating semipositive matrices and inverse positive matrices. Suppose F and G are two n-by-n matrices such that $G \geq F$, $G^{-1} \geq 0$, and assume there exists a vector $x > 0$ with $Fx > 0$. Then $F^{-1} > 0$. Given the conditions on F and G above, it is clear that $0 \leq G^{-1}(G - F) = I - G^{-1}F$. Thus $G^{-1}F$ must be a Z-matrix (see [HJ91]). Using the fact that there exists a vector $x > 0$ with $Fx > 0$, it follows that $G^{-1}Fx > 0$. Hence $G^{-1}F$ must be an M-matrix (see [HJ91, Sect. 2.5]). Thus $(G^{-1}F)^{-1} \geq 0$. Finally, since $G^{-1} \geq 0$, we have that $F^{-1} \geq 0$.

Now to complete the proof of the lemma, let $\tilde{B} = M$ and $\tilde{C} = F$ in the claim above. Let x be the unique vector that satisfies $\tilde{A}x = e$, where e is the vector of all ones. Then since $\tilde{A}$ is inverse positive, $x > 0$. Further we have $\tilde{C}x \geq \tilde{A}x > 0$. Hence the hypotheses of the claim are satisfied, and so we conclude that $\tilde{C} = S_n C S_n$ is inverse positive. Hence any matrix C, such that $A \overset{*}{\leq} C \overset{*}{\leq} B$, must be invertible and since A and B are TP, it follows that, in fact, $\det C > 0$.   □

We note here that this lemma also appeared in [Gar96], in which the proof contained a reference to an article on *monotone matrices*. A square matrix is monotone if and only if it is inverse nonnegative. The proof we presented above is an adaptation of the one used in [Gar96] and the subsequent reference, but since it is sufficiently different and requires fewer technical results, we decided to include it for completeness.

Theorem 3.2.4 *If $A, B, C \in M_n$, $A \overset{*}{\leq} C \overset{*}{\leq} B$, and A and B are TP, then C is TP.*

Proof. It suffices, by Theorem 3.1.4, to show that the initial minors of C are positive. But if C' is an initial minor of C and A', B' are the corresponding initial minors of A, B, respectively, then it follows that either $A' \overset{*}{\leq} C' \overset{*}{\leq} B'$ or $B' \overset{*}{\leq} C' \overset{*}{\leq} A'$ holds depending on the relative position of the initial minor. However, in either case Lemma 3.2.3 applies and $\det C' > 0$. Hence C is TP. □

Example 3.2.5 We note that, if TP is replaced by TN, then Theorem 3.2.4 no longer holds. For example,

$$\begin{bmatrix} 2 & 0 & 1 \\ 0 & 0 & 0 \\ 1 & 0 & 1 \end{bmatrix} \overset{*}{\leq} \begin{bmatrix} 3 & 0 & 4 \\ 0 & 0 & 0 \\ 4 & 0 & 5 \end{bmatrix} \overset{*}{\leq} \begin{bmatrix} 4 & 0 & 5 \\ 0 & 0 & 0 \\ 5 & 0 & 13 \end{bmatrix},$$

however, both end matrices are TN, while the middle matrix is not. The difficulty is that zero lines have been inserted in such a way that the $\overset{*}{\leq}$ partial order reverts to the usual partial order $\leq$.

In the intermediate case, it is not known whether Theorem 3.2.4 remains valid when TP is replaced by InTN.

3.3 EFFICIENT ALGORITHM FOR TESTING FOR TN

Returning now to sufficient sets of minors, we have, thus far, considered the TP case in detail. What about sets of minors whose nonnegativity is sufficient for TN? Unfortunately, the situation is rather different.

Example 3.3.1 For positive integers m and n and any index sets α, β, $|\alpha| = |\beta| = k \leq \min\{m, n\}$, there is an $A \in M_{m,n}$ such that every minor of A is nonnegative, except that $\det A[\alpha, \beta] < 0$. To see this choose a k-by-k

matrix for which all proper minors are nonnegative, but the determinant is negative. Place this matrix in the α, β positions of an m-by-n matrix all of whose remaining entries are 0. Such an A has the desired features.

If the rank of A is known, however, some reduction in the collection of minors needed to test for total nonnegativity is possible.

Theorem 3.3.2 *If $A \in M_{m,n}$ and $rank(A) = r$, then A is TN if and only if for all index sets α, β, such that $|\alpha| = |\beta|$ and $d(\beta) \leq n - r$, $\det A[\alpha, \beta] \geq 0$.*

As the proofs of this theorem are rather long and technical, we will only provide a brief discussion highlighting the key aspects of the arguments given in [And87, Thm. 2.1] and [Cry76, Thm. 1.3].

Both proofs rely on induction and both assume that all minors up to size $k - 1$ have been shown to be nonnegative. (Note that the base case, namely $k = 1$, follows since all 1-by-1 minors satisfy $d(\beta) = 0$.) Then the argument proceeds by assuming that there is a minor of size k that is negative. Moreover, such a k-by-k minor is chosen such that $d(\beta) = l$ is minimal, assuming β is the column set of this minor. Clearly, by hypothesis, $2 \leq k \leq r$ and $l = d(\beta) > n-r$. The latter inequality is essential for reaching a contradiction.

At this point both Ando and Cryer appeal to "Karlin's identity" (see Chapter 1 (1.6)) to link minors whose column sets have dispersion l to minors whose column sets have dispersion $l - 1$ (keeping in mind the minimality of l above).

Ando reaches a contradiction by employing Sylvester's determinantal identity and the inequality $l = d(\beta) > n - r$ to verify that $rank(A) > r$.

Cryer, on the other hand, obtains a contradiction by instead appealing to an auxiliary result (see [Cry76, Lemma 3.1]) on ranks of certain TN matrices.

We note that Theorem 3.3.2 could be applied to A^T to give the sufficiency of the minors $\det A[\alpha, \beta]$ with $d(\alpha) \leq n - r$.

Definition 3.3.3 A minor $\det A[\alpha, \beta]$ of $A \in M_{m,n}$ is *quasi-initial* if either $\alpha = \{1, 2, \ldots, k\}$ and β, $|\beta| = k$, is arbitrary or α, $|\alpha| = k$, is arbitrary, while $\beta = \{1, 2, \ldots, k\}$, $k = 1, 2, \ldots, \min\{m, n\}$.

In the case of triangular matrices the following result holds (see [And87, Cry76, GP93b]). Notice that only half of the quasi-initial minors are required given the triangular form.

Lemma 3.3.4 *Suppose $L \in M_n$ is invertible and lower triangular. Then L is InTN if and only if*

$$\det L[\alpha, \{1, 2, \ldots, k\}] \geq 0,$$

for all $\alpha \subseteq \{1, 2, \ldots, n\}$.

Proof. Suppose $\beta \subseteq \{1, 2, \ldots, n\}$ with $d(\beta) = 0$. Let $\alpha \subseteq \{1, 2, \ldots, n\}$ and assume that $\alpha = \{\alpha_1, \alpha_2, \ldots, \alpha_k\}$ and $\beta = \{\beta_1, \beta_2, \ldots, \beta_k\}$. If $\alpha_1 < \beta_1$, then $\det L[\alpha, \beta] = 0$, since L is lower triangular. If $\alpha_1 \geq \beta_1$ and $\beta_1 = 1$, then $\det L[\alpha, \beta] = \det L[\alpha, \{1, 2, \ldots, k\}] \geq 0$, by hypothesis. Finally, suppose $\alpha_1 \geq \beta_1$ and $\beta_1 > 1$. Let $\gamma = \{1, 2, \ldots, \beta_1 - 1\}$. Then

$$0 \geq \det L[\gamma \cup \alpha, \gamma \cup \beta],$$

by hypothesis. Moreover, since L is lower triangular

$$\det L[\gamma \cup \alpha, \gamma \cup \beta] = \left(\prod_{i=1}^{\beta_1 - 1} l_{ii} \right) \det L[\alpha, \beta].$$

Since L is invertible, $l_{ii} > 0$ for all i, hence $\det L[\alpha, \beta] \geq 0$. By Theorem 3.3.2, L is InTN. The converse is trivial. $\qquad \square$

The fact that all InTN matrices have factorizations into lower and upper triangular InTN matrices is needed along with the above fact to prove the next result for general InTN matrices.

Theorem 3.3.5 *If $A \in M_n$ is invertible, then A is InTN if and only if the leading principal minors of A are positive and all quasi-initial minors of A are nonnegative.*

Proof. If the leading principal minors of A are positive, it follows that A can be written as $A = LDU$, where D is a positive diagonal matrix and L (U) is a lower (upper) triangular matrix with unit main diagonal. Applying the Cauchy-Binet identity to the product LDU we have

$$0 \leq \det A[\alpha, \{1, 2, \ldots, k\}] = \sum_\gamma \det LD[\alpha, \gamma] \det U[\gamma, \{1, 2, \ldots, k\}]$$

$$= \det LD[\alpha, \{1, 2, \ldots, k\}] \det U[\{1, 2, \ldots, k\}]$$

$$= \det LD[\alpha, \{1, 2, \ldots, k\}].$$

Since D is a positive diagonal matrix, it follows that $\det L[\alpha, \{1, 2, \ldots, k\}] \geq 0$, for any $\alpha \subseteq \{1, 2, \ldots, n\}$. Hence, by Lemma 3.3.4, L is InTN. Similarly,

$$0 \leq \det A[\{1, 2, \ldots, k\}, \beta] = \sum_\gamma \det L[\{1, 2, \ldots, k\}, \gamma] \det DU[\gamma, \beta]$$

$$= \det L[\{1, 2, \ldots, k\}] \det DU[\{1, 2, \ldots, k\}, \beta]$$

$$= \det DU[\{1, 2, \ldots, k\}, \beta].$$

Again it follows that $\det U[\{1, 2, \ldots, k\}, \beta] \geq 0$ for any $\beta \subseteq \{1, 2, \ldots, n\}$. Hence, by Lemma 3.3.4, U is InTN. Thus if L, D, and U are all InTN, then $A = LDU$ must be InTN. $\qquad \square$

Recall that in [GK02] the term quasi-principal was used in conjunction with their investigations on criteria for a matrix to be oscillatory. Our definition is different from the one used in [GK02], and our intention is also different. This notion of quasi-principal was needed to verify that if an n-by-n matrix is oscillatory, then its $(n-1)$st power will be TP (see [GK02, Ch. 2]).

The sufficient sets of minors we have discussed are of theoretical value, and they do not lead to the most efficient (or simple) methods to check for total positivity or total nonnegativity, even when some advantage is taken in the overlap of calculation of different minors (as in [GM87]). Here, we restrict attention to the case of square matrices. Via calculation to produce an LU factorization, total positivity or total nonnegativity may assessed in $O(n^3)$ effort for $A \in M_n$. In fact, the SEB factorization may be calculated in $O(n^3)$ time if A is InTN.

Consider then the process of reducing a full matrix A to upper triangular form via Whitney elimination, beginning by eliminating the lower left entry with the one above, continuing up the column, then moving to the next column to the right, and so on. Once an upper triangular matrix is produced, the same process is applied to its transpose to reduce to diagonal form. At each step, a *multiple* of one entry is subtracted from that below it (to produce a zero). This multiplier will end up as the off-diagonal entry in one of the TN EB factors of the SEB factorization. If no entry prematurely becomes zero and all multipliers are positive, then the original A was TP. If A is invertible and no multiplier is negative, then A is InTN.

For singular TN matrices, it may happen that a zero entry appears before a nonzero. In this event, the entire row in which the zero lies must be zero and that row may be ignored and a nonzero row above it can be used in the elimination process (see also Chapter 2, Section 3). Again, allowing for zero rows, if no negative multiplier is encountered, or if a zero entry above a nonzero entry without the row being zero, the matrix is not TN.

We provide some illustrative examples to bring together the above discussion.

Example 3.3.6 The first illustration is meant to indicate that when reducing a singular TN matrix, a scan of the rows is necessary to detect whether the original matrix is TN. Consider the following sequence of eliminations up the first column.

$$\begin{bmatrix} 1 & 0 & 0 & 0 \\ 1 & 1 & 2 & 1 \\ 1 & 1 & 2 & 2 \\ 1 & 2 & 4 & 4 \end{bmatrix} \rightarrow \begin{bmatrix} 1 & 0 & 0 & 0 \\ 1 & 1 & 2 & 1 \\ 1 & 1 & 2 & 2 \\ 0 & 1 & 3 & 3 \end{bmatrix} \rightarrow \begin{bmatrix} 1 & 0 & 0 & 0 \\ 1 & 1 & 2 & 1 \\ 0 & 0 & 1 & 1 \\ 0 & 1 & 3 & 3 \end{bmatrix} \rightarrow \begin{bmatrix} 1 & 0 & 0 & 0 \\ 0 & 1 & 2 & 1 \\ 0 & 0 & 1 & 1 \\ 0 & 1 & 3 & 3 \end{bmatrix}.$$

At this point the algorithm is stopped, as we have encountered a zero (pivot) above a nonzero in the same column, and on further examination the corresponding row is not zero. If we had continued with the elimination, albeit not with TN bidiagonal matrices, we would end up with the InTN matrix

$$\begin{bmatrix} 1 & 0 & 0 & 0 \\ 0 & 1 & 2 & 1 \\ 0 & 0 & 1 & 1 \\ 0 & 0 & 0 & 1 \end{bmatrix}.$$

The second illustration is intended to demonstrate that reduction of singular TN matrices need not always produce zero rows (at least to triangular form).

$$\begin{bmatrix} 3 & 2 & 1 & 1 \\ 3 & 2 & 2 & 3 \\ 6 & 4 & 5 & 10 \\ 0 & 0 & 0 & 1 \end{bmatrix} \rightarrow \begin{bmatrix} 3 & 2 & 1 & 1 \\ 3 & 2 & 2 & 3 \\ 0 & 0 & 1 & 4 \\ 0 & 0 & 0 & 1 \end{bmatrix} \rightarrow \begin{bmatrix} 3 & 2 & 1 & 1 \\ 0 & 0 & 1 & 2 \\ 0 & 0 & 1 & 4 \\ 0 & 0 & 0 & 1 \end{bmatrix}.$$

Observe that in this case all the multipliers used in reducing this matrix to triangular form are positive.

The ideas developed thus far allow characterization of the zero/nonzero patterns possible in a TN matrix and, more generally, the possible arrangements of singular minors. In the event of an InTN matrix, the zero patterns, and so forth are more limited.

Chapter Four

Sign Variation of Vectors and TN Linear Transformations

4.0 INTRODUCTION

If the entries of a vector represent a sampling of consecutive function values, then a change in the sign of consecutive vector entries corresponds to the important event of the (continuous) function passing through zero. Since [Sch30], it has been known that as a linear transformation, a TP matrix cannot increase the number of sign changes in a vector (this has a clear meaning when both x and Ax are totally nonzero and will be given an appropriate general meaning later). The transformations that never increase the number of sign variations are of interest in a variety of applications, including approximation theory and shape preserving transforms (see, for example, [Goo96]) and are of mathematical interest as well. Here we develop an array of results about TP/TN matrices and sign variation diminution, along with appropriate converses.

4.1 NOTATION AND TERMS

A vector $x \in \mathbb{R}^n$ is called *totally nonzero* if no entry of x is zero. If $x \in \mathbb{R}^n$ is totally nonzero, then we may consider how the signs of consecutive entries of x vary. It is natural to total the number of times an entry differs in sign from the previous entry. Of course, when zero entries are present, a variety of measures of total sign variation may be imagined. Two have proven useful in this subject.

Definition 4.1.1 Let v be a function from totally nonzero vectors $x = [x_1, x_2, \ldots, x_n]^T \in \mathbb{R}^n$ into nonnegative integers defined by

$$v(x) = |\{i : x_i x_{i+1} < 0\}|, \text{ for } i = 1, 2, \ldots, n - 1$$

the *total sign variation of x* .

For general vectors $x \in \mathbb{R}^n$, $v_m(x)$ is the minimum value of $v(y)$ among totally nonzero vectors y that agree with x in its nonzero entries and $v_M(x)$ is the maximum value of $v(y)$ among all such vectors.

Of course in the totally nonzero case,

$$v_m(x) = v(x) = v_M(x),$$

and, in general, $v_m(x) \leq v_M(x) \leq n-1$. In case x has zero entries, $v_m(x)$ is also $v(x')$, in which x' is simply the result of deleting the zero entries from x. As a simple illustrative example, take $x = [1, 0, 1, -1, 0, 1]^T$. Then it follows that $v_m(x) = 2$ and $v_M = 4$.

We also let $x_m \in \mathbb{R}^n$ be a totally nonzero vector that agrees with x in its nonzero entries and for which $v(x_m) = v_m(x)$; x_M is similarly defined. The *initial* (*terminal*) sign of a vector x is that of the first (last) nonzero entry of x. The signs of the entries of an x_m or an x_M are not generally unique, but the signs of any entries of x_m or x_M occurring before the initial sign entry (after the terminal sign entry) of x are uniquely determined. If the jth position of x contains its initial sign, then in x_m all entries in positions before j have that same sign. If that entry is positive, then for each $i < j$ the ith entry of x_M has the sign of $(-1)^{j-i}$, and, if it is negative, the ith entry of x_M has the sign of $(-1)^{j-i+1}$. Corresponding statements may be made for signs after the terminal sign entry.

Entries in x_m or an x_M after the initial sign entry of x and before the terminal sign entry of x may or may not be sign determined. For zeros of x between consecutive, like-signed nonzero entries, the entries in x_m will have the same (like) sign while the entries of x_M must alternate insofar as possible. In the x_M case, if there is an odd number of zeros between two like-signed entries, x_M will exactly alternate beginning with the first of the like-signed entries (e.g., $+000+ \rightarrow +-+-+$). If there is an even number of zeros, there are multiple possibilities one pair of consecutive entries will have the same sign (e.g. $+00+ \rightarrow +-++$ or $++-+$ or $+--+$). So there are as many possibilities as the number of zeros plus one.

For zeros lying between consecutive, oppositely signed entries, x_m offers multiple possibilities, while x_M does as well for a odd number of zeros. For an even number of zeros, the intervening signs in x_M are uniquely determined as exact alternation is possible.

A very useful equation involving the quantities v_M and v_m of an arbitrary vector $u \in \mathbb{R}^n$ is given by

$$v_M(u) + v_m(Su) = n - 1, \tag{4.1}$$

where S is the alternating n-by-n signature matrix. This equation can be established by a standard induction argument and is omitted here. We remark that the equation above appeared in the study of the eigenvectors of TP matrices in [GK60, Chap. 2].

4.2 VARIATION DIMINUTION RESULTS AND EB FACTOR-IZATION

It is an extremely important fact that, viewed as a linear transformation, a TP matrix cannot increase total sign variation, in the strongest sense (see Theorem 4.3.5). But, there is actually a hierarchy of results for TN, InTN, and TP matrices, the broader classes enjoying necessarily weaker conclusions. In addition, there are three converses to the TP theorem, two of which will be of particular interest to us here (the third involving strictly sign regular matrices).

We begin with a simple observation about the effect of TN EB and GEB matrices on v, v_m and v_M.

Lemma 4.2.1 *(a) If A is an n-by-n TN GEB matrix, then*

$$v_m(Ax) \leq v_m(x),$$

for all $x \in \mathbb{R}^n$. Moreover, if $v_m(Ax) = v_m(x)$, then either x or Ax is 0 or the initial nonzero entries of x and Ax have the same sign.

(b) If A is an n-by-n TN EB matrix, then

$$v_m(Ax) \leq v_m(x) \text{ and } v_M(Ax) \leq v_M(x),$$

for all $x \in \mathbb{R}^n$. Moreover, if $v_M(Ax) = v_M(x)$, then either x is 0 or the signs of initial entries of x_M and $(Ax)_M$ coincide.

Proof. The first claim in (a) follows easily from the first claim in (b). The first two claims in (b) have essentially the same proof. Since A is a TN EB matrix, only one entry can change sign, and that one can change only in the direction of either the sign of the prior entry (in case A is lower EB) or the sign of the next entry (in case A is upper EB). In either event, even if the entry is changed to zero, neither v_m nor v_M can increase.

Note that, from above, the only way the initial (terminal) sign could change would be so as to decrease v_m or v_M. So, if equality occurs in either case and $x \neq 0$, then the initial (terminal) signs must remain the same. □

We note that, in general, if it is the case that $v_m(Ax) \leq v_m(x)$ for any square TN matrix and all $x \in \mathbb{R}^n$, then the non square case follows. If A is m-by-n and $m < n$, this is a triviality, as the image vector, which is on the lower side of the inequality, may be viewed as having been truncated. In case $m > n$, padding the nonsquare matrix to a square one by augmenting with zero columns verifies the claim.

Since any TN matrix may be expressed as a product of TN GEB matrices (see Theorem 2.3.2) we have

Theorem 4.2.2 *If A is an m-by-n TN matrix, then*

$$v_m(Ax) \leq v_m(x)$$

for all $x \in \mathbb{R}^n$.

Proof. Padding with zeros if necessary, the proof follows from repeated application of Lemma 4.2.1. □

Example 4.2.3 The conclusion of Theorem 4.2.2 cannot be strengthened (to $v_M(Ax) \leq v_M(x)$). Suppose $A = J$, the 4-by-4 matrix of all ones, and that $x = [1, -1, -1, 1]^T$. Then $v_M(x) = 2$ and $v_M(Ax) = v_M(0) = 3$.

Since $v_M(Ax) = v(Ax) = v_m(Ax)$ if Ax happens to be totally nonzero, then we have

Corollary 4.2.4 *If A is an m-by-n TN matrix, then*

$$v_M(Ax) \leq v_m(x)$$

whenever $x \in \mathbb{R}^n$ and $Ax \in \mathbb{R}^m$ is totally nonzero (in particular when both x and Ax are totally nonzero).

Proof. If Ax is totally nonzero, then $v_M(Ax) = v(Ax) = v_m(Ax) \leq v_m(x)$, by Theorem 4.2.2.   □

Since any InTN matrix is a product of TN EB matrices (and a matrix in D_+, see Theorem 2.2.2), we have

Theorem 4.2.5 *If A is an n-by-n InTN matrix, then*

$$v_m(Ax) \leq v_m(x)$$

for all $x \in \mathbb{R}^n$ and

$$v_M(Ax) \leq v_M(x)$$

for all $x \in \mathbb{R}^n$.

Proof. Both claims follow from several applications of Lemma 4.2.1, as A may be written as a product of TN EB matrices, according to Theorem 2.2.2.   □

Example 4.2.6 The conclusion of Theorem 4.2.5 cannot be improved (to $v_M(Ax) \leq v_m(x)$). Let A be any n-by-n tridiagonal InTN matrix (that is, A is a nonnegative P-matrix) and let $x = e_1$ the first standard basis vector. Then Ax has $n - 2$ zero entries, so it is clear that for $n \geq 3$, $v_M(Ax) > v_m(x) = 0$.

We have noted the validity of the strong conclusion: $v_M(Ax) \leq v_m(x)$ when A is TN and Ax is totally nonzero (Corollary 4.2.4). We may also reach the strong conclusion when x is totally nonzero and A is InTN.

Theorem 4.2.7 *If A is an n-by-n InTN matrix, then*

$$v_M(Ax) \leq v_m(x)$$

for all totally nonzero $x \in \mathbb{R}^n$.

Proof. We know from Corollary 4.2.4 that the claim is correct when Ax is totally nonzero. For a fixed totally nonzero vector x, consider a neighborhood $\{(z, Az)\} \subseteq \mathbb{R}^{2n}$ of x, Ax in which each component has the same nonzero sign as that of x, Ax, if the component of x, Ax is nonzero. Since A is invertible, there is a neighborhood $N(x)$ of x such that $\{N(x), AN(x)\} \subseteq \{(z, Az)\}$. Now,

$$v_M(Ax) = \max_{Az \in AN(x)} v(Az),$$

in which that maximum is taken over totally nonzero vectors z such that Az. Let $\tilde{z}$ be a maximizing vector. Then

$$v_M(Ax) \leq v_M(A\tilde{z}) = v(A\tilde{z}) \leq v(\tilde{z}) = v_m(x),$$

as desired.   □

Summarizing the above, we know that the strong inequality $(v_M(Ax) \leq v_m(x))$ holds whenever either x or Ax is totally nonzero, at least for InTN matrices. If both include 0 entries, the strong conclusion need not hold for any TN matrix that is not TP.

Lemma 4.2.8 *Suppose that A is an m-by-n TN matrix, but A is not TP. Then there is an $x \neq 0$, $x \in \mathbb{R}^n$ such that $v_M(Ax) > v_m(x)$.*

Proof. Since A is not TP, it must have a corner minor that is zero (see Theorem 3.1.10). Suppose, without loss of generality, that $\det A[\{1, 2, \ldots, k\}, \{n-k+1, n-k+2, \ldots, n\}] = 0$, and $x_1 \in \mathbb{R}^k$ is a nonzero null vector for this corner submatrix. Let

$$x = \begin{bmatrix} 0 \\ x_1 \end{bmatrix} \in \mathbb{R}^n.$$

Then the first k entries of Ax are 0, so that

$$v_M(Ax) \geq k > v_M(x_1) \geq v_m(x_1) = v_m(x),$$

as was claimed.   □

4.3 STRONG VARIATION DIMINUTION FOR TP MATRICES

Here, we will show that the strong conclusion does hold for TP matrices and *any* nonzero x. However, since this is not true for individual InTN EB factors, the proof must take into account something about all the nontrivial EB factors taken together. This requires an important fact about subspaces of vectors with bounded sign variation that will also be useful elsewhere.

Suppose that $u^{(1)}, \ldots, u^{(q)} \in \mathbb{R}^n$ are linearly independent and that $q < n$. When is v_M nontrivially bounded throughout that the subspace $\tilde{U} = Sp(\{u^{(1)}, \ldots, u^{(q)}\})$? Since $v_M(0) = n - 1$, the largest possible value, we exclude 0 from $\tilde{U}$ to get the set of vectors $\hat{U}$. Moreover, it is clear that values of v_M as high as $q - 1$ must occur in $\hat{U}$, so that we ask for which $\hat{U}$ is v_M never more than $q - 1$ on $\hat{U}$. There is a nice answer to this question, which we will use here and later on. Let U be the matrix with columns $u^{(1)}, \ldots, u^{(q)}$, so that U is n-by-q.

Theorem 4.3.1 *In the notation above, $v_M(u) \leq q - 1$ for all $u \in \hat{U}$ if and only if all the q-by-q minors formed by a selection of q rows from U are uniformly signed.*

Proof. For sufficiency, suppose that there is a $u \in \hat{U}$ such that $v_M(u) \geq q$. Append it to U on the right to obtain U'. Now, delete $n - q - 1$ rows from U' to obtain U'' so that the last column of U'' weakly alternates in sign and is nonzero. Now, U'' is nonsingular, as shown by the expansion of $\det U''$ down its last column, and using the (strong) uniformly signed hypothesis for U conveys to U'' with the last column deleted. This means that $\text{rank}(U') = q + 1$, contradicting that $u \in \tilde{U}$ and verifying sufficiency of the uniformly signed condition.

For necessity, suppose that the q-by-q minors of U are not all positive or all negative. If there is a q-by-q minor that is zero, then there is a nonzero vector y in the range of U that has at least q zero entries. No matter where these zeros are located in y, we have $v_M(y) \geq q$, a contradiction, from which we conclude that all q-by-q minors of U are nonzero. Thus, there are two minors opposite in sign. Without loss of generality, these may be assumed to lie in a $(q + 1)$-by-q submatrix

$$V = \begin{bmatrix} v_1^T \\ v_2^T \\ \vdots \\ v_{q+1}^T \end{bmatrix}$$

of U. Then V has a one-dimensional left null space, spanned by a nonzero vector z^T. Furthermore, the equation

$$z^T V = 0$$

implies that

$$z_i v_i^T = - \sum_{j \neq i} z_j v_j^T.$$

Set V_i to be the q-by-q submatrix of V obtained by deleting row i. We may apply Cramer's rule to the nonsingular linear system

$$z_i v_i^T = - \sum_{j \neq i} z_j v_j^T$$

with coefficient matrix V_i. From this, it follows that for each index $k \neq i$, we have

$$\frac{z_k}{z_i} = \frac{\det V_i}{\det V_k} = (-1)^{i-k+1},$$

in which the indices run from 1 to $q + 1$. Since the column space of V is the orthogonal complement of z^T, for each pair i, k, we may have an element of the column space with $m - 1$ zero entries and two nonzeros in positions i and k. If, for example, we take $i = q + 1$ and $k = 1$, we obtain a vector s in the column space of V with $v_M(s) = q$ and, then, via the same linear combination of the columns of U, a vector t in the column space of U with $v_M(t) \geq q$. This contradiction completes the proof. $\square$

As a consequence of Theorem 4.3.1, we have the fact that we will need here.

Corollary 4.3.2 *If A is an m-by-n TP matrix with $m > n$, then*

$$v_M(Ax) \leq n - 1$$

for each nonzero $x \in \mathbb{R}^n$.

Proof. The n-by-n minors of A are positive and thus uniformly signed. Thus, by Theorem 4.3.1, every nonzero vector z in the column space of A satisfies $v_M(z) \leq n - 1$, which gives the claim.   □

For the proof of our main result, one more idea, which is of independent interest, is needed. If several consecutive columns of a TP matrix are replaced by single column that is a nonnegative, nonzero linear combination of the replaced columns, then the resulting "column compressed" matrix is also TP. Suppose that A is an m-by-n TP matrix, $n' < n$, and $k_0 = 1 < k_1 < k_2 < \cdots < k_{n'} = n$ are indices. Suppose further that $c \in \mathbb{R}^n$, $c \geq 0$, and $c_{k_{i-1}+1} + \cdots + c_{k_i} > 0$, $i = 1, 2, \ldots, n'$.

Definition 4.3.3 In the notation above, with $A = [a_1, a_2, \ldots, a_n]$, let

$$a_i' = c_{k_{i-1}+1} a_{k_{i-1}+1} + \cdots + c_{k_i} a_{k_i},$$

$i = 1, 2, \ldots, n'$. Then the m-by-n matrix $A' = [a_1', a_2', \ldots, a_{n'}']$ is a *column compression* of A.

The key observation is that column compression preserves total positivity (nonnegativity).

Lemma 4.3.4 *If A is TP (resp. TN), then any column compression A' of A is also TP (resp. TN)*

Proof. Without loss of generality, we may suppose that $k_i - k_{i-1} > 1$ for only one i and that the components of c are 1 in all other positions. Successive application of this case gives the general one. Consider any minor of a' that includes column k_i; all others are the same as in A. Laplace expansion of that minor along column k_i reveals that the minor is a nonnegative, nonzero linear combination of minors of A. This verifies both claimed statements.   □

Our main result of the chapter is then the following.

Theorem 4.3.5 *If A is an m-by-n TP matrix and $x \in \mathbb{R}^n$ is nonzero, then*

$$v_M(Ax) \leq v_m(x).$$

Proof. Suppose that $n' = v_m(x) + 1$. We construct a totally nonzero, strictly alternating vector $x' \in \mathbb{R}^{n'}$ and an m-by-n' column compression A of A' such that $A'x' = Ax$ and $v_m(x') = v_m(x)$. The theorem then follows from Corollary 4.3.2. For the compression, let k_1 be the longest initial stretch of indices such that there is no sign variation in $x_1, x_2, \ldots, x_{k_1}$ and then $k_1 + 1, \ldots, k_2$ a maximal stretch without variation, and so on. This will end with $k_{n'} = n$ and each stretch will contain a nonzero coordinate. Now, let $c_j = |x_j|$, $j = 1, 2, \ldots, n$, for the compression and $x_j' = \pm 1$, $j = 1, 2, \ldots, n'$. The

x'_js strictly alternate, beginning with $+1$ or -1 depending on the nonzero signs among $x_1, \ldots, x_{k_1}$. We now have that the column compression A' is m-by-n'; A' is TP, $A'x' = Ax$, and $v_m(x') = v(x') = v_m(x) = n' - 1$. As $v_M(Ax) = v_M(A'x') \leq n' - 1$, by Corollary 4.3.2 and $v_m(x') = v_m(x) = n' - 1$, we conclude $v_M(Ax) \leq v_m(x)$, as claimed.   □

Some recent work has appeared on sign variation and total positivity that involves some in depth analysis of matrix factorization (see [DK08]).

4.4 CONVERSES TO VARIATION DIMINUTION

The major fact Theorem 4.3.5 has several converses. The first comes from Lemma 4.2.8.

Theorem 4.4.1 *Suppose that A is an m-by-n TN matrix. Then A is TP if and only if $v_M(Ax) \leq v_m(x)$ for all nonzero $x \in \mathbb{R}^n$.*

Proof. If A is TP, then the inequality follows from Theorem 4.3.5. If A is not TP, choose $x \neq 0$ according to Lemma 4.2.8 to verify the converse.   □

Armed with a slight refinement of Theorem 4.3.5, we will be able to give another converse. The next lemma was probably known to Karlin, as a version of this result can be found in [Kar64, Thm. 1.5, pg. 223]. For brevity, we let $\mathrm{sgn}(a)$ denote the sign of the nonzero real number a.

Lemma 4.4.2 *If A is an m-by-n TP matrix and $v_M(Ax) = v_m(x)$ for some nonzero $x \in \mathbb{R}^n$, then the initial entries of x_m and $(Ax)_M$ have the same sign ($+$ or $-$).*

Proof. We begin by proving a slightly weaker version of this, which will be used later in establishing the desired statement. Suppose x is as given, set $y = Ax$, and assume that $v_m(y) = v_m(x)$. Then we will verify that the initial entries of x_m and y_m have the same sign. Observe that we may assume x is a totally nonzero vector of maximum sign variation; otherwise we may apply a column compression to A. Thus, assume that $v_m(x) = v_m(y) = n - 1$. Hence there exists a sequence of indices $1 \leq i_1 < i_2 < \cdots < i_n \leq m$ such that $y_{i_j} y_{i_{j+1}} < 0$ for $j = 1, 2, \ldots, n - 1$. Using the equation $Ax = y$, and setting $B = A[\{i_1, i_2, \ldots, i_n\}, N]$, we have, by Cramer's rule, that

$$x_1 = \frac{\det \begin{bmatrix} y_{i_1} & | & | & | \\ y_{i_2} & | & | & | \\ \vdots & b_2 & \cdots & b_n \\ y_{i_n} & | & | & | \end{bmatrix}}{\det B},$$

where b_i is the ith column of B. Expanding along column one in the numerator gives

$$x_1 = \frac{1}{\det B} \sum_{s=1}^{n} (-1)^{s+1} y_{i_s} \det B(\{s\}, \{1\}).$$

Since $\det B > 0$, $\det B(\{s\},\{1\}) > 0$, and $(-1)^{s+1}y_{i_s}$ is of constant sign, it follows that the sign of x_1 is the same as the sign of y_{i_1}. Since $v_m(y) = v_m(x)$, it is evident that the sign of x_1 must coincide with the initial sign of y. (We could have chosen y_{i_i} to be the initial entry of y_m.)

To establish Lemma 4.4.2, assume that $v_M(y) = v_m(x) = n - 1$, that x has been chosen to be totally nonzero and of maximum sign variation, and that $Ax = y$. Thus, there exists a sequence of indices $1 \le i_1 < i_2 < \cdots < i_n \le m$ such that $(y_M)_{i_j}(y_M)_{i_{j+1}} < 0$ for $j = 1, 2, \ldots, n - 1$. As above, we may assume that the position of the first nonzero entry of y, say y_j is included in the list of indices $\{i_1, i_2, \ldots, i_n\}$. Moreover, if $j > 1$, then all the signs of the entries in y_M before position j are uniquely determined by $\operatorname{sgn}(y_M)_k = (-1)^{j-k}\operatorname{sgn}(y_j)$ for $k < j$. In this case, all the positions before j must also appear in this list of selected indices for y_M. The remainder of the proof mirrors the argument above, and therefore we may conclude by an application of Cramer's rule that the sign of x_1 must match the sign of $(-1)^{j+1}y_j$, and hence the signs of the initials entries of x_m and y_M are the same.   □

Remark. We note that, because of Lemma 4.2.1, if there is equality in the inequality of any of Theorem 4.2.2, Corollary 4.2.4, Theorem 4.2.5, or Theorem 4.2.7 for a particular x, then either $x = 0$, or there is an initial sign match involving x and Ax. For example, in the case of Theorem 4.2.2, we must have a match in the signs of the initial entries of x_m and $(Ax)_m$. In the case of Theorem 4.2.5, we must have a match in the signs of the initial entries of x_M and $(Ax)_M$.

The next converse is best viewed as a natural question within the topic of linear interpolation problems for TP matrices. This result appeared in [GJ04] in 2004 with a complete proof, which we only sketch below.

Theorem 4.4.3 *Suppose that $x \in \mathbb{R}^n$, $y \in \mathbb{R}^m$, and $x \ne 0$. Then there is an m-by-n TP matrix A such that $Ax = y$ if and only if*

(i) $v_M(y) \le v_m(x)$ *and*

(ii) *if $v_M(y) = v_m(x)$, then the signs of the initial entries of x_m and y_M coincide.*

The proof of Theorem 4.4.3 relies on three essential ingredients. It is clear that the necessity of the conditions (i) and (ii) follow from Theorem 4.3.5 and Lemma 4.4.2. To verify sufficiency, induction on n is employed along with two basic facts. The first fact is, if there exists an InTN matrix A such that $Ax = y$, then there exists a TP matrix A' so that $A'x = y$. The second observation is, for any pair of nonzero vectors x, y using InTN EB matrices, it follows that there exist InTN matrices B_m, B_M, C_m, C_M satisfying $x_m = B_m x$, $y_m = C_m y$, $x = B_M x_m$, and $y = C_M y_M$. Thus, if there is a TP matrix A' such that $A'x_m = x_M$, then setting $A = C_M A' B_m$ produces a TP matrix that satisfies $Ax = y$.

Finally, we mention a third converse to Theorem 4.3.5 that is known [Sch30] but outside of the range of things that we prove here.

Theorem 4.4.4 *Let A be an m-by-n matrix. Then $v_M(Ax) \leq v_m(x)$ for all nonzero $x \in \mathbb{R}^n$ if and only if A is (strictly) sign regular.*

Chapter Five

The Spectral Structure of TN Matrices

5.0 INTRODUCTION

By "spectral structure" we mean facts about the eigenvalues and eigenvectors of matrices in a particular class. In view of the well-known Perron-Frobenius theory that describes the spectral structure of general entrywise nonnegative matrices, it is not surprising that TN matrices have an important and interesting special spectral structure. Nonetheless, that spectral structure is remarkably strong, identifying TN matrices as a very special class among nonnegative matrices. All eigenvalues are nonnegative real numbers, and the sign patterns of the entries of the eigenvectors are quite special as well.

This spectral structure is most apparent in the case of IITN matrices (that is, the classical oscillatory case originally described by Gantmacher and Krein); the eigenvalues are all positive and distinct (i.e., the multiplicity of each eigenvalues is one): $\lambda_1 > \lambda_2 > \cdots > \lambda_n > 0$. Moreover, if $u^{(k)}$ is an eigenvector associated with λ_k, the kth largest eigenvalue, $v(u^{(k)}) = k - 1$, and there are strong relationships between the positions of the "sign changes" in successive eigenvectors (which are described below). The vector $u^{(k)}$ may have zero components when $1 < k < n$, but, then, $v_m(u^{(k)}) = v_M(u^{(k)})$, so that $v(u^{(k)})$ is always unambiguously defined.

As with Perron-Frobenius, where the spectral structure is weakened as we broaden the class of entrywise nonnegative matrices under consideration, when we relax to IrTN, multiple zero eigenvalues are possible, and nontrivial Jordan structure, associated with the eigenvalue zero may appear; what may happen is described below. However, even in IrTN, the positive eigenvalues remain distinct. Further relaxation yields predictable weakening; in the reducible case, multiple positive eigenvalues with nontrivial Jordan structure may occur.

Other spectral structure is also of interest: the relationship between eigenvalues of a matrix and those of a principal submatrix; between the eigenvalues and diagonal entries; and converses, such as an inverse eigenvalue problem, to results mentioned earlier. When the first or last row and column are deleted, there is conventional interlacing by the eigenvalues of the resulting TN matrix. Since, unlike Hermitian matrices, the TN matrices are not permutationally similarity invariant, deletion of interior rows and columns may yield something different, and does. In this event, the sort of interlacing that occurs is weaker. There is majorization between the eigenvalues and diagonal entries, which may be viewed as a consequence of interlacing, and it is relatively straightforward that any distinct positive numbers occur as the eigenvalues of a TP matrix (and thus of an IITN matrix). Other converses will also be discussed.

Finally, if A and B are n-by-n and TN, then so is AB. What relationships are there among the eigenvalues of A, B and their product AB? A complete answer is not known, but we give some inequalities.

5.1 NOTATION AND TERMS

Some specialized notions are useful in describing the remarkable structure occurring in the eigenvectors of an IITN matrix. First, imagine a piecewise linear function f on the interval $[0, n-1]$ determined by the (ordered) components of a vector $u \in \mathbb{R}^n$. Let $f(i-1) = u_i$ and between $i-1$ and i the graph of f be given by the line segment joining $(i-1, u_i)$ and (i, u_{i+1}), $i = 1, 2, \ldots, n-1$. We call the graph of f the u-profile. Unless u is uniformly signed ("unisigned"), f (or, informally its u-profile) has zeros. A zero may be identified with its real argument r, $0 < r < n-1$. We call these 0 nodes. The number and position of nodes will be important to us. In general, the number of nodes, when finite, for u lies between $v_m(u)$ and $v_M(u)$. In the event that $u \in \mathbb{R}^n$ has k nodes $r_1 < r_2 < \cdots < r_k$, and $w \in \mathbb{R}^n$ has $k+1$ nodes $s_1 < s_2 < \cdots < s_{k+1}$, and

$$s_1 \leq r_1 \leq s_2 \leq r_2 \leq \cdots \leq r_k \leq s_{k+1},$$

we say that the nodes of u and w alternate. (We avoid the term "interlace" to prevent confusion with a different situation in which that term is traditionally used.)

Example 5.1.1 If $u = (3, 1, -2, -3, -1)^T$ and $w = (5, -3, -5, -1, 2)^T$, then the profiles of u and v are presented in Figure 5.1, and the nodes of u and w alternate (in this case, strictly).

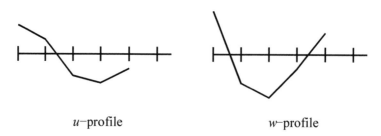

u–profile   w–profile

Figure 5.1 Vector profiles

Of course, the alternation of nodes may approximately be viewed as a kind of (weak) interlacing of the indices at which sign changes occur. In the above example, the first sign change in w occurs *before* the first sign change in u, and the second occurs after it. The notion of nodes is simply a way to make weak interlacing of the indices more precise, which will be especially needed in the case of eigenvectors of IITN matrices. We also note that similar notions were devised in [GK60] and in the survey paper [And87] for roughly the same purpose.

5.2 THE SPECTRA OF IITN MATRICES

Here, we consider only the eigenvalues of IITN matrices (i.e., the case of oscillatory matrices). Continuity considerations will then resolve similar issues for more general TN matrices.

We begin with a proof of the classical result (see [GK60] or [GK02]) on the eigenvalues of an IITN matrix.

Theorem 5.2.1 *Let A be an n-by-n IITN matrix. Then the eigenvalues of A are n distinct positive numbers.*

Proof. Assume that A is TP; otherwise we may replace A by A^k, in which A^k is TP for A in IITN. Suppose $\lambda_1, \lambda_2, \ldots, \lambda_n$ are the eigenvalues of A, arranged such that $|\lambda_1| \geq |\lambda_2| \geq \cdots \geq |\lambda_n|$. By Perron's theorem (see [HJ85]) it follows that λ_1 is real positive and strictly larger in modulus than any other eigenvalue of A. Hence λ_1 occurs as a simple eigenvalue of A. Suppose, to the contrary, that λ_j $(j \geq 2)$ is the first eigenvalue of A that is not a real number. It is clear that the product $\lambda_1\lambda_2 \cdots \lambda_j$ is an eigenvalue of $C_j(A)$ of maximum modulus, and hence, since $C_j(A)$ is an entrywise positive matrix, it follows from Perron's theorem that $\lambda_1\lambda_2 \cdots \lambda_j$ is a real positive number. Applying similar arguments to $C_{j-1}(A)$ implies that $\lambda_1\lambda_2 \cdots \lambda_{j-1}$ is a real positive number. Consequently, λ_j is a real positive number. Therefore, every eigenvalue of A is real and also positive.

Applying Perron's theorem to $C_j(A)$ yields,

$$\lambda_1\lambda_2 \cdots \lambda_j > \lambda_1\lambda_2 \cdots \lambda_{j-1}\lambda_{j+1},$$

which implies $\lambda_j > \lambda_{j+1}$. Thus the eigenvalues of A are distinct. $\square$

Because of Theorem 5.2.1, we may (and do) write the eigenvalues of an IITN matrix as

$$\lambda_1 > \lambda_2 > \cdots > \lambda_n > 0.$$

It is natural to ask whether there are any additional restrictions on the eigenvalues of an IITN matrix. The answer is no, as can be deduced from the following result that originally appeared in [BJ84].

Theorem 5.2.2 *If $\lambda_1 > \lambda_2 > \cdots > \lambda_n > 0$, then there is a a tridiagonal, n-by-n, IITN matrix A whose eigenvalues are $\lambda_1, \lambda_2, \ldots, \lambda_n$ and an n-by-n TP matrix B whose eigenvalues are $\lambda_1, \lambda_2, \ldots, \lambda_n$.*

Proof. Given the spectral data $\lambda_1 > \lambda_2 > \cdots > \lambda_n > 0$, it follows, according to [GW76], that there is a symmetric tridiagonal n-by-n matrix A with $a_{i,i+1} = a_{i+1,i} > 0$, $i = 1, 2, \ldots, n - 1$, whose eigenvalues are precisely $\lambda_1, \lambda_2, \ldots, \lambda_n$. The fact that A is also IITN follows from work in Section 1 of Chapter 0.

To verify the existence of a TP matrix with eigenvalues $\lambda_1, \lambda_2, \ldots, \lambda_n$, first set $\mu_i = \lambda_i^{1/(n-1)}$, for $i = 1, 2, \ldots, n$. Then $\mu_1 > \mu_2 > \cdots > \mu_n > 0$, and so there exists an IITN tridiagonal matrix C with eigenvalues $\mu_1, \mu_2, \ldots, \mu_n$. Setting $B = C^{n-1}$ it gives that C is TP and has eigenvalues $\lambda_1, \lambda_2, \ldots, \lambda_n$. $\square$

Because any TN matrix is a limit of TP matrices (see Chapter 2) and because a diagonal matrix with nonnegative diagonal entries is TN, we have the analogous results for TN matrices.

Corollary 5.2.3 *There is a TN matrix with eigenvalues $\lambda_1, \lambda_2, \ldots, \lambda_n$ if and only if $\lambda_1 \geq \lambda_2 \geq \cdots \geq \lambda_n \geq 0$.*

Corollary 5.2.4 *The n-by-n matrix A is similar to a TP matrix if and only if A has distinct positive eigenvalues.*

The analogous results for TN matrices are more subtle because of the possibility of nontrivial Jordan structure. However, one may say that the following is a result of Corollary 5.2.3.

Corollary 5.2.5 *An n-by-n diagonalizable matrix is similar to a TN matrix if and only if its eigenvalues are nonnegative real numbers.*

5.3 EIGENVECTOR PROPERTIES

In addition to eigenvalue properties, the Perron-Frobenius theory for general entrywise nonnegative matrices says that an identified eigenvector, that is associated with the spectral radius, is uniformly signed (usually taken to be positive), at least when the nonnegative matrix is irreducible. Other than that which follows from the principle of biorthogonality, no statement is made about any other eigenvector. For IITN matrices, very strong statements, both individually and relatively, may be made about every eigenvector. Because of the results of the prior section, we may suppose that the eigenvalues of the n-by-n IITN matrix A are

$$\lambda_1 > \lambda_2 > \cdots > \lambda_n > 0.$$

Associated with λ_k are a right eigenvector $u^{(k)} \in \mathbb{R}^n$ and a left eigenvector $w^{(k)} \in \mathbb{R}^n$, which, if we take the sign of the initial nonzero entry (which will be the first) to be positive and normalize, will be unique. Of course, $u^{(1)}$ ($w^{(1)}$) is entrywise positive by Perron-Frobenius, and again because of biorthogonality, no other $u^{(k)}$ $(w^{(k)})$, $k = 2, 3, \ldots, n$, can be positive. But the number of sign changes is predictable namely,

$$v_m(u^{(k)}) = v_M(u^{(k)}) = v(u^{(k)}) = k - 1, \ k = 1, 2, \ldots, n.$$

In particular, $u^{(1)}$ and $u^{(n)}$ are totally nonzero (the others may have zero entries), and $u^{(n)}$ strictly alternates in sign. Moreover, the nodes of $u^{(k)}$ and $u^{(k+1)}$, $k = 2, 3, \ldots, n - 1$, alternate. Of course, all of the above is valid for $w^{(k)}$, as well.

Two lemmas underlie much of the special structure of the eigenvectors. We refer to the first as the *eigenvector determinant conditions*. Similar conditions also appear, for natural reasons, in [GK60] and in [And87] in the context of eigenvalues and eigenvectors of TN matrices.

Lemma 5.3.1 *Let U be the matrix whose columns are the right eigenvectors of an IITN matrix, ordered as above (and W be the matrix whose rows are the left eigenvectors of an IITN matrix, ordered as above). Then, for $k = 1, 2, \ldots, n$, every k-by-k minor formed from the first k columns of U (first k rows of W) has the same (nonzero) sign.*

Proof. The idea of the proof is to establish that all such minors in question actually make up the coordinates of the eigenvector corresponding to the eigenvalue (Perron root) $\lambda_1 \lambda_2 \cdots \lambda_k$ for the positive matrix $C_k(A)$.

Without loss of generality, assume that A is TP, and let $\lambda_1 > \lambda_2 > \cdots > \lambda_n > 0$ be the eigenvalues of A with associated right eigenvectors $u^{(1)}, u^{(2)}, \ldots, u^{(n)} \in \mathbb{R}^n$. Hence, using U defined above, we know that $AU = UD$.

For each $k = 1, 2, \ldots, n - 1$ the kth compound $C_k(A)$ is entrywise positive with Perron root $\lambda_1 \lambda_2 \cdots \lambda_k$. In addition, this Perron root has a (unique) positive eigenvector.

Define the vector $\tilde{u}$ in $\mathbb{R}^{\binom{n}{k}}$ indexed by all k-subsets of N (conformally with $C_k(A)$), so the entry corresponding to $\alpha \subset N$ is given by $\det U[\alpha, \{1, 2, \ldots, k\}]$. Consider the α coordinate of the product $C_k(A)\tilde{u}$:

$$(C_k(A)\tilde{u})_\alpha = \sum_\beta \det A[\alpha, \beta] \det U[\beta, \{1, 2, \ldots, k\}]$$

$$= \det AU[\alpha, \{1, 2 \ldots, k\}], \text{ by Cauchy–Binet.}$$

Using the fact that $AU = UD$ we have

$$\det AU[\alpha, \{1, 2 \ldots, k\}] = \det UD[\alpha, \{1, 2 \ldots, k\}]$$

$$= \sum_\beta \det U[\alpha, \beta] \det D[\beta, \{1, 2, \ldots, k\}]$$

$$= \det U[\alpha, \{1, 2, \ldots, k\}] \det D[\{1, 2, \ldots, k\}]$$

$$= (\lambda_1 \lambda_2 \cdots \lambda_k) \det U[\alpha, \{1, 2 \ldots, k\}]$$

$$= (\lambda_1 \lambda_2 \cdots \lambda_k) \tilde{u}_\alpha.$$

Putting together the equations above, we deduce that $\tilde{u}$ is an eigenvector of $C_k(A)$ corresponding to the eigenvalue $\lambda_1 \lambda_2 \cdots \lambda_k$. Hence $\tilde{u}$ is a totally nonzero vector all of whose entries are either positive or negative. This completes the proof of the claim.   □

Our next result along these lines involves key inequalities on the sign variation of vectors in the span of the right and left eigenvectors of an IITN matrix.

Lemma 5.3.2 *For an n-by-n IITN matrix A with the above notation, if u is any nontrivial linear combination of $u^{(p)}, u^{(p+1)}, \ldots u^{(q)}$ (w of $w^{(p)}, w^{(p+1)}, \ldots, w^{(q)}$), then*

$$p - 1 \leq v_m(u) \leq v_M(u) \leq q - 1,$$

$$(p - 1 \leq v_m(w) \leq v_M(w) \leq q - 1).$$

Proof. The inequality $v_M(u) \leq q-1$ follows from Theorem 4.3.1 and Lemma 5.3.1. To demonstrate the inequality $p-1 \leq v_m(u)$, we will instead show that $v_M(Su) \leq n - p$, and then use equation (4.1), which states that $v_M(Su) + v_m(u) = n - 1$, to verify the desired inequality. If we set $u^{*(s)} = Su^{(s)}$ for each right eigenvector $u^{(s)}$, then $u^{*(s)}$ is the right eigenvector for the eigenvalue of the unsigned adjugate matrix of A lying in position $n - s + 1$. This follows since the eigenvalues of this matrix are the reciprocals of the eigenvalues of A. Since the unsigned adjugate matrix of A is also TP, we may apply Theorem 4.3.1 and Lemma 5.3.1 again to deduce that $v_M(Su) \leq (n - p + 1) - 1 = n - p$, since Su is a linear combination of eigenvectors ranging up to the eigenvalue in position $n - p + 1$. Hence by equation (4.1), we have that $v_m(u) \geq n - 1 - (n - p) = p - 1$, which completes the proof. □

We may now state the main facts about eigenvectors of IITN matrices.

Theorem 5.3.3 *If A is an n-by-n IITN matrix and $u^{(k)}$ ($w^{(k)}$) is the right (left) eigenvector associated with λ_k, the kth largest eigenvalue of A, then*

$$v(u^{(k)}) = k - 1 \ (v(w^{(k)}) = k - 1),$$

for $k = 1, 2, \ldots, n$. Moreover, the first and last components of $u^{(k)}$ ($w^{(k)}$) are nonzero, and $u^{(1)}$ and $w^{(n)}$ ($w^{(1)}$ $w^{(n)}$) are totally nonzero.

Proof. Using Lemma 5.3.2 we may take $p = q = k$ and deduce that

$$k - 1 \leq v_m(u^{(k)}) \leq v_M(u^{(k)}) \leq k - 1,$$

which of course implies that $v(u^{(k)}) = k - 1$. The fact that $v_m(u^{(k)}) = v_M(u^{(k)})$ implies that the initial and terminal entries of $u^{(k)}$ are both nonzero. The final claim follows from Perron's theorem for entrywise positive matrices, applied to both A and the unsigned adjugate of A.   □

Example 5.3.4 Consider the 3-by-3 IITN matrix

$$A = \begin{bmatrix} 1 & 1 & 0 \\ 1 & 3 & 1 \\ 0 & 1 & 1 \end{bmatrix},$$

with eigenvalues: $2 + \sqrt{3} > 1 > 2 - \sqrt{3} > 0$. Then an eigenvector associated with $\lambda_2 = 1$ is

$$u^{(2)} = \begin{bmatrix} 1 \\ 0 \\ -1 \end{bmatrix},$$

so that $v(u^{(2)}) = 1$, in spite of the fact that $u^{(2)}$ has a zero-component (in an interior position).

In addition to the sign pattern requirements on each eigenvector, associated with the relative position of its eigenvalue, there are also relative restrictions on the position at which the "sign changes" occur. These are conveniently described in terms of the nodes of the eigenvectors defined earlier. Note that the proof described below is essentially the proof provided in [GK02].

Theorem 5.3.5 *If $A \in IITN$ is n-by-n and $u^{(k)}$ $(w^{(k)})$ is the right (left) eigenvector associated with λ_k, the kth largest eigenvalue of A, then the k nodes of $u^{(k+1)}$ $(w^{(k+1)})$ alternate with the $k - 1$ nodes of $u^{(k)}$ $(w^{(k)})$, $k = 2, 3, \ldots, n - 1$.*

Proof. Recall that $v(u^{(k)}) = k - 1$ and $v(u^{(k+1)}) = k$, by Theorem 5.3.3, and if $u = cu^{(k)} + du^{(k+1)}$ $(c^2 + d^2 > 0)$, then Lemma 5.3.2 gives

$$k - 1 \leq v_m(u) \leq v_M(u) \leq k.$$

To reach a contradiction, let us assume that between some two consecutive nodes (α, β) of the $u^{(k+1)}$-profile there are no nodes of the $u^{(k)}$-profile. Then it follows, from the definition of profiles, that the $u^{(k)}$-profile and $u^{(k+1)}$-profile are different from zero and have constant sign (possibly different) on the interval (α, β). Up to multiplication by a scalar, we may assume that both profiles are positive on (α, β). For even further simplicity, we assume that $d = -1$, so that $u = cu^{(k)} - u^{(k+1)}$. The first claim to verify is that the $u^{(k)}$-profile is nonzero at both α and β. If not, suppose the $u^{(k)}$-profile is zero at α. Then for any c, the u-profile is zero at α. Now choose a number γ so that

$$\alpha < \gamma < \min\{\beta, \lfloor \alpha + 1 \rfloor\}.$$

Then set $c = \frac{u^{(k+1)}(\gamma)}{u^{(k)}(\gamma)}$ (here $u^{(\cdot)}(\gamma)$ means the image of the $u^{(\cdot)}$-profile at the point γ). For each such point γ we have $u(\gamma) = 0$. Thus a nontrivial segment of the u-profile is zero, which must imply that two consecutive coordinates of u must be zero. In this case, we then have $v_M(u) \geq v_m(u) + 2$, which contradicts the fact that $k - 1 \leq v_m(u) \leq v_M(u) \leq k$, and verifies the desired claim. Note that this claim also establishes that the $u^{(k)}$-profile and $u^{(k+1)}$-profile do not have any common nodes.

Since $u = cu^{(k)} - u^{(k+1)}$, and both the $u^{(k)}$-profile and the $u^{(k+1)}$-profile are positive on the interval (α, β), it follows that for large enough $c > 0$ this u-profile is positive on $[\alpha, \beta]$. Therefore we may choose a value for c, call it c_0, at which $u_0(x) := c_0 u^{(k)}(x) - u^{(k+1)}(x)$ first vanishes at some δ (i.e., $u_0(\delta) = 0$) with $\alpha \leq \delta \leq \beta$. Evidently, $c_0 > 0$, and $\alpha < \delta < \beta$. Hence the u_0-profile lies on one side (positive side) of the axis with a zero at δ. If a nontrivial segment of u_0 is zero, then we reach a contradiction as above, and if the u_0-profile is positive before δ, zero at δ, and positive after δ (in a small neighborhood of δ), then the vector u_0 has a zero coordinate surrounded by two positive coordinates. Thus $v_M(u_0) \geq v_m(u_0) + 2$, a contradiction.

So, there is at least one node of the $u^{(k)}$-profile between two successive nodes of the $u^{(k+1)}$-profile. Since $v(u^{(k)}) = k - 1$ and $v(u^{(k+1)}) = k$, there can only be exactly one node of the $u^{(k)}$-profile between two successive nodes of the $u^{(k+1)}$-profile, which completes the proof. □

We remark here that the proof included above is very similar to the proof provided in [GK02].

Theorems 5.3.3 and 5.3.5 do not fully characterize the eigenvectors of an IITN matrix. First of all, the left eigenvectors are precisely linked to the right eigenvectors via

$$A = UDU^{-1},$$

so that the rows of U^{-1} are the left eigenvectors if the columns of U are the right eigenvectors (with the order in U^{-1} being top-to-bottom, relative to the right-to-left order in U). Second, the eigenvector sign conditions (in Lemma 5.3.2 and Theorem 5.3.3) ultimately follow from the more precise determinantal conditions of Lemma 5.3.1, and, finally, the mentioned inversion linkage between the left and right eigenvectors means, for example, that there are conditions on the right eigenvectors not explicitly mentioned in Theorems 5.3.3 and 5.3.5.

Example 5.3.6 Suppose

$$U = \begin{bmatrix} 1 & a & x \\ 1 & b & -y \\ 1 & -c & z \end{bmatrix}$$

represents the left eigenvectors of a 3-by-3 TP matrix (we are implicitly assuming that all variables are positive). As written, the columns of U satisfy the eigenvector sign conditions mentioned above; to ensure that the determinantal conditions hold, we must additionally assume that $a > b$ (so that all minors in the first two columns are negative). In this case, observe (by Laplace expansion along the 3rd column) that the determinant of U is necessarily negative. Moreover,

$$U^{-1} = \frac{1}{\det U} \begin{bmatrix} bz - yc & -(az + xc) & -(ay + bx) \\ -(z + y) & z - x & y + x \\ -(c + b) & c + a & b - a \end{bmatrix}.$$

From the eigenvector-eigenvalue equations, the rows of U^{-1} are the corresponding left eigenvectors, so they in turn must also satisfy the eigenvector sign conditions and the determinantal conditions. By inspection, U^{-1} satisfies these conditions only if $a > b$ (already assumed above) and $bz < yc$

(that is, the bottom right 2-by-2 minor of U is negative). We then note that this second inequality $(bz < yc)$ is an additional constraint (not explicitly covered in the results above) which must be satisfied by the entries of U.

By reverting to Lemma 5.3.1, however, we may offer the following comprehensive characterization regarding the eigenvectors of an IITN matrix.

Theorem 5.3.7 *Let column vectors* $u^{(1)}, u^{(2)}, \ldots, u^{(n)} \in \mathbb{R}^n$ *and row vectors* $w^{(1)T}, w^{(2)T}, \ldots, w^{(n)T} \in \mathbb{R}^n$ *be given, such that* $WU = I$, *with*

$$U = \begin{bmatrix} u^{(1)} & u^{(2)} & \cdots & u^{(n)} \end{bmatrix} \text{ and } W = \begin{bmatrix} w^{(1)T}, \\ w^{(2)T} \\ \vdots \\ w^{(n)T} \end{bmatrix}.$$

Then the following statements are equivalent.

(1) *For* $1 \le p \le q \le n$, $p - 1 \le v_m(u) \le v_M(u) \le q - 1$, *for* $u \ne 0$ *in the span of* $\{u^{(p)}, u^{(p+1)}, \ldots, u^{(q)}\}$ *and* $p - 1 \le v_m(w) \le v_M(w) \le q - 1$, *for* $w \ne 0$ *in the span of* $\{w^{(p)}, w^{(p+1)}, \ldots, w^{(q)}\}$,

(2) *For each* $q = 1, 2, \ldots, n$, $v_M(u) \le q - 1$, *for* $u \ne 0$ *in the span of* $\{u^{(1)}, u^{(2)}, \ldots, u^{(q)}\}$ *and* $v_M(w) \le q - 1$, *for* $w \ne 0$ *in the span of* $\{w^{(1)}, w^{(2)}, \ldots, w^{(q)}\}$.

(3) *The eigenvector determinantal conditions (of Lemma 5.3.1) hold for* $u^{(1)}, u^{(2)}, \ldots, u^{(n)}$ *and for* $w^{(1)T}, w^{(2)T}, \ldots, w^{(n)T}$,

(4) *There exist thresholds* $r_1, r_2, \ldots, r_{n-1} > 1$ *such that whenever* $r_i \lambda_{i+1} < \lambda_i$ *with* $\lambda_n > 0$ *and* $D = \text{diag}(\lambda_1, \ldots, \lambda_n)$, *then* $A = UDW$ *is TP.*

(5) *If* $D = \text{diag}(\lambda_1, \ldots, \lambda_n)$, *with* $\lambda_1 > \lambda_2 > \cdots > \lambda_n > 0$, *and* $A = UDW$, *then there is a positive power of* A *that is TP,*

(6) *There is a TP matrix with right eigenvectors* $u^{(1)}, u^{(2)}, \ldots, u^{(n)}$ *and left eigenvectors* $w^{(1)}, w^{(2)}, \ldots, w^{(n)}$ *arranged in standard order.*

(7) *There is an IITN matrix with right eigenvectors* $u^{(1)}, u^{(2)}, \ldots, u^{(n)}$ *and left eigenvectors* $w^{(1)}, w^{(2)}, \ldots, w^{(n)}$ *arranged in standard order.*

Proof. First we demonstrate that conditions (1), (2), and (3) are equivalent; we then show that statements (3) through (7) are equivalent.

Using Theorem 4.3.1 and Lemma 5.3.2, we know that $(3) \Rightarrow (1)$. The implication $(1) \Rightarrow (2)$ is trivial, and Theorem 4.3.1 also proves that $(2) \Rightarrow (3)$.

$(3) \Rightarrow (4)$: Assume that the determinantal conditions are satisfied by U and W (as according to Lemma 5.3.1). Consider n positive numbers $\lambda_1, \lambda_2, \ldots, \lambda_n$, and let $D = \text{diag}(\lambda_1, \ldots, \lambda_n)$. Define the n-by-n matrix $A = UDW$.

The key equation to examine is the following. For any $\alpha, \beta \subset N$, with $|\alpha| = |\beta| = k$,

$$\det A[\alpha, \beta] = \det U[\alpha, \{1, 2, \ldots, k\}](\lambda_1 \lambda_2 \cdots \lambda_k) \det W[\{1, 2, \ldots, k\}, \beta]$$

$$+ \sum_{\substack{\gamma = \{i_1, i_2, \ldots, i_k\} \\ \gamma \neq \{1, \ldots, k\}}} \det U[\alpha, \gamma](\lambda_{i_1} \lambda_{i_2} \cdots \lambda_{i_k}) \det W[\gamma, \beta]. \tag{5.1}$$

An important item to note is that since $WU = I$ and (3) is assumed, it follows that for any pair α, β, $\det U[\alpha, \{1, 2, \ldots, k\}]$ and $\det W[\{1, 2, \ldots, k\}, \beta]$ are nonzero and of the same sign. Suppose that there are real numbers $r_1, r_2, \ldots, r_{n-1} > 1$ so that $r_i \lambda_{i+1} < \lambda_i$. Then it follows that

$$\frac{\lambda_{i_1} \lambda_{i_2} \cdots \lambda_{i_k}}{\lambda_1 \lambda_2 \cdots \lambda_k} \leq \left(\frac{1}{\prod_{j=1}^{i_1 - 1} r_j} \right) \left(\frac{1}{\prod_{j=2}^{i_2 - 1} r_j} \right) \cdots \left(\frac{1}{\prod_{j=k}^{i_k - 1} r_j} \right) \tag{5.2}$$

for any k-subset $\{i_1, i_2 \ldots, i_k\}$ of N not equal to $\{1, 2, \ldots, k\}$.

Hence to verify that $\det A[\alpha, \beta] > 0$, it is enough (keeping in mind (5.1)) to choose r_i larger than one so that

$$1 > \sum_{\gamma \neq \{1, \ldots, k\}} \left| \frac{\det U[\alpha, \gamma]}{\det U[\alpha, \{1, 2, \ldots, k\}]} \right| \left| \frac{(\lambda_{i_1} \lambda_{i_2} \cdots \lambda_{i_k})}{(\lambda_1 \lambda_2 \cdots \lambda_k)} \right| \left| \frac{\det W[\gamma, \beta]}{\det W[\{1, 2, \ldots, k\}, \beta]} \right|.$$

Using (5.2), it is clear that each r_i can be chosen large enough to ensure that the finite sum above is less than one. Furthermore, they can be chosen even larger, if necessary, so that this sum is less than one for each pair of index sets α and β. Thus if the determinantal conditions are satisfied by the matrices U and W, then such thresholds do indeed hold. (Observe that the thresholds do not depend on λ but do depend on U and W.)

$(4) \Rightarrow (5)$: Suppose the thresholds conditions are satisfied: There exist (positive) numbers $r_1, r_2, \ldots, r_{n-1} > 1$ such that whenever $r_i \lambda_{i+1} < \lambda_i$ with $\lambda_n > 0$ and $D = \text{diag}(\lambda_1, \ldots, \lambda_n)$, then $A = UDW$ is TP.

Suppose $D = \text{diag}(\lambda_1, \ldots, \lambda_n)$, with $\lambda_1 > \lambda_2 > \cdots > \lambda_n > 0$ are given and let $A = UDW$. Then for every positive integer k, $A^k = UD^kW$. Thus for each $i = 1, 2, \ldots n - 1$, since $\lambda_i / \lambda_{i+1} > 1$, there exists a $k_i > 1$ so that

$$\left(\frac{\lambda_i}{\lambda_{i+1}} \right)^{k_i} > r_i > 1,$$

for any fixed r_i. Choosing $k = \max(k_1, k_2, \ldots, k_n)$, it follows that the eigenvalues of A^k meet the threshold conditions. Thus A^k is TP.

The implications $(5) \Rightarrow (6)$ and $(6) \Rightarrow (7)$ are trivial, and Lemma 5.3.1 shows that $(7) \Rightarrow (3)$. This completes the proof of the theorem. $\square$

5.4 THE IRREDUCIBLE CASE

Relaxing IITN matrices to IrTN does indeed allow for more general types of Jordan or spectral structure. For instance, not all IrTN matrices are diagonalizable, as can be seen by considering the following basic, yet informative, example,

$$A = \begin{bmatrix} 3 & 3 & 2 & 1 \\ 2 & 2 & 3 & 2 \\ 1 & 1 & 2 & 3 \\ 1 & 1 & 2 & 3 \end{bmatrix}.$$

Then A can be shown to be IrTN and the Jordan canonical form of A is given by

$$\begin{bmatrix} \begin{array}{cc|cc} a & 0 & & \\ 0 & b & & \\ \hline & & 0 & 1 \\ & & 0 & 0 \end{array} \end{bmatrix},$$

where $a = 7.8284$ and $b = 2.1716$. We do note that the positive eigenvalues of A are still distinct, which is not a coincidence.

If A is an m-by-n matrix, then recall that the *rank* of A, denoted by rank(A), is the size of its largest invertible square submatrix. It is well known that $n - \text{rank}(A)$ is equal to the geometric multiplicity of the zero eigenvalue (i.e., the nullity of A).

Similarly, the *principal rank* of an n-by-n matrix A, denoted by p-rank(A), is the size of the largest invertible principal submatrix of A. Evidently, $0 \leq$ p-rank(A) $\leq$ rank(A) $\leq \min(m, n)$. Furthermore, when A is assumed to be TN, $n-\text{p-rank}(A)$ is equal to the algebraic multiplicity of the zero eigenvalue of A. To see this, observe that the characteristic polynomial of A can be written as

$$\begin{aligned} f(x) &= x^n - c_1 x^{n-1} + \cdots \pm c_p x^{n-p} \\ &= x^{n-p}(x^p - c_1 x^{p-1} + \cdots \pm c_p), \end{aligned}$$

where p is the principal rank of A. In this case, $c_p \neq 0$ since c_p represents the sum of all p-by-p principal minors of A, and at least one of them is positive while the remaining are nonnegative (no accidental cancellation). As a consequence, if A is TN, then p-rank(A) is equal to the number of nonzero (or in this case positive) eigenvalues of A. This connection does not hold for general matrices, as can be seen from the following simple example,

$$\begin{bmatrix} 1 & 1 & 1 \\ 1 & 2 & 3 \\ 1 & 1 & 1 \end{bmatrix},$$

which has rank and p-rank equal to two, but only one nonzero eigenvalue.

In summary we have that, if k is the smallest positive integer such that rank(A^k) = p-rank(A), then k is equal to the size of the largest Jordan block corresponding to the eigenvalue zero (under the assumption that A is TN).

Consider the following illustrative example (compare with above).

Example 5.4.1 Let

$$A = \begin{bmatrix} 3 & 2 & 1 \\ 2 & 3 & 2 \\ 1 & 2 & 3 \end{bmatrix}.$$

Then A is a 3-by-3 TP matrix.

Consider the 4-by-4 irreducible TN matrix

$$B = \left[\begin{array}{c|ccc} 3 & 3 & 2 & 1 \\ 2 & 2 & 3 & 2 \\ 1 & 1 & 2 & 3 \\ \hline 1 & 1 & 2 & 3 \end{array} \right].$$

Then rank(B) = rank(A) = 3 and p-rank(B) = p-rank$(A) - 1 = 2$, from which it follows that B has one 2-by-2 Jordan block corresponding to zero and two distinct positive eigenvalues.

In general, this "asymmetric" bordering of a TN matrix preserves the rank but may change the principal rank. Observe that if we border the matrix B above in a similar manner, then the resulting TN matrix has the same rank and principal rank as B.

Before we come to our main observations regarding the spectral structures of IrTN matrices, we need the following additional definition, which first appeared in [FGJ00].

Definition 5.4.2 Let A and B be two square matrices, not necessarily of the same size. Then we say that A and B have the *same nonzero Jordan structure* if the distinct nonzero eigenvalues of A and B can be put into 1-1 correspondence so that each corresponding pair has the same number and sizes of Jordan blocks. Furthermore, if A and B are the same size, we say that A and B have the same *qualitative Jordan structure* if they have the same nonzero Jordan structure and if the number and sizes of the Jordan blocks corresponding to zero coincide.

For example, if

$$A = \left[\begin{array}{cc|c|cc|cc} a & 1 & & & & & \\ & a & & & & & \\ \hline & & b & & & & \\ \hline & & & c & & & \\ & & & & c & & \\ \hline & & & & & 0 & 1 \\ & & & & & & 0 \end{array} \right] \quad \text{and } B = \left[\begin{array}{c|cc|cc|cc} x & & & & & & \\ \hline & y & 1 & & & & \\ & & y & & & & \\ \hline & & & x & & & \\ & & & & z & & \\ \hline & & & & & 0 & \\ & & & & & & 0 \end{array} \right].$$

Then A and B have the same nonzero Jordan structure, but not the same qualitative Jordan structure. Note that, two matrices that are similar necessarily have the same qualitative Jordan structure.

To establish that the positive eigenvalues of an IrTN matrix are distinct, we make use of a key reduction lemma (see Lemma 5.4.3), from which the main result will follow by a sequential application (see [FGJ00] for original reference).

In [RH72] a similar "reduction-type" result was given to show that, if A is InTN, then there exist an InTN matrix S and a tridiagonal InTN matrix T such that $TS = SA$. This fact was later extended to singular TN matrices [Cry76].

Since we are concerned with IrTN matrices, the reduction result needed here is different in desired outcome than the one provided in [RH72]. We state and prove this reduction result now; see also [FGJ00].

Lemma 5.4.3 (basic lemma) *Suppose that $A = [a_{ij}]$ is an n-by-n irreducible TN matrix such that for some fixed $i < j < n$, $a_{lm} = 0$ for all $l < i$, $m \geq j$, $a_{i,j+1} > 0$, and $a_{it} = 0$ for all $t > j+1$. Then there exists an irreducible TN matrix A' such that*

(i) *$(A')_{lm} = 0$ for all $l < i$, $m \geq j$, $(A')_{i,j+1} = 0$, and*

(ii) *A' either is n-by-n and similar to A or is $(n-1)$-by-$(n-1)$ and has the same nonzero Jordan structure as A.*

Proof. By our assumptions, $a_{lj} = a_{l,j+1} = 0$ for $l < i$. Also, since A is irreducible and $a_{i,j+1} > 0$, a_{ij} is also positive. Use column j to eliminate $a_{i,j+1}$ via the elementary upper triangular bidiagonal nonsingular matrix $U_j = I - \alpha e_j e_{j+1}^T$. Consider the matrix $A' = U_j^{-1} A U_j$. From Chapter 2, we have shown that AU_j is TN, and since S^{-1} is TN, we have $A' = U_j^{-1} A U_j$ is TN. Clearly the $(i, j+1)$ entry of A' is zero. Observe that A' will be in double echelon form, unless A' contains a zero column, which necessarily must be column $j+1$. Assume for now that column $j+1$ of A' is nonzero. Then we must show that A' is irreducible. Note that, by construction, and the irreducibility of A, $(A')_{k,k+1} = a_{k,k+1} > 0$ for $k \neq j$, $(A')_{k,k-1} = a_{k,k-1} > 0$ for $k \neq j, j+1$, and $(A')_{j,j-1} = a_{j,j-1} + \alpha a_{j+1,j-1} > 0$ for $k \neq j$. Thus, by Corollary 1.6.6, we only need to show that $(A')_{j,j+1} > 0$ and $(A')_{j+2,j+1} > 0$. Since $(A')_{j,j+2} = a_{j,j+2} + \alpha a_{j+1,j+2}$ is positive, so is $(A')_{j,j+1}$ (recall that A' is in double echelon form). Now consider $(A')_{j+2,j+1} = a_{j+2,j+1} - \alpha a_{j+2,j}$. Then either $a_{j+2,j} = 0$, and therefore $(A')_{j+2,j+1} > 0$, or both $(A')_{j+2,j} = a_{j+2,j}$ and $(A')_{j+2,j+2} = a_{j+2,j+2}$ are positive, from which the positivity of $(A')_{j+2,j+1}$ follows from the double echelon form of A'. Thus, A' is irreducible. Finally, suppose the $(j+1)$th column of A' is zero. Then (as in [RH72] and [Cry76]) consider the matrix obtained by deleting the $(j+1)$th row and column of A', and denote it by A''. It is not hard to see that A' is similar to a matrix $\begin{bmatrix} A'' & 0 \\ * & 0 \end{bmatrix}$. Therefore, the nonzero Jordan structure of A'' is the same as A', which in turn is the same as A. Moreover, since A'' is a submatrix of A', A'' is TN. The only point remaining to prove is that A'' is irreducible. To this end, it suffices to show that $(A')_{j+2,j}$ and $(A')_{j,j+2}$ are positive. But $(A')_{j,j+2} = a_{j,j+2} + \alpha a_{j+1,j+2} > 0$, as A is irreducible. For the same reason, since $(A')_{j+2,j+1} = a_{j+2,j+1} - \alpha a_{j+2,j} = 0$, we have $(A')_{j+2,j} = a_{j+2,j} > 0$ and the proof is complete.   □

We may now demonstrate that the positive eigenvalues of an IrTN matrix are distinct. As a culmination we will generalize the original result in [GK60] on the eigenvalues of an IITN matrix (see [FGJ00]).

Theorem 5.4.4 *Let A be an irreducible TN matrix. Then there exists an irreducible tridiagonal TN matrix T (not necessarily of the same size as A) with the same nonzero Jordan structure as A. Moreover, T is obtained from A by a sequence of similarity transformations and projections.*

Proof. Sequential application of the basic lemma (Lemma 5.4.3) results in a k-by-k ($k \leq n$) irreducible lower Hessenberg TN matrix L with the same nonzero Jordan structure as A. Consider the matrix $U = L^T$. Observe that U is upper Hessenberg. Since this property is preserved under similarity transformations by upper triangular matrices, if a zero column is produced via the procedure described in the proof of Lemma 5.4.3, it must be the last column. In this case, deleting the last column and row then produces an upper Hessenberg matrix (see Example 5.4.7 below for definition) of smaller dimension with the same nonzero Jordan structure as U. Applying Lemma 5.4.3 repeatedly, we obtain an irreducible TN tridiagonal matrix T^T with the same nonzero Jordan structure as A. Then, T satisfies all the conditions of the theorem.   □

The next result represents a fundamental result on the positive eigenvalues of an IrTN matrix, and may be viewed as a generalization of the distinctness of the positive eigenvalues of IITN matrices, as IITN $\subset$ IrTN (see [FGJ00]).

Theorem 5.4.5 *Let A be an n-by-n irreducible TN matrix. Then the positive eigenvalues of A are distinct.*

Proof. By the previous theorem, there exists an irreducible TN tridiagonal matrix T with the same nonzero Jordan structure as A. By Lemma 0.1.1 the positive eigenvalues of T are distinct; hence the positive eigenvalues of A are distinct.   □

In the event A is IITN, then as a consequence of the proof of Theorem 5.4.4, it follows that the tridiagonal matrix T produced is, in fact, similar to A, which gives

Corollary 5.4.6 *The eigenvalues of an IITN matrix are real, positive, and distinct.*

The size of the tridiagonal matrix (T) obtained in Theorem 5.4.5 is either the number of nonzero eigenvalues of A or this number plus one (in the event T is singular).

Observe that the number of nonzero eigenvalues of A will play a central role in our analysis of the qualitative Jordan structures of IrTN matrices.

Example 5.4.7 A matrix $B = [b_{ij}]$ is said to be a *lower Hessenberg matrix* if $b_{ij} = 0$ for all i, j with $i + 1 < j$. Consider the n-by-n lower Hessenberg

(0,1)-matrix

$$H = \begin{bmatrix} 1 & 1 & 0 & \cdots & 0 \\ 1 & 1 & 1 & \ddots & \vdots \\ \vdots & \vdots & \ddots & \ddots & 0 \\ 1 & 1 & \cdots & 1 & 1 \\ 1 & 1 & \cdots & 1 & 1 \end{bmatrix}.$$

Then H is an irreducible TN matrix with rank $n - 1$. (Observe that H is singular and that $H[\{1, 2, \ldots, n - 1\}, \{2, 3, \ldots, n\}]$ is a nonsingular lower triangular matrix.) In fact, p-rank$(H) = \lceil \frac{n}{2} \rceil$. To see this, observe that if n is odd (even), then $H[\{1, 3, 5, \ldots, n\}]$ ($H[\{2, 4, 6, \ldots, n\}]$) is a nonsingular lower triangular matrix. Hence p-rank$(H) \geq \lceil \frac{n}{2} \rceil$. To show p-rank$(H) \leq \lceil \frac{n}{2} \rceil$, suppose there exists an index set α such that $|\alpha| > \lceil \frac{n}{2} \rceil$ and $\det H[\alpha] > 0$. Then α must contain at least one consecutive pair of indices, say i and $i + 1$, where $1 < i < n - 1$. Since H is lower Hessenberg and $\det H[\alpha] > 0$, it follows that index $i + 2 \in \alpha$. Continuing in this manner will show that $H[\alpha]$ is singular, since either both indices $n - 1$ and n will be in α, or the maximum index in α is less than n, in which case there will exist a pair of indices $k, k + 1$ in α and $k + 2$ not in α. Thus p-rank$(H) = \lceil \frac{n}{2} \rceil$.

Much more is known about the relationships between the rank, p-rank, and overall Jordan canonical form of a given IrTN matrix, and we encourage the reader to consult the work [FGJ00] as a primer into this area.

For example, one topic of interest is characterizing all the triples

$$(n, \text{rank}(A), \text{p} - \text{rank}(A)),$$

where n is the size of a square TN matrix A.

The fact that any TN matrix has a bidiagonal factorization (see Chapter 2) is an extremely useful tool when considering Jordan structures of TN matrices. In particular, the quantities rank and principal rank of a given matrix and irreducibility can all be interpreted via the diagram associated with a particular bidiagonal factorization of a TN matrix. For example, from a given diagram we can verify whether the associated TN matrix is irreducible by determining whether there exists a path in this diagram from any index i to each of $i - 1$, i, and $i + 1$ (ignoring $i - 1$ when $i = 1$, and $i + 1$ when $i = n$), by Corollary 1.6.6. Similarly, since rank and principal rank are defined in terms of nonsingular submatrices, it follows that rank and principal rank can be interpreted as the largest collection of vertex disjoint paths beginning on the left and terminating on the right, in the diagram, and the largest collection of vertex disjoint paths that begin and terminate in the same index set, respectively.

We highlight some of the key observations along these lines, and we encourage the reader to consult [FGJ00] for additional details. We begin by studying the triples $(n, \text{rank}, \text{p-rank})$ among the class of irreducible TN matrices. First, observe that the triples $(n, 1, 1)$ and (n, n, n) certainly exist, for all $n \geq 1$, by considering the matrix J of all ones, and any n-by-n TP matrix, respectively. Along these lines, we say that a triple (n, r, p) is *realizable* if there exists an n-by-n IrTN matrix A with rank r and p-rank p.

A preliminary result on these triples, with fixed principal rank, highlights the relationship between rank and p-rank of an IrTN matrix (particularly given extreme values or p-rank).

Proposition 5.4.8 *The triple* $(n, k, 1)$ *is realizable by an n-by-n irreducible TN matrix if and only if $k = 1$.*

Similarly, when the principal rank is fixed to be two, we have

Proposition 5.4.9 *Suppose the triple* $(n, k, 2)$ *is realizable by an n-by-n irreducible TN matrix. Then $2 \leq k \leq \lceil \frac{n+1}{2} \rceil$. Moreover, each such k is realizable.*

Recall that the $(0,1)$ lower Hessenberg matrix in Example 5.4.7 has rank equal to $n - 1$ and principal rank equal to $\lceil \frac{n}{2} \rceil$. The next result shows that $\lceil \frac{n}{2} \rceil$ is the smallest possible value for principal rank in the case when rank is $n - 1$ (see [FGJ00]).

Proposition 5.4.10 *Suppose the triple* $(n, n - 1, k)$ $(n \geq 2)$ *is realizable by an n-by-n irreducible TN matrix. Then $\lceil \frac{n}{2} \rceil \leq k \leq n - 1$. Moreover, for each such k, the triple is realizable.*

If A is an n-by-n matrix and D is a positive diagonal matrix, then the Jordan structure of A and AD can be vastly different. (For example, it is known that if A is a matrix with positive principal minors, then there exists a positive diagonal matrix D such that the eigenvalues of AD are positive and distinct, even though A may not even be diagonalizable.) However, in the case of irreducible TN matrices, it turns out that the Jordan structure of A and AD (D a positive diagonal matrix) coincide.

We begin with the following lemma, with the proof from [FGJ00] included.

Lemma 5.4.11 *Let A be an n-by-n TN matrix and suppose D is an n-by-n positive diagonal matrix. Then*

$$\text{rank}((AD)^k) = \text{rank}(A^k)$$

and

$$\text{p–rank}((AD)^k) = \text{p–rank}(A^k),$$

where $k \geq 1$.

Proof. Let $C_j(A)$ denote the jth compound of A. Since D is a positive diagonal matrix, it follows that $C_j(D) = D'$ is a positive diagonal matrix for all j. Hence $C_j(AD) = C_j(A)C_j(D) = C_j(A)D'$, where the first equality follows from the Cauchy-Binet identity for determinants. Since D' is a positive diagonal matrix, the zero/nonzero patterns of $C_j(AD)$ and $C_j(A)$ are the same. Moreover, since $C_j(A)$ and $C_j(AD)$ are entrywise nonnegative matrices and $C_j(A^k) = (C_j(A))^k$, it follows that the zero/nonzero pattern of each $C_j(A^k)$ is completely determined by $C_j(A)$. Since the zero/nonzero patterns of $C_j(AD)$ and $C_j(A)$ are the same, it follows that the zero/nonzero patterns of $C_j(A^k)$ and $C_j((AD)^k)$ are the same. Observe that the rank and the principal rank of a given matrix are given by the largest j, such that the jth compound is nonzero, and the largest j such that the jth compound has a nonzero diagonal, respectively. Hence it follows that $\text{rank}((AD)^k) = \text{rank}(A^k)$ and p-rank$((AD)^k) = $ p-rank(A^k), where $k \geq 1$. This completes the proof.   □

From this we may show that the Jordan structure of A and AD are the same, whenever A is TN and irreducible.

Theorem 5.4.12 *Suppose A is an n-by-n irreducible TN matrix and D is a positive diagonal matrix. Then A and AD have the same qualitative Jordan structure.*

Proof. Since A is irreducible (and hence AD is) and since p-rank$(AD) = $ p-rank(A), we have that the number of distinct positive eigenvalues of A and AD are equal. Moreover, since the number and sizes of the Jordan blocks corresponding to zero are completely determined by the ranks of powers, it follows that A and AD have the same qualitative Jordan structure, since $\text{rank}((AD)^k) = \text{rank}(A^k)$, for $k \geq 1$ (by Lemma 5.4.11).   □

The assumption of irreducibility in the above result is necessary as seen by the following example. Let

$$A = \begin{bmatrix} 1 & 1 & 0 \\ 0 & 1 & 1 \\ 0 & 0 & 1 \end{bmatrix}.$$

Then A is TN and is itself a Jordan block, and hence is not diagonalizable. However, if $D = \text{diag}(1, 2, 3)$, then

$$AD = \begin{bmatrix} 1 & 2 & 0 \\ 0 & 2 & 3 \\ 0 & 0 & 3 \end{bmatrix},$$

which has distinct eigenvalues and hence is diagonalizable. Thus A and AD do not have the same qualitative Jordan structure.

Some interesting consequences to the above results are now mentioned; see [FGJ00].

Corollary 5.4.13 *Suppose A is an n-by-n irreducible TN matrix partitioned as follows: $A = \begin{bmatrix} A_{11} & a_{12} \\ a_{21} & a_{22} \end{bmatrix}$, where A_{11} is $(n-1)$-by-$(n-1)$, and a_{22} is a scalar. Define the $(n+1)$-by-$(n+1)$ matrix B as follows*

$$B = \left[\begin{array}{cc|c} A_{11} & a_{12} & a_{12} \\ a_{21} & a_{22} & a_{22} \\ \hline a_{21} & a_{22} & a_{22} \end{array} \right].$$

Then B is an irreducible TN matrix with $\mathrm{rank}(B) = \mathrm{rank}(A)$ *and* p-rank$(B) =$ p-rank(A), *and the Jordan structure of B is the same as A, except B has one more 1-by-1 Jordan block associated with the eigenvalue zero.*

Proof. The fact that B is TN is trivial, and since $a_{22} > 0$ (because A is irreducible), B is irreducible. Also by the symmetry of the bordering scheme, it follows that $\mathrm{rank}(B) = \mathrm{rank}(A)$ and p-rank$(B) = $ p-rank(A). Let $L = L_n(-1)$; then an easy calculation reveals that

$$LBL^{-1} = \begin{bmatrix} A_{11} & 2a_{12} & a_{12} \\ a_{21} & 2a_{22} & a_{22} \\ 0 & 0 & 0 \end{bmatrix} = \begin{bmatrix} AD & \begin{bmatrix} a_{12} \\ a_{22} \end{bmatrix} \\ 0 & 0 \end{bmatrix},$$

where $D = I \oplus [2]$. Observe that $\mathrm{rank}(B^k) = \mathrm{rank}(LB^k L^{-1}) = \mathrm{rank}((LBL^{-1})^k)$. Since

$$(LBL^{-1})^k = \begin{bmatrix} (AD)^k & (AD)^{k-1} \begin{bmatrix} a_{12} \\ a_{22} \end{bmatrix} \\ 0 & 0 \end{bmatrix},$$

and $\begin{bmatrix} a_{12} \\ a_{22} \end{bmatrix}$ is in the span of AD, it follows that

$$\mathrm{rank}\left(\begin{bmatrix} (AD)^k & (AD)^{k-1} \begin{bmatrix} a_{12} \\ a_{22} \end{bmatrix} \\ 0 & 0 \end{bmatrix} \right) = \mathrm{rank}((AD)^k).$$

By Theorem 5.4.12, we have $\mathrm{rank}(B^k) = \mathrm{rank}((AD)^k) = \mathrm{rank}(A^k)$. The result now follows easily.   □

Corollary 5.4.14 *If the triple* (n, k, p) *is realizable by an irreducible TN matrix, then the triple* $(n + 1, k, p)$ *is also realizable by an irreducible TN matrix.*

Summarizing the work thus far, it is clear that for IrTN, the relationship between rank and principal rank is quite strong. What is even more surprising is the restriction principal rank places on the Jordan structure associated to zero (in addition to being the size minus the algebraic multiplicity of zero). We have already demonstrated that when the principal rank on an IrTN matrix is one, then so is the rank. Hence, in this case, all the Jordan blocks associated with the eigenvalue zero must have size one. This type of constraint persists when the principal rank is 2, as is seen by the next fact. See [FGJ00] for the original reference. We include the proof from [FGJ00] for completeness.

Theorem 5.4.15 *Let A be an irreducible TN matrix with principal rank equal to two. Then the size of the largest Jordan block corresponding to zero is at most two.*

Proof. As before we assume that A has a bidiagonal factorization and an associated diagram. Let P be a shortest path from index n on the left to index

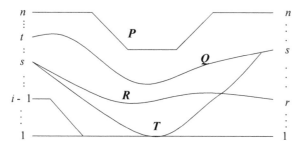

Figure 5.2 Principal rank two

n on the right. Since p-rank(A) = 2, it follows that this path P does not intersect the bottom of this diagram. Suppose P drops to level i, that is, does not use any edges of the diagram induced by the set $\{1, 2, \ldots, i-1\}$; then (as in the case of Proposition 5.4.9) any path from any index $2, 3, \ldots, i-1$, disjoint from P, must intersect the bottom row, otherwise p-rank(A) > 2. To show that the size of the largest Jordan block is at most two, we will show rank(A^2) = 2. To prove this, it is enough to show that any path from any of the indices $\{i, i+1, \ldots, n-1\}$ to $\{1, 2, \ldots, n\}$ must either intersect P or the bottom row, or terminate among the indices $\{1, 2, \ldots, i-1\}$. Suppose there exists a path Q originating at some index $t \in \{i, i+1, \ldots, n-1\}$ and terminating at $s \in \{i, i+1, \ldots, n-1\}$ without intersecting P or the bottom row. Since Q does not intersect P, it must drop below level i, as P was a shortest path. Assume $t \geq s$ (the argument for $s \leq t$ is similar). Since A is irreducible, there exists a path from s to $s + 1$, but in this case such a path must intersect Q. We claim that any path from s, disjoint from P, must intersect the bottom level. To see this, suppose there exists a path R from s that does not intersect the bottom level (see also Figure 5.2). Recall that any path T from s to s, disjoint from P, must intersect the bottom level (since p-rank(A) = 2), and hence any such path must intersect R. Thus there exists a path from s to s that does not intersect P and is disjoint from the bottom level: take R until the intersection of R and T; then follow T until it intersects Q (which may be at s), and then proceed to s. This contradicts that fact that p-rank(A) = 2. Therefore any path originating from $\{i, i+1, \ldots, n-1\}$ must satisfy one of the following: (1) intersects P; (2) intersects the bottom row; (3) terminates in $\{1, 2, \ldots, i-1\}$; or (4) if it terminates at $s \geq i$, then any path beginning at s that is disjoint from P must intersect the bottom row. (We note that these cases are not mutually exclusive.) It now follows that the rank of A^2 is two. Certainly, rank(A^2) $\geq$ 2, as p-rank(A) = 2. Suppose there exist at least three vertex-disjoint paths constituting the rank of A^2. Since P was chosen to be a shortest such path, at most one of these paths can intersect P. Moreover, since these paths are vertex-disjoint, at most one can terminate among the vertices $\{1, 2, \ldots, i-1\}$ (which also includes that case of a path intersecting the bottom level). Thus the only possibility left is case (4). But in this case, any path beginning from s that is disjoint from P must intersect the bottom level. Hence these paths cannot be disjoint for the diagram representing A^2 (which is obtained simply by concatenating two diagrams associated with A). This completes the proof.   □

An immediate consequence is then,

Corollary 5.4.16 *Let A be an n-by-n irreducible TN matrix with* p-rank$(A) = 2$. *Then* rank$(A^2) = 2$.

Recall that the *index* of an eigenvalue λ corresponding to a matrix A is the smallest positive integer k such that rank$((A - \lambda I)^k) = $ rank$((A - \lambda)^{k+1})$. Equivalently, the index of an eigenvalue is equal to the size of the largest Jordan block corresponding to that eigenvalue. It was observed in [FGJ00] that in many examples of IrTN matrices, the size of the largest Jordan block corresponding to zero does not exceed the principal rank. Based on the evidence, this relation was conjectured in [FGJ00] to be true in general.

Since for IrTN matrices the number of positive eigenvalues is equal to the principal rank, the claim that the index of zero does not exceed the principal rank is equivalent to

$$\text{rank}(A^{\text{p}-\text{rank}(A)}) = \text{p-rank}(A)$$

for any IrTN matrix A. The above equality was established in [FG05], and we present it now with proof, as in [FG05].

Theorem 5.4.17 *Let A be an n-by-n irreducible TN matrix with a principal rank equal to k with $1 \le k < n$. Then* rank$(A^k) = k$. *In particular, the size of the largest Jordan block corresponding to zero is at most k.*

Proof. We will use induction on k and n. The case of $k = 1$ and arbitrary n was treated in Proposition 5.4.8. Suppose that A is an n-by-n irreducible TN matrix with principal rank equal to k. Let $\mathcal{D}$ be an irreducible diagram that represents A, and let P be the highest path from n to n in $\mathcal{D}$ (see Figure 5.3). Choose the largest possible vertex m ($m < n$) such that there

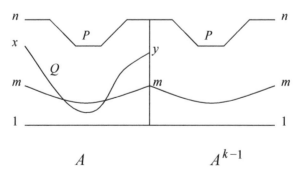

Figure 5.3 Planar diagram $\mathcal{D}$

exists a path from m to m that does not intersect P. Observe that such a vertex exists because the principal rank of A is greater than 1. Define a subdiagram $\mathcal{D}'$ of $\mathcal{D}$ in the following way. First, delete all edges that belong to P or have a common vertex with P. Then, $\mathcal{D}'$ is a union of all paths from i to j $(1 \leq i, j \leq m)$ in the resulting diagram. Note that $\mathcal{D}'$ is irreducible since $\mathcal{D}$ is. Furthermore, by the maximality of m, the principal rank of $\mathcal{D}'$ is equal to $k - 1$. (A matrix A' represented by $\mathcal{D}'$ should not be confused with an m-by-m principal submatrix of A.)

Suppose $m < n - 1$, and let x be any vertex such that $m < x < n$. Then one the following must hold: Any path Q from x must either: (i) intersect P, or (ii) terminate at a sink y, where $y \leq m$.

Suppose neither of these cases hold; that is, assume Q begins at x and terminates at y with $y > m$, and does not intersect P. Then if $x < y$, it follows that there exists a path T from x to x that does not intersect P. Indeed, since A is irreducible there exists a path R from x to x, and by the maximality of m, this path R must intersect P. Thus R must intersect Q. To construct T (a path from x that does not intersect P), follow R and Q until they part, then follow Q until they meet up again, and then follow R until x (note that R and Q may meet at x). On the other hand, if $y < x$, then, applying similar reasoning, it follows that there exists a path from y to y that does not intersect P. Since both $x, y > m$, both cases contradict the maximality of m.

Observe that an immediate consequence of the above claim is that any two paths that begin and end in $\{m + 1, \ldots, n\}$ must intersect. Furthermore, if there is a path in $\mathcal{D}$ that starts (resp. terminates) in $\{m + 1, \ldots, n - 1\}$ and does not intersect P, then there is no path that terminates (resp. starts) in $\{m + 1, \ldots, n - 1\}$ and does not intersect P, for if both such paths exist, they have to intersect (since both intersect a path from m to m) and thus, can be used to produce a path that starts and terminates in $\{m + 1, \ldots, n - 1\}$ and does not intersect P. Therefore, we can assume that $\mathcal{D}$ does not contain a path that terminates in $\{m+1, \ldots, n-1\}$ and does not intersect P (otherwise one can deal with the transpose of A instead of A).

Next, consider a matrix A^k. It can be represented by the diagram $\mathcal{D}^k$, obtained by a concatenation of k copies of $\mathcal{D}$. Let r be the rank of A^k. Then there exists a family F of r vertex-disjoint paths in $\mathcal{D}^k$. Since P was chosen to be the highest path from n to n in $\mathcal{D}$, at most one of these paths can intersect a path P^k defined in $\mathcal{D}^k$ in an obvious way. Without loss of generality, we may assume that $P^k \in F$. Let Q be any other path in F, and let Q_i be the part of Q that lies in the ith copy of $\mathcal{D}$ in $\mathcal{D}^k$ (counting from left to right). Since none of the Q_i intersect P, by the previous discussion, the end points of Q_i must belong to the index set $\{1, 2, \ldots, m\}$. This means that the path obtained by gluing $Q_2, \ldots, Q_k$ belongs to $(\mathcal{D}')^{k-1}$. In other words, F gives rise to a family of $r - 1$ vertex-disjoint paths in $(\mathcal{D}')^{k-1}$ and, therefore,

$$r = \operatorname{rank}(A^k) \leq 1 + \operatorname{rank}((A')^{k-1}) .$$

By the induction assumption, $\operatorname{rank}((A')^{k-1}) = \text{p-rank}(A') = k - 1$ and we obtain

$$k = \text{p-rank}(A) = \text{p-rank}(A^k) \leq \text{rank}(A^k) \leq k,$$

which completes the proof.   □

The next result is an immediate consequence of Theorem 5.4.17, and represents an interesting inequality involving p-rank and rank of IrTN matrices (see [FGJ00] and [FG05]).

Theorem 5.4.18 *Let A be an n-by-n irreducible TN matrix. Then*

$$\text{p-rank}(A) \geq \left\lceil \frac{n}{n - \text{rank}(A) + 1} \right\rceil.$$

5.5 OTHER SPECTRAL RESULTS

As was demonstrated in the previous sections, the eigenvalues of TN/TP matrices enjoy a number of striking properties. For example, all the eigenvalues are nonnegative real numbers. Since all the eigenvalues are real (a property shared with Hermitian matrices), it is natural to explore other potential important properties, such as interlacing and majorization (also satisfied by positive semidefinite matrices). We have also encountered, on a number of occasions, connections between general TN/TP matrices and tridiagonal TN/TP matrices and noted the similarities between their spectral structures. Such a strong connection is suggestive of a spectral comparison between the classes of TN matrices and positive semidefinite matrices. For example, it has been shown here (but first noted in [RH72]), that if A is TP, then there exist an InTN matrix S and an IITN tridiagonal matrix T such that $AS = TS$. In particular, A and T share the same eigenvalues, and since T is diagonally similar to a symmetric tridiagonal matrix (i.e., a positive semidefinite matrix), it readily follows that the eigenvalues of A are all real, positive, and distinct.

This relationship was also exploited to some degree by Gantmacher and Krein. For instance, they began the study of the spectral theory of IITN matrices by first performing a careful analysis of tridiagonal IITN matrices, which they referred to as Jacobi matrices.

5.5.1 Eigenvalue Interlacing

We begin with the topic of the interlacing of eigenvalues, and with it comes our first definition.

Definition 5.5.1 Let $a_1 \geq a_2 \geq \cdots \geq a_n$ and $b_1 \geq b_2 \geq \cdots \geq b_{n-1}$ arranged in decreasing order. Then we say the sequence $\{b_i\}$ *interlaces* the sequence $\{a_i\}$ if $a_n \leq b_{n-1} \leq a_{n-1} \cdots \leq b_1 \leq a_1$. Further, if all of the above inequalities can be taken to be strict, we say the sequence $\{b_i\}$ *strictly interlaces* the sequence $\{a_i\}$.

It has long been known that for any Hermitian matrix A, the eigenvalues of A and $A(i)$ interlace. In [GK60] (and [GK02, pp. 107–108]) a similar result was shown to hold for TN matrices whenever $i = 1$ or n. Later in [Fri85] a number of interesting observations were added on possible extensions, including exhibiting examples in which interlacing need not hold in general when $1 < i < n$ (see also the example below).

We present the eigenvalue interlacing result from [GK60], although the proof technique used is adapted from [LM02] and is reminiscent of the idea employed in the proof of Theorem 5.4.4.

Theorem 5.5.2 *If A is an n-by-n IITN matrix, then the eigenvalues of A strictly interlace the eigenvalues of the two principal submatrices of order $n - 1$, $A(1)$ or $A(n)$, obtained from A by deleting the first row and column or the last row and column.*

Proof. From Theorem 5.4.4 we know that A is similar, via an IITN matrix S, to an IITN tridiagonal matrix T. Upon a more careful inspection of the matrix S conveying the similarity, we note that S can be written as

$$S = [1] \oplus S',$$

where S' is $(n - 1)$-by-$(n - 1)$. This essentially follows since the first and last column are never used to annihilate the (1,2) or (2,1) entry of A. Given this, we observe that A is similar to T and $A(1)$ is similar to $T(1)$. Thus strict interlacing between the eigenvalues of A and $A(1)$ is a consequence of the strict interlacing between the eigenvalues of T and $T(1)$ (see Section 0.1). The fact that we may replace n with 1 follows from Theorem 1.4.1. $\square$

A number of interesting consequences follow rather nicely from this fact. In general, the concept of interlacing can be extended inductively to smaller principal submatrices, which can be stated as follows. Let $a_1 \geq a_2 \geq \cdots \geq a_n$ and $b_1 \geq b_2 \geq \cdots \geq b_{n-k}$ be arranged in decreasing order. Then we say the sequence $\{b_i\}$ *interlaces* the sequence $\{a_i\}$ if $a_{i+k} \leq b_i \leq a_i$, for $i = 1, 2, \ldots, n - k$ (see [HJ85]).

By applying Theorem 5.5.2 successively, it can be shown that

Corollary 5.5.3 *If A is an n-by-n IITN matrix, then the eigenvalues of A and $A[\alpha]$ strictly interlace whenever α is a contiguous subset of N.*

On the other hand, since TN matrices may be approximated by TP matrices the next result follows easily by continuity.

Corollary 5.5.4 *If A is an n-by-n TN matrix, then the eigenvalues of A and $A[\alpha]$ interlace, whenever α is a contiguous subset of N.*

Comparing TN matrices and positive semidefinite matrices, one is immediately led to the fact that positive semidefinite matrices are closed under permutation similarity, while TN is not. Thus, a natural question to ask is whether interlacing holds for arbitrary $(n - 1)$-by-$(n - 1)$ principal submatrices. It is not difficult to observe that the answer is no in general.

Example 5.5.5 Let

$$A = \begin{bmatrix} 2 & 1 & 0 \\ 1 & 2 & 1 \\ 1 & 2 & 2.5 \end{bmatrix}.$$

Then A is IITN with (approximate) eigenvalues $4.0678, 1.8266, 0.6056$. Now $A(2)$ has eigenvalues $2, 2.5$. Evidently, the eigenvalues of $A(2)$ do not interlace the eigenvalues of A.

However, recently a complete "interlacing" picture for arbitrary $(n-1)$-by-$(n-1)$ principal submatrices of a TN/TP matrix has been worked out. We omit the proof here, but encourage the reader to consult [Pin98].

Theorem 5.5.6 *Let A be an n-by-n TP matrix. Suppose the eigenvalues of A are given by $\lambda_1 > \lambda_2 > \cdots > \lambda_n > 0$, and let the eigenvalues of $A(k)$, $1 < k < n$, be $\mu_1 > \mu_2 > \cdots > \mu_{n-1} > 0$. Then*

$$\lambda_{j-1} > \mu_j > \lambda_{j+1}$$

for $j = 1, 2, \ldots, n-1$ (here $\lambda_0 := \lambda_1$).

For TN matrices, a similar statement holds; however, the inequalities above need not be strict.

The subtle difference between the conclusion in Theorem 5.5.6 and standard interlacing is the left inequality; for usual interlacing λ_{j-1} above would be replaced with λ_j.

We close this section by a remarking that certain refinements of eigenvalue interlacing for TN matrices have appeared in part motivated by an important problem on biorthogonal polynomials (see [FW07] for details on the result below and for other related work).

Theorem 5.5.7 *Suppose $A = [a_{ij}]$ is an n-by-n TN matrix and assume that $A(1)$ (resp. $A(n)$) is oscillatory. Then strict interlacing holds between the eigenvalues of A and of $A(1)$ (resp. $A(n)$) if and only if A is irreducible.*

5.5.2 Eigenvalues and majorization

As with the topic of eigenvalue interlacing for TN/TP matrices, it is natural to consider connections between the eigenvalues and the main diagonal entries of a TN or TP matrix. Clearly, they each have the same sum (namely, the trace) and, as a consequence of Perron's theorem (or also Theorem 5.5.6), it follows the λ_1 is at least the maximum of the diagonal entries of A.

Moreover, it is well known that for positive semidefinite (even Hermitian) matrices, the eigenvalues majorize the main diagonal entries (see [HJ85]). A similar form of majorization holds for TN matrices (again it may be viewed as a consequence of interlacing), but because we are not able to arbitrarily permute rows and columns of a TN matrix, the implication is a bit more subtle.

We begin by recalling the definition of majorization.

Definition 5.5.8 Let $a = (a_1, a_2, \ldots, a_n)$ and $b = (b_1, b_2, \ldots, b_n)$ be two sequences of real numbers arranged in decreasing order. We say the sequence a *majorizes* the sequence b, and write $a \succ b$, if

$$\sum_{i=1}^{k} b_i \leq \sum_{i=1}^{k} a_i, \quad k = 1, 2, \ldots, n-1, \quad \text{and} \quad \sum_{i=1}^{n} b_i = \sum_{i=1}^{n} a_i.$$

We say a strictly majorizes b if all the inequalities above are strict.

The next fact, which can be found in [Gar82b], states that the main diagonal entries of a TN/TP matrix are majorized by its eigenvalues (as is the case for positive semidefinite matrices). For positive semidefinite matrices, this type of majorization is essentially a direct consequence of eigenvalue interlacing after arranging the main diagonal entries. In contrast, a more careful argument is required in the TN case, but as such, it still relies heavily on eigenvalue interlacing. The proof presented here mirrors the one given in [Gar82b].

Theorem 5.5.9 *Let $n \geq 2$ and $A = [a_{ij}]$ be an IITN matrix. Suppose $\lambda = (\lambda_1, \ldots, \lambda_n)$ are the eigenvalues of A arranged in decreasing order and that $d = (d_1, \ldots, d_n)$ are the main diagonal entries of A arranged in decreasing order. Then λ strictly majorizes d.*

Proof. The proof proceeds by induction on n, and observe that the case $n = 2$ follows from interlacing (Theorem 5.5.2). Assume that such a majorization holds for any $(n-1)$-by-$(n-1)$ IITN matrix. Since A is IITN, it follows easily that both $B := A(1)$ and $C := A(n)$ are IITN as well (see Corollary 2.6.7). Since order is necessary here, suppose $a_{11} = d_s$ and $a_{nn} = d_t$, and we may assume that $s < t$ (i.e., $a_{11} \geq a_{nn}$); otherwise we may reserve the rows and columns of A. For ease of notation, we let $\lambda(B) = (\lambda_1(B), \ldots, \lambda_{n-1}(B))$ and $\lambda(C) = (\lambda_1(C), \ldots, \lambda_{n-1}(C))$ denote the ordered list of eigenvalues for the matrices B and C, respectively. Similarly, we let $d(B) = (d_1(B), \ldots, d_{n-1}(B))$ and $d(C) = (d_1(C), \ldots, d_{n-1}(C))$ denote the respective ordered list of main diagonal entries for B and C. Applying the induction hypothesis to both B and C, we conclude that for each $k = 1, 2, \ldots, t-1$

$$\sum_{i=1}^{k} d_i(A) = \sum_{i=1}^{k} d_i(B) \leq \sum_{i=1}^{k} \lambda_i(B),$$

and that for $k = 2, \ldots, n-1$

$$\sum_{i=k}^{n-1} d_i(C) > \sum_{i=k}^{n-1} \lambda_i(C).$$

Furthermore, by Theorem 5.5.2, we have

$$\sum_{i=1}^{k} \lambda_i(B) < \sum_{i=1}^{k} \lambda_i(A),$$

for each $k = 1, 2, \ldots, n - 1$. At this stage, we have verified the theorem for each k with $1 \leq k \leq t - 1$.

Now suppose that for some l, with $t \leq l \leq n - 1$, the corresponding inequality in majorization failed for A, namely,

$$\sum_{i=1}^{l} d_i(A) \geq \sum_{i=1}^{l} \lambda_i(A).$$

On the other hand, we have already shown that for each $k = s + 1, \ldots, n$

$$\sum_{i=k}^{n} d_i(A) = \sum_{i=k-1}^{n-1} d_i(C) > \sum_{i=k-1}^{n-1} \lambda_i(C) > \sum_{i=k}^{n} \lambda_i(A). \qquad (5.3)$$

Using (5.3), we may conclude that

$$\operatorname{tr}(A) = \sum_{i=1}^{l} d_i(A) + \sum_{i=l+1}^{n} d_i(A) > \sum_{i=1}^{l} \lambda_i(A) + \sum_{i=l+1}^{n} \lambda_i(A) = \operatorname{tr}(A),$$

clearly reaching a contradiction.   □

A direct consequence of Theorem 5.5.9, using continuity is the following.

Corollary 5.5.10 *Let A be an TN matrix. Suppose $\lambda = (\lambda_1, \ldots, \lambda_n)$ are the eigenvalues of A arranged in decreasing order, and that $d = (d_1, \ldots, d_n)$ are the main diagonal entries of A arranged in decreasing order. Then λ majorizes d.*

5.5.3 Eigenvalue inequalities and products of TN matrices

A simple consequence of the interlacing inequalities is that if A is Hermitian, then

$$\lambda_{\min}(A) \leq \lambda_{\min}(A[\alpha]), \qquad (5.4)$$

for any nonempty index set α. Here $\lambda_{\min}(A)$ denotes the smallest eigenvalue of A, and is well defined only if A has real eigenvalues. We define $\lambda_{\max}(A)$ similarly. In fact, a product analog of (5.4), namely,

$$\lambda_{\min}(AB) \leq \lambda_{\min}(A[\alpha]B[\alpha]), \qquad (5.5)$$

is also known to hold for $A, B \in M_n$, Hermitian and positive semidefinite (see [FJM00]). Since $\lambda_{\min}$ is well defined for TN matrices and even arbitrary products of TN matrices, not only are (5.4) and (5.5) valid for special index sets for TN matrices, but so is the generalization of (5.5) to arbitrary products of TN matrices $A_1, A_2, \ldots, A_k$, namely,

$$\lambda_{\min}(A_1 A_2 \cdots A_k) \leq \lambda_{\min}(A_1[\alpha]A_2[\alpha] \cdots A_k[\alpha]), \qquad (5.6)$$

in which α is a nonempty (contiguous) index set. Similar inequalities were established for positive semidefinite and M-matrices in [FJM00].

Before we come to our main results concerning (5.5) and (5.6), we need some notation and background. Let A and B be two m-by-n real matrices. We say $A \overset{(t)}{\geq} B$ if and only if

$$\det A[\alpha, \beta] \geq \det B[\alpha, \beta]$$

for all $\alpha, \beta \subseteq N$ with $|\alpha| = |\beta|$. In other words, $A \overset{(t)}{\geq} B$ if and only if $C_k(A) \geq C_k(B)$ (entrywise) for every $k = 1, 2, \ldots, \min\{m, n\}$ (see [And87]). Thus $A \overset{(t)}{\geq} 0$ means A is TN. Observe that if $A \overset{(t)}{\geq} B \overset{(t)}{\geq} 0$ and $C \overset{(t)}{\geq} D \overset{(t)}{\geq} 0$, then $AC \overset{(t)}{\geq} BD \overset{(t)}{\geq} 0$, which follows easily from the Cauchy-Binet identity for determinants (assuming the products exist). Unfortunately, $A \overset{(t)}{\geq} B$ does not enjoy some of the useful properties that the positive definite or entrywise orderings possess. For example, $A \overset{(t)}{\geq} B$ does not imply $A - B \overset{(t)}{\geq} 0$, and if in addition $A \overset{(t)}{\geq} B \overset{(t)}{\geq} 0$, then it is not true in general that $A/A[\alpha] \overset{(t)}{\geq} B/B[\alpha]$ or $\tilde{T}B^{-1}\tilde{T} \overset{(t)}{\geq} \tilde{T}A^{-1}\tilde{T}$, for $\tilde{T} = \operatorname{diag}(1, -1, \cdots, \pm 1)$, in the event B, and hence A, is invertible. The following lemma is proved in [And87, Thm. 3.7].

Lemma 5.5.11 *If A is an n-by-n TN matrix, and $\alpha = \{1, 2, \ldots, k\}$ or $\alpha = \{k, k+1, \ldots, n\}$, then*

$$A[\alpha] \overset{(t)}{\geq} A/A[\alpha^c] \overset{(t)}{\geq} 0,$$

provided that $A[\alpha^c]$ is invertible.

We now prove the following result, which can also be found in [FJM00] with proof.

Lemma 5.5.12 *Let A and B be n-by-n TN matrices. Then*

$$\lambda_{\min}(AB) \leq \lambda_{\min}(A[\alpha]B[\alpha]),$$

if $\alpha = \{1, 2, \ldots, k\}$ or $\alpha = \{k, k+1, \ldots, n\}$ for some k, $1 \leq k \leq n$.

Proof. If either A or B is singular, then the inequality is trivial. Thus we assume both A and B are nonsingular. We prove this lemma in the case when $\alpha = \{1, 2, \ldots, k\}$, the proof in the other case is similar. Since A (B) is invertible, it follows that $A[\alpha]$ ($B[\alpha]$) is invertible by the interlacing of eigenvalues. As before, the inequality $\lambda_{\min}(AB) \leq \lambda_{\min}(A[\alpha]B[\alpha])$ is equivalent to the inequality $\lambda_{\max}(B^{-1}A^{-1}) \geq \lambda_{\max}(B[\alpha]^{-1}A[\alpha]^{-1})$. Let $C = \tilde{T}B^{-1}\tilde{T}$ and $D = \tilde{T}A^{-1}\tilde{T}$, where $\tilde{T} = \operatorname{diag}(1, -1, \ldots, \pm 1)$. Then C and D are both TN. Now

$$\begin{aligned}
\lambda_{\max}(B^{-1}A^{-1}) &= \lambda_{\max}(CD) \\
&\geq \lambda_{\max}(CD[\alpha]) \\
&\geq \lambda_{\max}(C[\alpha]D[\alpha]),
\end{aligned}$$

and both inequalities follow from the monotonicity of the Perron root. By Lemma 5.5.11 and the fact that both $C/C[\alpha^c]$ and $D/D[\alpha^c]$ are TN, we have

$$\lambda_{\max}(C[\alpha]D[\alpha]) \geq \lambda_{\max}(C/C[\alpha^c] \cdot D/D[\alpha^c]).$$

Using the classical formula for the partitioned form of the inverse of a matrix, it follows that $C/C[\alpha^c] = (C^{-1}[\alpha])^{-1} = (\tilde{T}B\tilde{T}[\alpha])^{-1} = \tilde{T}[\alpha](B[\alpha])^{-1}\tilde{T}[\alpha]$. Hence

$$\lambda_{\max}(C/C[\alpha^c] \cdot D/D[\alpha^c]) = \lambda_{\max}((B[\alpha])^{-1}(A[\alpha])^{-1}),$$

which implies

$$\lambda_{\max}(B^{-1}A^{-1}) \geq \lambda_{\max}(B[\alpha]^{-1}A[\alpha]^{-1}),$$

as desired.   □

This leads to

Theorem 5.5.13 *Let A and B be n-by-n TN matrices. Then*

$$\lambda_{\min}(AB) \leq \lambda_{\min}(A[\alpha]B[\alpha])$$

for any contiguous index set $\alpha \subseteq N$.

Proof. The proof follows from at most two applications of Lemma 5.5.12. For example, if $\alpha = \{i, i+1, \ldots, i+k\} \subseteq N$, then

$$\lambda_{\min}(AB) \leq \lambda_{\min}(A[\{i, \ldots, n\}]B[\{i, \ldots, n\}])$$
$$\leq \lambda_{\min}(A[\alpha]B[\alpha]),$$

as desired.   □

Observe that if $A_1, A_2, \ldots, A_k, B_1, B_2 \ldots, B_k \in M_{m,n}$ and $A_i \overset{(t)}{\geq} B_i \overset{(t)}{\geq} 0$, for $i = 1, 2, \ldots, k$, then

$$A_1 A_2 \cdots A_k \overset{(t)}{\geq} B_1 B_2 \cdots B_k \overset{(t)}{\geq} 0.$$

The above fact is needed in the proof of the next result, see also [FJM00].

Theorem 5.5.14 *Let $A_1, A_2, \ldots, A_k$ be n-by-n TN matrices. Then*

$$\lambda_{\min}(A_1 A_2 \cdots A_k) \leq \lambda_{\min}(A_1[\alpha]A_2[\alpha] \cdots A_k[\alpha]),$$

for any contiguous index set $\alpha \subseteq N$.

Proof. As was the case in the proof of Theorem 5.5.13, it is enough to prove this result in the case in which all A_i are nonsingular and $\alpha = \{1, 2, \ldots, k\}$. For each i, let $B_i = \tilde{T}A_i^{-1}\tilde{T}$, so that B_i is TN. Then

$$\lambda_{\max}(A_k^{-1} \cdots A_1^{-1}) = \lambda_{\max}(B_k \cdots B_1)$$
$$\geq \lambda_{\max}(B_k \cdots B_1[\alpha])$$
$$\geq \lambda_{\max}(B_k[\alpha] \cdots B_1[\alpha]),$$

in which both inequalities follow from the monotonicity of the Perron root. By Lemma 5.5.11, the fact that each $B_i/B_i[\alpha^c]$ is TN, and the remark preceding the statement of Theorem 5.5.14, we have

$$\lambda_{\max}(B_k[\alpha]\cdots B_1[\alpha]) \geq \lambda_{\max}(B_k/B_k[\alpha^c]\cdots B_1/B_1[\alpha^c]).$$

As before,

$$\lambda_{\max}(B_k/B_k[\alpha^c]\cdots B_1/B_1[\alpha^c]) = \lambda_{\max}((A_k[\alpha])^{-1}\cdots(A_1[\alpha])^{-1}),$$

from which the desired inequality follows. $\square$

Since the class of TN matrices is not closed under simultaneous permutation of rows and columns, it is not surprising that these results hold for only certain index sets. Moreover, in the case in which α is not a contiguous subset, consider the following counterexample (see [FJM00]).

Example 5.5.15 Let $A = \begin{bmatrix} 1 & 0 & 0 \\ 1.3 & 1.3 & 0 \\ 1 & 1 & 1 \end{bmatrix}$ and $B = A^T$. Then both A

and B are nonsingular TN matrices. Suppose $\alpha = \{1,3\}$. Then $A[\alpha]B[\alpha] = \begin{bmatrix} 1 & 1 \\ 1 & 2 \end{bmatrix}$. A simple computation reveals

$$\lambda_{\min}(AB) \approx .387 > .381 \approx \lambda_{\min}(A[\alpha]B[\alpha]).$$

Lemma 5.5.12, Theorems 5.5.13, and 5.5.14 can be extended in the nonsingular case to include (as an intermediate inequality) the smallest eigenvalue of a product of TN matrices involving Schur complements. For example, in the proof of Lemma 5.5.12 we actually proved the following,

$$\lambda_{\min}(AB) \leq \lambda_{\min}(A/A[\alpha^c]B/B[\alpha^c])$$
$$\leq \lambda_{\min}(A[\alpha]B[\alpha])$$

for $\alpha = \{1, 2, \ldots, k\}$ or $\alpha = \{k, k+1, \ldots, n\}$ and A and B nonsingular TN matrices. Consequently, it follows that Theorem 5.5.14 may also be extended to include

$$\lambda_{\min}(A_1 A_2 \cdots A_k) \leq \lambda_{\min}(A_1/A_1[\alpha^c]A_2/A_2[\alpha^c]\cdots A_k/A_k[\alpha^c])$$
$$\leq \lambda_{\min}(A_1[\alpha]A_2[\alpha]\cdots A_k[\alpha])$$

for any contiguous set α, and for nonsingular TN matrices $A_1, A_2, \ldots, A_k$.

Finally, we extend Theorems 5.5.13 and 5.5.14 to the product of the jth smallest eigenvalues of AB (see [FJM00] for the details).

Corollary 5.5.16 *Let A and B be n-by-n TN matrices. Then for any contiguous index set $\alpha \subseteq N$*

$$\prod_{i=1}^{j} \lambda_{n-i+1}(AB) \leq \prod_{i=1}^{j} \lambda_{|\alpha|-i+1}(A[\alpha]B[\alpha]),$$

for $j = 1, 2, \ldots, |\alpha|$.

Also we note the Corollary 5.5.16 can easily be extended to the case of an arbitrary finite product of TN matrices.

5.5.4 Certain Types of Inverse Eigenvalue Problems for TN Matrices

In general, an inverse eigenvalue problem asks: given a collection of numbers $\{\lambda_1, \ldots, \lambda_n\}$ (repeats allowed), find an n-by-n matrix A having some desired property and eigenvalues $\lambda_1, \ldots, \lambda_n$. There are many such versions of this inverse eigenvalue problem, each asking the matrix A to satisfy a particular property of interest. Perhaps the most famous version is the "non-negative inverse eigenvalue problem," which asks: given complex numbers $\{\lambda_1, \ldots, \lambda_n\}$ (repeats allowed), find an n-by-n entry-wise nonnegative matrix A with eigenvalues $\lambda_1, \ldots, \lambda_n$. This problem is extremely complicated and to date is still unresolved in general.

For the class of TN matrices, a natural problem along these lines is whether, given a collection of nonnegative numbers $\{\lambda_1, \ldots, \lambda_n\}$, there exist an n-by-n TN matrix A with eigenvalues $\lambda_1, \ldots, \lambda_n$. As stated, the answer to this question is certainly in the affirmative–simply let A be the diagonal matrix $\text{diag}(\lambda_1, \ldots, \lambda_n)$. Now, suppose we modify the problem as follows: given a collection of distinct positive numbers $\{\lambda_1, \ldots, \lambda_n\}$, does there exist an n-by-n TP matrix A with eigenvalues $\lambda_1, \ldots, \lambda_n$? As we saw in Theorem 5.2.2, such a matrix A can always be constructed without any additional conditions on the collection $\{\lambda_1, \ldots, \lambda_n\}$, and relies heavily on the classical fact that a IITN tridiagonal matrix B can be constructed with given eigenvalues $\mu_i := \lambda_i^{1/n}$.

Other variants of an inverse eigenvalue problem that pertain to TN/TP matrices have appeared in the literature, and have highlighted other spectral properties enjoyed by this class. We mention just a few examples in passing as illustrations. For example, we can slightly extend the TP inverse eigenvalue problem as follows (see [FG05]). Recall that the notions of irreducibility, rank, and principal rank have natural interpretations in terms of the planar diagrams associated with EB factorizations of TN matrices.

Corollary 5.5.17 *Let $\mathcal{D}$ be irreducible planar diagram of order n and of principal rank p. Then, for every $0 < \lambda_1 < \lambda_2 < \cdots < \lambda_p$, there is a TN matrix A with associated diagram $\mathcal{D}$ and with positive eigenvalues $\lambda_1, \ldots, \lambda_p$.*

Another example may be viewed as a type of converse to majorization for TN/TP matrices. In [Gar85], it was verified that majorization of a vector x (properly ordered) of nonnegative numbers by some other vector y (properly ordered) of nonnegative numbers is not enough to guarantee the existence of a TN matrix A with eigenvalues y and diagonal entries x. What extra conditions are required is not known at present.

Along similar lines it was shown in [LM02] that an analogous converse to eigenvalue interlacing need not hold in general. However, it is interesting to note that the following statement is indeed valid (see [LM02]). Suppose $k = 1, 2, n - 2$, or $n - 1$, and let $\lambda_1 \geq \lambda_2 \geq \cdots \geq \lambda_n$ and $\mu_1 \geq \mu_2 \geq \cdots \geq \mu_{n-k}$ arranged in decreasing order that satisfy $\lambda_{i+k} \leq \mu_i \leq \lambda_i$ for

$i = 1, 2, \ldots, n - k$. Then there exists a TN matrix A with an $(n - k)$-by-$(n-k)$ contiguous principal submatrix B such that A and B have eigenvalues $\lambda_1, \lambda_2, \ldots, \lambda_n$ and $\mu_1, \mu_2, \ldots, \mu_{n-k}$, respectively.

In addition, a similar result is shown to hold for TP matrices (assuming, of course, that all inequalities are strict), when $k = 1$ or $n - 1$; however, the cases $k = 2, n - 2$ remain open currently. For the remaining values of k ($2 < k < n - 2$), such a result does not hold for either TN or TP (see [LM02]).

We close this section with another type of inverse eigenvalue problem related to the Jordan canonical form of IrTN matrices. We have already seen that the Jordan structure of such matrices are very much controlled by the parameters rank and principal rank. As it turns out, this is basically all that is required to determine the Jordan structure. The next result appeared in [FG05], which we state without proof.

Theorem 5.5.18 *Let A be an irreducible n-by-n TN matrix. Then*

(1) *A has at least one non-zero eigenvalue. Furthermore, the non-zero eigenvalues of A are positive and distinct.*

(2) *Let p be the number of non-zero eigenvalues of A. For each $i \geq 1$, let $m_i = m_i(A)$ denote the number of Jordan blocks of size i corresponding to the zero eigenvalue in the Jordan canonical form of A. Then the numbers m_i satisfy the following conditions:*

   (a) $\sum i m_i = n - p$.

   (b) $m_i = 0$ for $i > p$.

   (c) $\sum m_i \geq \frac{n-p}{p}$.

(3) *Conversely, let $n \geq 2$, let $1 \leq p \leq n$, let $\lambda_1, \ldots, \lambda_p$ be arbitrary distinct positive numbers and let $m_1, m_2, \ldots$ be nonnegative integers satisfying conditions in 2.*

   Then there exists an irreducible TN matrix A having non-zero eigenvalues and such that $m_i(A) = m_i$ for all $i \geq 1$.

Chapter Six

Determinantal Inequalities for TN Matrices

6.0 INTRODUCTION

In this chapter we begin an investigation into the possible relationships among the principal minors of TN matrices. A natural and important goal is to examine all the inequalities among the principal minors of TN matrices. More generally, we also study families of determinantal inequalities that hold for all TN matrices.

In 1893, Hadamard noted the fundamental determinantal inequality

$$\det A \leq \prod_{i=1}^{n} a_{ii} \tag{6.1}$$

holds whenever $A = [a_{ij}]$ is a positive semidefinite Hermitian matrix. Hadamard observed that it is not difficult to determine that (6.1) is equivalent to the inequality

$$|\det A| \leq \sqrt{\prod_{i=1}^{n} \left(\sum_{j=1}^{n} |a_{ij}|^2 \right)}, \tag{6.2}$$

where $A = [a_{ij}]$ is an arbitrary n-by-n matrix.

In 1908, Fischer extended Hadamard's inequality (6.1) by demonstrating that

$$\det A \leq \det A[S] \cdot \det A[N \setminus S], \tag{6.3}$$

holds for any positive semidefinite Hermitian matrix A and for any index set $S \subset N = \{1, 2, \ldots, n\}$. Recall that $A[S]$ denotes the principal submatrix of A lying in rows and columns S, and for brevity we will let S^c denote the complement of S relative to N (that is, $S^c = N \setminus S$). As usual we adopt the convention that $\det A[\phi] := 1$. Evidently, Hadamard's inequality (6.1) follows from repeated application of Fischer's inequality by beginning with $S = \{1\}$. For a detailed survey on various principal minor inequalities of the standard positivity classes the reader is directed to the paper [FJ00].

Around the time of World War II, Koteljanskiĭ was interested in many aspects of positivity in linear algebra, and, among his many interesting and important contributions was a unifying theory of a certain class of determinantal inequalities involving principal minors, which now bear his name. Formally, Koteljanskiĭ generalized Fischer's inequality (6.3) within the class of positive semidefinite matrices, and, perhaps more importantly, he extended the class of matrices (beyond positive semidefinite) for which this inequality is valid (see [Kot53, Kot63a]). As Koteljanskiĭ inequalities play a fundamental role in the field of determinantal inequalities, we expend some effort to carefully explain this seminal result.

For an n-by-n matrix A, and any two index sets $S, T \subseteq N$, Koteljanskiĭ's inequality states:

$$\det A[S \cup T] \cdot \det A[S \cap T] \leq \det A[S] \cdot \det A[T]. \qquad (6.4)$$

It is apparent, for any positive semidefinite matrix, that Koteljanskiĭ's inequality implies Fischer's inequality (bearing in mind the convention $\det A[\phi] = 1$), and hence is a generalization of both Fischer's and Hadamard's inequalities.

In addition, Koteljanskiĭ observed that (6.4) holds for a larger class of matrices, which includes the positive semidefinite matrices. Before we can derive this class, we should recall some key definitions.

Recall that a matrix is called a P_0-matrix if all its principal minors are nonnegative. Consider the subclass of P_0-matrices A that satisfy

$$\det A[U, V] \det A[U, V] \geq 0, \qquad (6.5)$$

for any two index sets $S, T \subseteq N$ with the properties:

(1) $|U| = |V|$ and

(2) $|U \cup V| = |U \cap V| + 1$.

We note that a minor of the form $\det A[U, V]$, where U and V satisfy the conditions $|U| = |V|$ and $|U \cup V| = |U \cap V| + 1$, has been called an *almost principal minor* (in the sense that the index sets U, V differ only in precisely one index). Note that condition (6.5) requires that all symmetrically placed almost principal minors have the same sign (ignoring zeros), and it is obvious that not all P_0-matrices satisfy (6.5). It is clear that any positive semidefinite matrix satisfies (6.5), and it is also clear that any TN also satisfies (6.5). A slightly more general result is that (6.5) holds for both M-matrices and inverse M-matrices as well (see, for example, [Car67, FJ00]) for more details.

Perhaps an even more useful property is that if A is an invertible P_0-matrix and satisfies (6.5), then A^{-1} is P_0 and satisfies (6.5), see [FJ00].

The main observation [Kot53, Kot63a] is the following fact.

Theorem 6.0.1 *Suppose A is an n-by-n P_0-matrix. Then A satisfies (6.4) for all index sets U, V satisfying (1) and (2) above if and only if A satisfies (6.5) for all index sets S, T.*

For simplicity of notation, if A is a P_0-matrix that satisfies (6.4) for all index sets S, T, then we call A a *Koteljanskiĭ matrix*, and denote the class of all Koteljanskiĭ matrices by $\mathcal{K}$.

The punch line is then that both TN matrices and positive semidefinite (as well as M and inverse M) matrices satisfy Koteljanskiĭ's determinantal inequality, and therefore they also satisfy both Fischer's inequality and

Hadamard's inequality. Moreover, all these determinantal inequalities lie naturally within a general family of multiplicative minor inequalities which we survey further throughout the remainder of this chapter. In fact, the study of determinantal inequalities has seen a reemergence in modern research among linear algebraists (see, for example [BJ84] and subsequent works [FJ00, FJ01, FGJ03]), where an important theme emerged; namely, to describe all possible inequalities that exist among products of principal minors via certain types of set-theoretic conditions. Employing this completely combinatorial notion led to a description of the necessary and sufficient conditions for all such inequalities on three or fewer indices, and also verified other inequalities for cases of more indices, all with respect to positive semidefinite matrices.

Our intention here is to employ these ideas and similar ones more pertinent to theory of TN matrices to derive a similar description. One notable difference between the class of TN matrices and the positive semidefinite matrices is that TN matrices are not in general closed under simultaneous permutation of rows and columns, while positive semidefinite matrices are closed under such an operation.

6.1 DEFINITIONS AND NOTATION

Recall from Chapter 0 that for a given n-by-n matrix, and $S, T \subseteq N$, $A[S, T]$ denotes the submatrix of A lying in rows S and columns T. For brevity, $A[S, S]$ will be shortened to $A[S]$.

More generally, if $\alpha = \{\alpha_1, \alpha_2, \ldots, \alpha_p\}$ is a collection of index sets (i.e., $\alpha_i \subset N$), and $A \in M_n$, we let

$$\alpha(A) = \det A[\alpha_1] \det A[\alpha_2] \cdots \det A[\alpha_p]$$

denote the product of the principal minors $\det A[\alpha_i]$, $i = 1, 2, \ldots, p$. For example, if $\alpha = \{S \cup T, S \cap T\}$ for any two index sets S, T, then

$$\alpha(A) = \det A[S \cup T] \det A[S \cap T].$$

If, further, $\beta = \{\beta_1, \beta_2, \ldots, \beta_q\}$ is another collection of index sets with $\beta_i \subset N$, then we write $\alpha \leq \beta$ with respect to a fixed class of matrices $\mathcal{C}$ if

$$\alpha(A) \leq \beta(A)$$

for each $A \in \mathcal{C}$. Thus, for instance, if $\alpha = \{S \cup T, S \cap T\}$ and $\beta = \{S, T\}$ for any two index sets S, T, then the statement $\alpha \leq \beta$ with respect to the class $\mathcal{K}$ is a part of Koteljanskiĭ's theorem (Theorem 6.0.1).

We shall also find it convenient to consider ratios of products of principal minors. For two such collections α and β of index sets, we interpret α/β as both a numerical ratio $\alpha(A)/\beta(A)$ for a given matrix, and also as a formal ratio to be manipulated according to natural rules. When interpreted numerically, we obviously need to assume that $\beta(A) \neq 0$ for A, and we note that if A is InTN, then $\det A[S] > 0$, for all index sets S. Consequently, when discussing ratios of products of principal minors, we restrict ourselves to InTN (or even TP) matrices.

Since, by convention, $\det A[\phi] = 1$, we also assume, without loss of generality, that in any ratio α/β both collections α and β have the same number of index sets, since if there is a disparity between the total number of index sets in α and β, the one with fewer sets may be padded out with copies of ϕ. Also, note that either α or β may include repeated index sets (which count).

When considering ratios of principal minors, we may also be interested in determining all pairs of collections α and β such that

$$\frac{\alpha(A)}{\beta(A)} \leq K, \tag{6.6}$$

for some constant K (which may depend on n), and for all A in InTN. We may also write $\alpha/\beta \leq K$, with respect to InTN.

As a matter of exposition, if given two collections α and β of index sets, and $\alpha \leq \beta$ with respect to TN, then we have determined a (determinantal) inequality, and we say the ratio $\frac{\alpha}{\beta}$ is *bounded* by 1. For example, if $\alpha = \{S \cup T, S \cap T\}$ and $\beta = \{S, T\}$ for any two index sets S, T, then the ratio α/β is bounded by 1 with respect to InTN (and invertible matrices in $\mathcal{K}$). If, further, $\alpha/\beta \leq K$ with respect to InTN, then we say α/β is *bounded by the constant K* with respect to InTN.

6.2 SYLVESTER IMPLIES KOTELJANSKIĬ

As suggested in the previous section, the work of Koteljanskiĭ on determinantal inequalities and the class of matrices $\mathcal{K}$ is so significant in the area of determinantal inequalities that we continue to study this topic, but from a different angle. In this section, we demonstrate how Sylvester's determinantal identity can be used to verify Koteljanskiĭ's theorem (Theorem 6.0.1). In some sense, Sylvester's identity is implicitly used in Koteljanskiĭ's original work, so this implication is not a surprise, but nonetheless it is still worth discussing.

Recall that a special instance of Sylvester's identity (1.5) is

$$\det A \det A[S]$$
$$= \det A[S \cup \{i\}] \det A[S \cup \{j\}]$$
$$\quad - \det A[S \cup \{i\}, S \cup \{j\}] \det A[S \cup \{j\}, S \cup \{i\}],$$

where $S \subset N$, with $|S| = n - 2$ and $i, j \notin S$. Thus for any matrix in $\mathcal{K}$, we can immediately deduce that

$$\det A \det A[S] \leq \det A[S \cup \{i\}] \det A[S \cup \{j\}],$$

as $\det A[S \cup \{i\}, S \cup \{j\}] \det A[S \cup \{j\}, S \cup \{i\}] \geq 0$ for all such S (cf. (6.5)). Observe that the inequality

$$\det A \det A[S] \leq \det A[S \cup \{i\}] \det A[S \cup \{j\}]$$

is an example of a Koteljanskiĭ inequality. In fact, it is a very special example of such an inequality, and we refer to it as a *basic Koteljanskiĭ inequality*. To be more precise, a Koteljanskiĭ inequality is called *basic* if it is of the form,

$$\det A[S \cup T] \det A[S \cap T] \le \det A[S] \det A[T],$$

in which $|S| = |T|$, and $|S \cap T| = |S| - 1 = |T| - 1$. In this case, it is evident that $|S \cup T| = |S| + 1 = |T| + 1$. Such conditions are intimately connected with the almost principal minors that were discussed in the previous section. From the above analysis, it is clear that Sylvester's identity can be used to verify basic Koteljanskiĭ inequalities for all matrices in $\mathcal{K}$.

To make the exposition simpler, we define a notion of "gap." For two index sets $S, T \subseteq N$, we define the *gap of S and T* as

$$\mathrm{gap}(S, T) = |S \cup T| - |S \cap T|.$$

Observe that for all $S, T \subset N$, $\mathrm{gap}(S, T) \ge 0$, and if $\mathrm{gap}(S, T) = 0$ or 1, then the corresponding Koteljanskiĭ inequality is trivial. Summarizing parts of the discussion above essentially establishes that if $\mathrm{gap}(S, T) = 2$, then Koteljanskiĭ's holds for sets S and T assuming that A was in $\mathcal{K}$. However, we outline the argument in the gap two case for completeness.

Proposition 6.2.1 *Suppose S and T are two index sets such that* $\mathrm{gap}(S, T) = 2$. *Then*

$$\det A[S \cup T] \det A[S \cap T] \le \det A[S] \det A[T],$$

for any A satisfying (6.5).

Proof. If $\mathrm{gap}(S, T) = 2$ for two index sets S and T, then one of the following cases must hold: (1) $|S| = |T|$, or (2) $|S| = |T| \pm 2$. If case (2) occurs, then the corresponding Koteljanskiĭ inequality is trivial, and, if (1) holds, then $|S \cup T| = |S| + 1 = |T| + 1$, and the corresponding Koteljanskiĭ inequality is basic, and thus valid. □

The idea for the proof of the main result is to reduce the gap by one and apply induction, using Proposition 6.2.1 as a base case. We state the next result in terms of ratios, so we are implicitly assuming the use of invertible matrices. The general case then follows from continuity considerations. For brevity, we may let (S) denote $\det A[S]$.

Proposition 6.2.2 *Suppose S and T are two index sets for which the associated Koteljanskiĭ inequality holds and $\mathrm{gap}(S, T) > 2$. Then for any invertible A satisfying (6.5) we have*

$$\frac{\det A[S \cup T] \det A[S \cap T]}{\det A[S] \det A[T]}$$
$$= \frac{\det A[U \cup V] \det A[U \cap V]}{\det A[U] \det A[V]} \cdot \frac{\det A[W \cup X] \det A[W \cap X]}{\det A[W] \det A[X]},$$

such that

$$\mathrm{gap}(S, T) > \{\mathrm{gap}(U, V), \mathrm{gap}(W, X)\}.$$

Proof. Without loss of generality, assume that $|S| \leq |T|$. Since $\mathrm{gap}(S,T) > 2$ (implying a nonbasic inequality), we know that $|S \cap T| < |T| - 1$, and hence there must exist an index $i \in T$, so that $i \notin S \cap T$. Consider the set-theoretic factorization

$$\frac{(S \cup T)(S \cap T)}{(S)(T)} = \frac{(S \cup T)((S \cup i) \cap T)}{(S \cup i)(T)} \cdot \frac{(S \cup i)(S \cap T)}{((S \cup i) \cap T)(S)}.$$

Observe that both factors represent (in a natural way) Koteljanskiĭ ratios and that the gap for each factor on the right is less than $\mathrm{gap}(S,T)$, by construction. Applying a simple induction argument will complete the proof.
□

Proposition 6.2.2 gives rise to two interesting consequences, both of which are useful to note and follow below. The other item we wish to bring attention to is the proof technique used for establishing Proposition 6.2.2: namely, connecting the verification of a numerical ratios by extending this notion to a factorization (or product) of ratios of index sets. This idea will appear often in the next few sections.

Corollary 6.2.3 *For matrices in $\mathcal{K}$, basic Koteljanskiĭ inequalities may be viewed (set-theoretically) as a generating set for all Koteljanskiĭ inequalities. That is, any valid Koteljanskiĭ inequality may be rewritten as a product of basic Koteljanskiĭ inequalities.*

Corollary 6.2.4 *Let A be an n-by-n matrix satisfying (6.5). Then Sylvester's determinantal identity can be used to verify that Koteljanskiĭ's inequality*

$$\det A[S \cup T] \det A[S \cap T] \leq \det A[S] \det A[T]$$

holds for all pairs of index sets S, T.

6.3 MULTIPLICATIVE PRINCIPAL MINOR INEQUALITIES

One of the central problems in this area is to characterize, via set-theoretic conditions, all pairs of collection of index sets such that

$$\frac{\alpha(A)}{\beta(A)} \leq K,$$

over all InTN matrices A. It is clear that since TP is the closure of TN (and hence of InTN), that the bounded ratios with respect to InTN and with respect to TP are the same.

This general issue was resolved in [FHJ98] for the case of M-matrices and inverse M-matrices. In fact, the bound K can always be chosen to be one for both the M- and the inverse M-matrices. The same problem has received much attention for positive semidefinite matrices (see, for example, [BJ84]), but a general resolution in this case seems far off.

We now turn our attention to the case of InTN matrices, and we begin with an illustrative example, that in many ways was the impetus of the investigation that began in [FGJ03].

Example 6.3.1 Let $\alpha = \{\{1,2\},\{3\}\}$ and $\beta = \{\{1,3\},\{2\}\}$. Then $\alpha(A) \leq \beta(A)$, for any InTN matrix A, which can be seen by

$$\det A[\{1,3\}]\det A[\{2\}] = (a_{11}a_{33} - a_{13}a_{31})a_{22}$$

$$= a_{11}a_{33}a_{22} - a_{13}a_{31}a_{22}$$

$$\geq a_{11}a_{33}a_{22} - a_{12}a_{23}a_{31} \quad \text{since } a_{12}a_{23} \geq a_{13}a_{22}$$

$$\geq a_{11}a_{33}a_{22} - a_{12}a_{21}a_{33} \quad \text{since } a_{21}a_{33} \geq a_{23}a_{31}$$

$$= (a_{11}a_{22} - a_{12}a_{21})a_{33}$$

$$= \det A[\{1,2\}]\det A[\{3\}].$$

On the other hand, if $\alpha = \{\{1,2\},\{3\}\}$ and $\beta = \{\{1,3\},\{2\}\}$, then $\alpha(A) \leq \beta(A)$ is not a valid inequality in general, since if

$$A = \begin{bmatrix} 1 & 2 & 1 \\ t+1 & 2t+3 & t+2 \\ 1 & 3 & 3 \end{bmatrix},$$

then $\alpha(A) = 4t + 6$ and $\beta(A) = 3$, where $t > 0$ is arbitrary.

These basic examples illustrate two particular points regarding determinantal inequalities for TN/TP matrices. First, arbitrary permutation of the indices in the collections α and β of an established bounded ratio need not produce a bounded ratio, and, second, verification of a bounded ratio can be intricate and subtle.

Let $\alpha = \{\alpha_1, \alpha_2, \ldots, \alpha_p\}$ be any collection of index sets from N. For $i \in N$, we define $f_\alpha(i)$ to be the number of index sets in α that contain the element i. In other words, $f_\alpha(i)$ counts the number of occurrences of the index i among the index sets $\alpha_1, \ldots, \alpha_p$ in α.

The following proposition, which represents a basic necessary condition, demonstrates that in order for any ratio α/β to be bounded with respect to the TN/TP matrices, the multiplicity of each index in α and β must coincide (see [FGJ03]).

Proposition 6.3.2 *Let α and β be two collections of index sets. If $\frac{\alpha}{\beta}$ is bounded with respect to the class of TN matrices, then $f_\alpha(i) = f_\beta(i)$, for every $i = 1, 2, \ldots, n$.*

Proof. Suppose there exists an index i for which $f_\alpha(i) \neq f_\beta(i)$. For $k > 0$ let $D_i(k) = \text{diag}(1, \ldots, 1, k, 1, \ldots, 1)$, where the number k occurs in the (i,i)th entry of $D_i(k)$. Then $D_i(k)$ is an invertible TN matrix for every such value of k. Evaluation of α/β at $D_i(k)$ gives

$$\frac{\alpha(D_i(k))}{\beta(D_i(k))} = k^{(f_\alpha(i)-f_\beta(i))}.$$

Hence $\frac{\alpha}{\beta}$ is unbounded, since we may select $k > 1$ or $k < 1$ as according to the sign of $f_\alpha(i) - f_\beta(i)$. $\square$

If $\frac{\alpha}{\beta}$ is a given ratio that satisfies the condition $f_\alpha(i) = f_\beta(i)$, for every $i = 1, 2, \ldots, n$, then we say that $\frac{\alpha}{\beta}$ satisfies condition *set-theoretic zero*, ST0 for short. It is clear that STO is not a sufficient condition for a ratio to be bounded with respect to InTN; see the previous example. Furthermore, since the closure of the TP matrices is the TN matrices, STO is also a basic necessary condition for any ratio to be bounded with respect to TP.

The fact that any TN (InTN, TP) matrix has an elementary bidiagonal factorization proves to be very useful for verifying the boundedness of a given ratio.

Suppose we are presented with a specific ratio $\frac{\alpha}{\beta}$. To verify that this ratio is unbounded, say with respect to InTN, we need to show that there exists choices for the nonnegative parameters $\{l_i\}$ and $\{u_j\}$ and choices for the positive parameters $\{d_i\}$, such that $\alpha(A)$ increases without bound faster than $\beta(A)$, when A is written as in (2.12).

Now an important item to keep in mind is that when A is presented via its elementary bidiagonal factorization, then both $\alpha(A)$ and $\beta(A)$ are subtraction-free (polynomial) expressions in the nonnegative variables $\{l_i\}$ and $\{u_j\}$ and the positive variables $\{d_i\}$. Along these lines, if we are able to identify a subcollection of these variables for which the degree with respect to this subcollection in $\alpha(A)$ exceeds the corresponding degree in $\beta(A)$, then we may assign all the variables in this subcollection the single variable $t > 0$, and set all remaining variables be equal to 1. Under this assignment, both $\alpha(A)$ and $\beta(A)$ are polynomials in the single variable t with $\deg(\alpha(A)) > \deg(\beta(A))$. Thus, letting $t \to \infty$ implies that $\frac{\alpha}{\beta}$ is not a bounded ratio. Consider the following ratio (written set-theoretically):

$$\frac{\alpha}{\beta} = \frac{\{1,3\}\{2\}}{\{1,2\}\{3\}}.$$

Let A be an arbitrary InTN matrix, and suppose A is written in terms of its SEB factorization:

$$A = L_3(l_3)L_2(l_2)L_3(l_1)U_3(u_1)U_2(u_2)U_3(u_3).$$

Then, it is easily verified that

$$\alpha(A) = (1 + l_2u_2)(1 + (l_1 + l_3)(u_1 + u_3)),$$

and

$$\beta(A) = (1)(1 + (l_1 + l_3)(u_1 + u_3) + l_3l_2u_3u_2).$$

If, for example, we set $l_2 = l_3 = u_2 = u_1 = t$, and all remaining variables to one, then $\alpha(A)$ is a polynomial in t of degree 4, whereas $\beta(A)$ is a polynomial in t of degree 3. Consequently, $\frac{\alpha(A)}{\beta(A)}$ increases without bound as $t \to \infty$, and therefore $\frac{\alpha}{\beta}$ is an unbounded ratio with respect InTN.

On the other hand, the ratio

$$\frac{\alpha}{\beta} = \frac{\{1,2\}\{3\}}{\{1,3\}\{2\}}$$

is, indeed, a bounded ratio with respect to InTN. To see this, observe that the difference (which is a subtraction-free expression)

$$\beta(A) - \alpha(A) = l_2 u_2 (l_3 u_1 + l_1 u_1 + l_1 u_3) + l_2 u_2 \geq 0,$$

for all nonnegative values of the parameters (here we are assuming A is presented as above). Hence $\frac{\alpha}{\beta} \leq 1$ with respect to InTN.

A more general necessary condition for a proposed ratio to be bounded is known, and makes use of a notion of majorization (see [FGJ03]).

Given two nondecreasing sequences $M_1 = \{m_{1j}\}_{j=1}^n, M_2 = \{m_{2j}\}_{j=1}^n$ of nonnegative integers such that $\sum_{j=1}^\infty m_{1j} = \sum_{j=1}^\infty m_{2j} < \infty$, we say that M_1 majorizes M_2 (denoted $M_1 \succeq M_2$) if the following inequalities hold:

$$\sum_{j=1}^k m_{1j} \geq \sum_{j=1}^k m_{2j}, \qquad k = 1, 2, \ldots, n.$$

Conjugate sequences $M_j^* = \{m_{jk}^*\}_{k=1}^\infty, j = 1, 2$ are defined by

$$m_{jk}^* = |\{m_{ij} \in M_i : m_{ij} \geq k\}|. \tag{6.7}$$

It is well known that $M_1 \succeq M_2$ if and only if $M_2^* \succeq M_1^*$.

Now let $\alpha = \{\alpha_1, \ldots, \alpha_p\}$, $\beta = \{\beta_1, \ldots, \beta_p\}$ be two collections of index sets. Let $L = \{i, \ldots, i+m\}$ be any contiguous subset of N. We call such an index set L an interval. Define $m(\alpha, L)$ to be a sequence of numbers $|\alpha_i \cap L|$ rearranged in a nondecreasing order.

Given a ratio $\frac{\alpha}{\beta}$, we say that this ratio satisfies condition (M) if

$$m(\alpha, L) \succeq m(\beta, L) \quad \text{for every interval} \quad L \subseteq N; \tag{6.8}$$

see also [FGJ03] with proof.

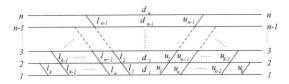

Figure 6.1 A general n-by-n diagram

Theorem 6.3.3 If the ratio $\frac{\alpha}{\beta}$ is bounded, then it satisfies condition (M).

Proof. First, note that the condition $\sum_{j=1}^p m_j(\alpha, L) = \sum_{j=1}^p m_j(\beta, L)$ follows from Proposition 6.3.2. Next, let $L = N$. For $l = 1, \ldots, n$, consider a TP matrix $A_{nl}(t)$ that corresponds to the diagram in Figure 6.1 with weights $d_1, \ldots, d_l$ equal to t and the rest of the weights equal to 1. Then, for any $S \subseteq N$, the degree of $\det A_{nl}(t)[S]$ as a polynomial in t is equal to $\min(l, |S|)$. It follows that $\deg \alpha(A_{nl}(t)) = \sum_i \min(l, |\alpha_i|) = \sum_{j=1}^l m_j^*(\alpha, N)$, where $m^*(\alpha, N)$ is a sequence conjugate to $m(\alpha, N)$ defined as in (6.7). Thus conditions $\deg(\alpha(A_{nl}(t))) \leq \deg(\beta(A_{nl}(t)))$ $(l = 1, \ldots, n)$ necessary for $\frac{\alpha}{\beta}$ to be bounded translate into inequalities $\sum_{j=1}^{l} m_j^*(\alpha, N) \leq \sum_{j=1}^{l} m_j^*(\beta, N)$ ($l = 1, \ldots, n$). Therefore, $m^*(\beta, N) \succeq m^*(\alpha, N)$ or, equivalently, $m(\alpha, N) \succeq m(\beta, N)$. Now, to prove (6.8) for an arbitrary interval $L = \{i, \ldots, i+m\}$, it is sufficient to apply the above argument to InTN matrices of the form $I_{i-1} \oplus A_{m+1,l} \oplus I_{n-m-i}$ for $l = 1, \ldots, m+1$.   □

Corollary 6.3.4 *For $J \subseteq N$, let $f_\alpha(J)$ (resp. $f_\beta(J)$) denote the number of index sets in α (resp. β) that contain J. If the ratio $\frac{\alpha}{\beta}$ is bounded, then $f_\alpha(L) > f_\beta(L)$ for every interval $L \subseteq N$.*

Proof. It suffices to notice that all elements of $m(\alpha, L)$ are not greater than $|L|$ and exactly $f_\alpha(L)$ of them are equal to $|L|$.   □

Remark. In [BJ84], it is shown that a ratio $\frac{\alpha}{\beta}$ is bounded with respect to the class of positive definite matrices, if $f_\alpha(J) \geq f_\beta(J)$, for every subset $J \subseteq N$ (the converse need not hold). In [FHJ98], part of one of the main results can be stated as follows: a ratio $\frac{\alpha}{\beta}$ is bounded with respect to the class of M-matrices if and only if it satisfies: $f_\alpha(\{i\}) = f_\beta(\{i\})$, for all $i = 1, 2, \ldots, n$, and $f_\alpha(J) \geq f_\beta(J)$, for every subset $J \subseteq N$. We note here that the condition $f_\alpha(J) \geq f_\beta(J)$ for arbitrary J is neither necessary nor sufficient (see [Fal99]) for a ratio to be bounded with respect to the InTN matrices.

6.3.1 Characterization of Principal Minor Inequalities for $n \leq 5$

A collection of bounded ratios with respect to InTN is referred to a *generator* if and only if any bounded ratio with respect to InTN can be written as products of positive powers of ratios from this collection. The idea employed is to assign the weights $0, 1, t, t^{-1}$, in which t is a nonnegative variable that we make arbitrarily large, to the variables $\{l_i\}$ $\{u_j\}$. For such a weighting, each principal minor will then be a function in terms of t. For a ratio of principal minors to be bounded, the degree of t in the denominator must be greater than or equal to the degree of t in the numerator. As before, we may let (S) denote $\det A[S]$.

We consider the 3-by-3 case in full detail, which can also be found in [FGJ03]. Every ratio on 3 indices can be written in the following way:

$$(1)^{x_1} (2)^{x_2} (3)^{x_3} (12)^{x_{12}} (13)^{x_{13}} (23)^{x_{23}} (123)^{x_{123}}$$

where x_{α_i} is the degree of (α_i) in the ratio. Let y_{α_i} be the degree of t in (α_i). Then the expression

$$y_1 x_1 + y_2 x_2 + y_3 x_3 + y_{12} x_{12} + y_{13} x_{13} + y_{23} x_{23} + y_{123} x_{123},$$

represents the degree of t in this ratio. By assumption, we set $(1) = (12) = (123) = 1$. Then, $y_1 = y_{12} = y_{123} = 0$, and the simplified expression has the form $y_2 x_2 + y_3 x_3 + y_{13} x_{13} + y_{23} x_{23}$. For this ratio to be bounded, $y_2 x_2 + y_3 x_3 + y_{13} x_{13} + y_{23} x_{23} \leq 0$. Using diagrams with different weightings, we produce a list of conditions, each of the form $y_2x_2 + y_3x_3 + y_{13}x_{13} + y_{23}x_{23} \leq 0$. Condition ST0 (for $n = 3$) is equivalent to $x_1 + x_{12} + x_{13} + x_{123} = 0$, $x_2 + x_{12} + x_{23} + x_{123} = 0$, and $x_3 + x_{13} + x_{23} + x_{123} = 0$.

Example 6.3.5 Let A be a 3-by-3 invertible TN matrix,

$$A = \begin{bmatrix} 1 & t^{-1} & 0 \\ 2 & 1+2t^{-1} & t \\ 1 & 1+t^{-1} & 1+t \end{bmatrix}.$$

For the matrix A, a simple computation yields $(2) = 1 + 2t^{-1}$, $(3) = 1 + t$, $(13) = 1 + t$, and $(23) = 2 + 2t^{-1}$. Thus $y_2 = 0$, $y_3 = 1$, $y_{13} = 1$, and $y_{23} = 0$. Therefore, the expression $x_3 + x_{13}$ represents the total degree of t in any ratio. In order for a ratio to be bounded with respect to InTN, it must satisfy the condition $x_3 + x_{13} \leq 0$.

In the 3-by-3 case, consider the following four necessary conditions (see Figure 6.2) that are used to define the cone of bounded ratios. Because the edges of the left side of each diagram are all weighted 1, it is sufficient to consider the right side only (see [FGJ03]).

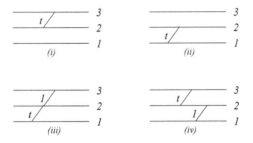

Figure 6.2 Necessary diagrams for 3-by-3 case

Each of the graphs in Figure 6.2 gives rise to a halfspace

$$(6.2.i)\ \ x_3 + x_{13} \leq 0, \quad (6.2.ii)\ \ x_2 + x_{23} \leq 0,$$

$$(6.2.iii)\ \ x_2 + x_3 + x_{23} \leq 0, \quad (6.2.iv)\ \ x_3 + x_{13} + x_{23} \leq 0.$$

This list of inequalities can be written as a matrix inequality $Ax \leq 0$, in which A is a 4-by-4 real matrix and we can define an associated polyhedral cone $C = \{x \in \mathbb{R}^4 : Ax \leq 0\}$. In the 3-by-3 case we then have the following characterization of all bounded ratios (see [FGJ03] for a complete proof).

Theorem 6.3.6 *Suppose $\frac{\alpha}{\beta}$ is a ratio of principal minors on three indices. Then $\frac{\alpha}{\beta}$ is bounded with respect to the invertible TN matrices if and only if $\frac{\alpha}{\beta}$ can be written as a product of positive powers of the bounded ratios*

$$\frac{(13)(\emptyset)}{(1)(3)}, \ \frac{(123)(2)}{(12)(23)}, \ \frac{(12)(3)}{(13)(2)}, \ \frac{(23)(1)}{(13)(2)}.$$

Proof. The four diagrams in Figure 6.2 give rise to the following four neces-sary conditions: (a) $x_3 + x_{13} \leq 0$; (b) $x_2 + x_{23} \leq 0$; (c) $x_2 + x_3 + x_{23} \leq 0$; and (d) $x_3 + x_{13} + x_{23} \leq 0$. Recall that the necessary conditions (ST0) are equiva-lent to the three equations $x_1 + x_{12} + x_{13} + x_{123} = 0$, $x_2 + x_{12} + x_{23} + x_{123} = 0$, and $x_3 + x_{13} + x_{23} + x_{123} = 0$. For simplicity of the analysis we use the equa-tions above to convert the four inequalities to "greater than" inequalities, namely, (a) $x_{23} + x_{123} \geq 0$; (b) $x_{12} + x_{123} \geq 0$; (c) $x_{12} + x_{13} + x_{23} + 2x_{123} \geq 0$; and (d) $x_{123} \geq 0$. Fix $x_{123} = k \geq 0$. Then the intersection given by the linear inequalities: $x_{23} \geq -k$, $x_{12} \geq -k$, and $x_{12} + x_{13} + x_{23} \geq -2k$, forms a polyhedral cone. Translate the variables $(x_{12}, x_{23}, x_{13})^T$ to $(x_{12} + k, x_{23} + k, x_{13})^T = (z_{12}, z_{23}, z_{13})^T$ so that the corresponding inequal-ities, in terms of z, are translated to the origin. Observe that the inter-section of $z_{12} = z_{23} = 0$ is the ray given by $\{(0, 0, t)^T : t \geq 0\}$; the re-maining extreme rays are determined similarly. In this case, the polyhedral cone C formed by the intersection of the above hyperplanes is given by $C = \{t_1(0, 0, 1)^T + t_2(0, 1, -1)^T + t_3(1, 0, -1)^T : t_i \geq 0, \}$. Therefore any vector $v \in C$ may be written as $v = (z_{12}, z_{23}, z_{13})^T = (t_3, t_2, t_1 - t_2 - t_3)^T$. Writing these equations in terms of x gives $x_{12} = t_3 - k$, $x_{23} = t_2 - k$, and $x_{13} = t_1 - t_2 - t_3$. Using the equations given by (ST0), we may now solve for the remaining variables, for example, $x_1 = -x_{12} - x_{13} - x_{123} = t_2 - t_1$. Similarly, it follows that $x_2 = k - t_2 - t_3$, and $x_3 = t_3 - t_1$. Finally, we substitute the above x-values back into the original ratio, that is,

$$(123)^{x_{123}}(12)^{x_{12}}(13)^{x_{13}}(23)^{x_{23}}(1)^{x_1}(2)^{x_2}(3)^{x_3}$$

$$= (123)^k(12)^{t_3-k}(13)^{t_1-t_2-t_3}(23)^{t_2-k}(1)^{t_2-t_1}(2)^{k-t_2-t_3}(3)^{t_3-t_1}$$

$$= \left[\frac{(123)(2)}{(12)(23)}\right]^k \left[\frac{(13)(\phi)}{(1)(3)}\right]^{t_1} \left[\frac{(23)(1)}{(13)(2)}\right]^{t_2} \left[\frac{(12)(3)}{(13)(2)}\right]^{t_3},$$

where $k, t_1, t_2, t_3 \geq 0$. This completes the proof. $\qquad\square$

For the cases $n = 4$ and 5, the analysis is similar but not surprisingly more involved. The number of necessary conditions increase, the number of variables increase, and the number of generators (or extremals) increases (see [FGG98] and [FGJ03]). We summarize the main results from these works below.

Theorem 6.3.7 *Suppose $\frac{\alpha}{\beta}$ a ratio of principal minors on four indices. Then $\frac{\alpha}{\beta}$ is bounded with respect to the invertible TN matrices if and only if $\frac{\alpha}{\beta}$ can be written as a product of positive powers of the following bounded ratios:*

$$\frac{(14)(\emptyset)}{(1)(4)}, \frac{(2)(124)}{(12)(24)}, \frac{(3)(134)}{(13)(34)}, \frac{(23)(1234)}{(123)(234)}, \frac{(12)(3)}{(13)(2)}, \frac{(1)(24)}{(2)(14)}, \frac{(2)(34)}{(3)(24)},$$

$$\frac{(4)(13)}{(3)(14)}, \frac{(12)(134)}{(13)(124)}, \frac{(13)(234)}{(23)(134)}, \frac{(34)(124)}{(24)(134)}, \frac{(24)(123)}{(23)(124)}, \frac{(14)(23)}{(13)(24)}.$$

Theorem 6.3.8 *Suppose $\frac{\alpha}{\beta}$ a ratio of principal minors on five indices. Then $\frac{\alpha}{\beta}$ is bounded with respect to the invertible TN matrices if and only if $\frac{\alpha}{\beta}$ can be written as a product of positive powers of the following bounded ratios:*

$$\frac{(15)(\emptyset)}{(1)(5)}, \; \frac{(2)(125)}{(12)(25)}, \; \frac{(3)(135)}{(13)(35)}, \; \frac{(4)(145)}{(14)(45)}, \; \frac{(23)(1235)}{(235)(123)}, \; \frac{(24)(1245)}{(124)(245)}, \; \frac{(34)(1345)}{(134)(345)},$$

$$\frac{(234)(12345)}{(1234)(2345)}, \; \frac{(12)(3)}{(2)(13)}, \; \frac{(13)(4)}{(3)(14)}, \; \frac{(14)(5)}{(15)(4)}, \; \frac{(1)(25)}{(2)(15)}, \; \frac{(2)(35)}{(3)(25)}, \; \frac{(3)(45)}{(4)(35)}, \; \frac{(12)(135)}{(13)(125)},$$

$$\frac{(13)(235)}{(23)(135)}, \; \frac{(14)(245)}{(24)(145)}, \; \frac{(13)(145)}{(14)(135)}, \; \frac{(23)(245)}{(24)(235)}, \; \frac{(123)(24)}{(124)(23)}, \; \frac{(124)(25)}{(24)(125)}, \; \frac{(34)(124)}{(24)(134)},$$

$$\frac{(35)(134)}{(34)(135)}, \; \frac{(45)(135)}{(35)(145)}, \; \frac{(35)(125)}{(25)(135)}, \; \frac{(24)(345)}{(34)(245)}, \; \frac{(123)(1245)}{(124)(1235)}, \; \frac{(124)(1345)}{(134)(1245)},$$

$$\frac{(134)(2345)}{(234)(1345)}, \; \frac{(235)(1234)}{(234)(1235)}, \; \frac{(245)(1235)}{(235)(1245)}, \; \frac{(345)(1245)}{(245)(1345)}, \; \frac{(14)(23)}{(13)(24)}, \; \frac{(15)(24)}{(14)(25)},$$

$$\frac{(25)(34)}{(24)(35)}, \; \frac{(125)(134)}{(124)(135)}, \; \frac{(135)(234)}{(134)(235)}, \; \frac{(145)(235)}{(135)(245)}.$$

Note that the generators in the list (38 in total) all consist of two sets over two sets just as in the $n = 3$ and $n = 4$ cases.

Since, by Corollary 6.3.14 (to follow), each of the ratios in Theorems 6.3.6, 6.3.7, and 6.3.8 is bounded by one, we obtain the following.

Corollary 6.3.9 *For $n \leq 5$, any ratio $\frac{\alpha}{\beta}$ of principal minors is bounded with respect to the invertible TN matrices if and only if it is bounded by one.*

The above essentially represents the current state, in terms of complete characterizations of all of the principal minor inequalities with respect to InTN. It is clear that this technique could extend, however, the computation becomes prohibitive. One of the important properties of determinantal inequalities for InTN seems to be the underlying connection with subtraction-free expression (in this case with respect to the parameters from the SEB factorization).

6.3.2 Two-over-Two Ratio Result: Principal Case

As seen in the previous characterizations, ratios consisting of two index sets over two index sets seem to play a fundamental role. In fact, for $n \leq 5$, all the generators for the bounded ratios of products of principal minors have this form. Also, Koteljanskiĭ's inequality can be interpreted as a ratio of two sets over two sets.

Before our main observations of this section, we first consider the following very interesting and somewhat unexpected ratio (see [FGJ03]). We include a proof for completeness and to highlight the concept of subtraction-free expressions and their connections to determinantal inequalities for InTN.

Proposition 6.3.10 *For any 4-by-4 invertible TN matrix the following inequality holds,*

$$\frac{(14)(23)}{(13)(24)} \leq 1.$$

Proof. Consider an arbitrary 4-by-4 InTN matrix presented via its elementary bidiagonal factorization. Then it is straightforward to calculate each of the four 2-by-2 minors above. For instance, $(14) = 1 + u_3 u_2 l_3 l_2 + u_3 l_3$, $(23) = 1 + u_6 l_4 + u_6 l_1 + u_4 l_6 + u_4 l_4 + u_6 l_6 + u_1 l_6 + u_1 l_4 + u_1 l_1 + u_4 l_1 + u_6 l_6 u_5 l_5 + u_6 l_6 u_5 l_2 + u_6 l_6 u_2 l_5 + u_6 l_6 u_2 l_2 + u_6 l_4 u_5 l_2 + u_6 l_4 u_2 l_2 + u_4 l_6 u_2 l_5 + u_4 l_6 u_2 l_2 + u_4 l_4 u_2 l_2$, $(13) = 1 + u_5 l_5 + u_5 l_2 + u_2 l_5 + u_2 l_2$, and finally $(24) = 1 + u_6 l_4 + u_3 l_3 + u_6 l_1 + u_4 l_6 + u_4 l_4 + u_6 l_6 + u_1 l_6 + u_1 l_4 + u_1 l_1 + u_4 l_1 + u_6 l_6 u_3 l_3 + u_6 l_4 u_3 l_3 + u_6 l_1 u_3 l_3 + u_4 l_6 u_3 l_3 + u_4 l_4 u_3 l_3 + u_4 l_1 u_3 l_3 + u_1 l_6 u_3 l_3 + u_1 l_4 u_3 l_3 + u_1 l_1 u_3 l_3 + u_6 l_6 u_3 u_2 l_3 l_2 + u_6 l_4 u_3 u_2 l_3 l_2 + u_4 l_6 u_3 u_2 l_3 l_2 + u_4 l_4 u_3 u_2 l_3 l_2$.

Next we compute the difference between the expressions for the denominator and the numerator, namely, $(13)(24) - (14)(23) = u_5 l_5 + u_5 l_2 + u_2 l_5 + u_2 l_2 + u_5 u_4 l_5 l_4 + u_5 u_4 l_1 l_5 + u_5 u_4 l_1 l_2 + u_5 u_1 l_5 l_4 + u_5 u_1 l_1 l_5 + u_5 u_1 l_1 l_2 + u_2 u_1 l_5 l_4 + u_2 u_1 l_1 l_5 + u_2 u_1 l_1 l_2 + u_6 l_4 u_5 l_5 + u_6 l_4 u_2 l_5 + u_6 l_1 u_5 l_5 + u_6 l_1 u_5 l_2 + u_6 l_1 u_2 l_5 + u_6 l_1 u_2 l_2 + u_4 l_6 u_5 l_5 + u_4 l_6 u_5 l_2 + u_4 l_4 u_5 l_2 + u_4 l_4 u_2 l_5 + u_4 l_1 u_2 l_5 + u_4 l_1 u_2 l_2 + u_1 l_6 u_5 l_5 + u_1 l_6 u_5 l_2 + u_1 l_6 u_2 l_5 + u_1 l_6 u_2 l_2 + u_1 l_4 u_5 l_2 + u_1 l_4 u_2 l_2 + u_5 l_5 u_3 l_3 + u_5 l_2 u_3 l_3 + u_2 l_5 u_3 l_3 + u_5 l_5 u_6 l_4 u_3 l_3 + u_5 l_5 u_6 l_1 u_3 l_3 + u_5 l_5 u_4 l_6 u_3 l_3 + u_5 l_5 u_4 l_4 u_3 l_3 + u_5 l_5 u_4 l_1 u_3 l_3 + u_5 l_5 u_1 l_6 u_3 l_3 + u_5 l_5 u_1 l_4 u_3 l_3 + u_5 l_5 u_1 l_1 u_3 l_3 + u_5 l_5 u_6 l_4 u_3 u_2 l_3 l_2 + u_5 l_5 u_4 l_6 u_3 u_2 l_3 l_2 + u_5 l_5 u_4 l_4 u_3 u_2 l_3 l_2 + u_5 l_2 u_6 l_1 u_3 l_3 + u_5 l_2 u_4 l_6 u_3 l_3 + u_5 l_2 u_4 l_4 u_3 l_3 + u_5 l_2 u_4 l_1 u_3 l_3 + u_5 l_2 u_1 l_6 u_3 l_3 + u_5 l_2 u_1 l_4 u_3 l_3 + u_5 l_2 u_1 l_1 u_3 l_3 + u_5 l_2{}^2 u_4 l_6 u_3 u_2 l_3 + u_5 l_2{}^2 u_4 l_4 u_3 u_2 l_3 + u_2 l_5 u_6 l_4 u_3 l_3 + u_2 l_5 u_6 l_1 u_3 l_3 + u_2 l_5 u_4 l_4 u_3 l_3 + u_2 l_5 u_4 l_1 u_3 l_3 + u_2 l_5 u_1 l_6 u_3 l_3 + u_2 l_5 u_1 l_4 u_3 l_3 + u_2 l_5 u_1 l_1 u_3 l_3 + u_2{}^2 l_5 u_6 l_4 u_3 l_3 l_2 + u_2{}^2 l_5 u_4 l_4 u_3 l_3 l_2$. Observe that the above expression is a subtraction free expression in nonnegative variables and hence is nonnegative. This completes the proof. $\square$

We now move on to the central result of this section; consult [FGJ03] for complete details of the facts that follow.

Theorem 6.3.11 *Let $\alpha_1, \alpha_2, \beta_1$ and β_2 be subsets of $\{1, 2, \ldots, n\}$. Then the ratio $\frac{(\alpha_1)(\alpha_2)}{(\beta_1)(\beta_2)}$ is bounded with respect to the invertible TN matrices if and only if:*

(1) this ratio satisfies (ST0); and

(2) $\max(|\alpha_1 \cap L|, |\alpha_2 \cap L|) \geq \max(|\beta_1 \cap L|, |\beta_2 \cap L|)$, for every interval $L \subseteq \{1, 2, \ldots, n\}$.

The proof is divided into several steps. First, observe that the necessity of the conditions follows immediately from Theorem 6.3.3. Indeed, it is not hard to see that these conditions are equivalent to condition (M) for $\frac{\alpha}{\beta}$ with $\alpha = (\alpha_1, \alpha_2)$, $\beta = (\beta_1, \beta_2)$.

Suppose γ is a subset of $\{1, 2, \ldots, n\}$ and L is a given interval of $\{1, 2, \ldots, n\}$. Then we let $g(\gamma, L) = \max(|\gamma \cap L|, |\gamma^c \cap L|)$. Finally, suppose that $i_1 < i_2 < \cdots < i_k$ and $j_1 < j_2 < \cdots < j_l$ are indices of $\{1, 2, \ldots, n\}$, so that $k - 1 \leq l \leq k + 1$. Then we say the *sequence* $\{i_t\}$ *interlaces the sequence* $\{j_t\}$ if one of following cases occur: (1) $l = k + 1$ and $j_1 \leq i_1 \leq j_2 \leq i_2 \leq \cdots \leq j_k \leq i_k \leq j_l$; (2) $l = k$ and $i_1 \leq j_1 \leq i_2 \leq j_2 \leq \cdots \leq j_{k-1} \leq i_k \leq j_l$; or (3) $l = k - 1$ and $i_1 \leq j_1 \leq i_2 \leq j_2 \leq \cdots \leq j_l \leq i_k$. In the event $\gamma = \{i_1, i_2, \ldots, i_k\}$ with $i_j < i_{j+1}$ and $\delta = \{j_1, j_2, \ldots, j_l\}$ with $j_i < j_{i+1}$ $(k - 1 \leq l \leq k + 1)$, and the sequence $\{i_t\}$ interlaces the sequence $\{j_t\}$, then we say that γ *interlaces* δ. The next proposition follows immediately from the definitions above.

Proposition 6.3.12 *Let γ and δ be two nonempty index sets of $\{1, 2, \ldots, n\}$. If δ interlaces δ^c, then $g(\gamma, L) \geq g(\delta, L)$ for every interval L.*

We have already established the following inequalities with respect to the TN matrices: $(1, 2)(\phi) \leq (1)(2)$ (Koteljanskiĭ), $(1, 2)(3) \leq (1, 3)(2)$, and $(1, 4)(2, 3) \leq (1, 3)(2, 4)$ (Proposition 6.3.10). These inequalities will serve as the base cases for the next result, which uses induction on the number of indices involved in a ratio (see [FGJ03]).

Theorem 6.3.13 *Let γ and δ be two nonempty index sets of N. Then the ratio $\frac{(\gamma)(\gamma^c)}{(\delta)(\delta^c)}$ is bounded with respect to the invertible TN matrices if and only if $g(\gamma, L) \geq g(\delta, L)$ for every interval L of N.*

Observe that Theorem 6.3.13 establishes that main result in this special case of the index sets having no indices in common.

Proof of Theorem 6.3.11: Note that the ratio $\frac{(\alpha_1)(\alpha_2)}{(\beta_1)(\beta_2)}$ is bounded with respect to the invertible TN matrices if and only if $\frac{(S)(S^c)}{(T)(T^c)}$ is bounded with respect to the invertible totally nonnegative matrices, for some S and T obtained from α_1, α_2 and β_1, β_2, respectively, by deleting common indices and shifting (as these operations preserve boundedness, see [FGJ03]).

There are many very useful consequences to Theorems 6.3.11 and 6.3.13, which we include here.

The proof of the above result follows directly from Theorem 6.3.13 and the fact that the base case ratios are each bounded by one; see [FGJ03] for more details.

Corollary 6.3.14 *Let $\alpha_1, \alpha_2, \beta_1$ and $\beta_2 \subseteq N$. Then the ratio $\frac{(\alpha_1)(\alpha_2)}{(\beta_1)(\beta_2)}$ is bounded with respect to the invertible TN matrices if and only if it is bounded by one.*

For the next consequence we let (odds) $= (1, 3, 5, \ldots)$ (i.e., the minor consisting of all the odd integers of N) and (evens) $= (2, 4, 6, \ldots)$ (the minor consisting of all the even integers of N). Then the next result follows directly from Theorem 6.3.13.

Corollary 6.3.15 *For $n \geq 2$, and $\gamma \subseteq \{1, 2, \ldots, n\}$,*

$$(1, 2, 3 \ldots, n)(\phi) \leq (\gamma)(\gamma^c) \leq (\text{odds})(\text{evens}),$$

for any TN matrix.

Of course the first inequality is just Fischer's inequality.

6.3.3 Inequalities Restricted to Certain Subclasses of InTN/TP

As part of an ongoing effort to learn more about all possible principal minor inequalities over all TN/TP matrices, it is natural to restrict attention to important subclasses. Along these lines two subclasses have received some significant attention: the tridiagonal InTN matrices and a class known as STEP 1.

Suppose A is a tridiagonal InTN matrix; then A is similar (via a signature matrix) to an M-matrix. As a result, we may apply the characterization that is known for all multiplicative principal minor inequalities with regard to M-matrices. In doing so, we have the following characterization (see [FJ01]). We begin with the following definition, and note that the condition presented has a connection to condition (M) restricted to ratios consisting of two index sets.

Let $\alpha = \{\alpha_1, \alpha_2, \ldots, \alpha_p\}$ be a given collection of index sets. We let $f_\alpha(L)$ denote the number of index sets in α that contain L, where $L \subseteq N$. We say the ratio ratio α/β satisfies (ST1)′ if

$$f_\alpha(L) \geq f_\beta(L) \tag{6.9}$$

holds for all *contiguous* subsets L of N. This condition, in a more general form (namely, the condition holds for all subsets of N), has appeared in the context of determinantal inequalities for both positive semidefinite and M-matrices (see [BJ84, FHJ98] and the survey paper [FJ00]).

Now, for tridiagonal matrices, the next property is key for the characterization that follows. If A is a tridiagonal matrix, then

$$A[\{1, 2, 4, 6, 7\}] = A[\{1, 2\}] \oplus A[\{4\}] \oplus A[\{6, 7\}],$$

where $\oplus$ denotes direct sum. In particular,

$$\det A[\{1, 2, 4, 6, 7\}] = \det A[\{1, 2\}] \det A[\{4\}] \det A[\{6, 7\}].$$

It is clear that to compute a principal minor of a tridiagonal matrix it is enough to compute principal minors based on associated contiguous index sets.

The main result regarding the multiplicative principal minor inequalities for tridiagonal matrices (see [FJ01]) is then

Theorem 6.3.16 *Let $\frac{\alpha}{\beta}$ be a given ratio of index sets. Then the following are equivalent:*

(i) $\frac{\alpha}{\beta}$ satisfies (ST0) and (ST1)′;

(ii) $\frac{\alpha}{\beta}$ *is bounded with respect to the tridiagonal TN matrices;*

(iii) $\frac{\alpha}{\beta}$ *with respect to the tridiagonal TN matrices.*

We also bring to the attention of the reader that there is a nice connection between the determinantal inequalities that hold with respect to the tridiagonal TN matrices and Koteljanskiĭ's inequality (see [FJ01]).

Another subclass that has been studied in this context has been referred to as the class STEP 1. As a consequence of the SEB factorization of any InTN matrix, we know that other orders of elimination are possible when factoring an InTN matrix into elementary bidiagonal factors. For instance, any InTN matrix A can be written as

$$A = (L_2(l_k))(L_3(l_{k-1})L_2(l_{k-2}))\cdots(L_n(l_{n-1})\cdots L_3(l_2)L_2(l_1))D$$
$$(U_2(u_1)U_3(u_2)\cdots U_n(u_{n-1}))\cdots(U_2(u_{k-2})U_3(u_{k-1}))(U_2(u_k)),(6.10)$$

where $k = \binom{n}{2}$ $l_i, u_j \geq 0$ for all $i,j \in \{1,2,\ldots,k\}$; and D is a positive diagonal matrix. Further, given the factorization above (6.10), we introduce the following notation:

$$\tilde{L}_1 = (L_n(l_{n-1})\cdots L_3(l_2)L_2(l_1)),$$

$$\tilde{U}_1 = (U_2(u_1)U_3(u_2)\cdots U_n(u_{n-1})),$$

$$\vdots$$

$$\tilde{L}_{n-2} = (L_3(l_{k-1})L_2(l_{k-2})), \ \tilde{U}_{n-2} = (U_2(u_{k-2})U_3(u_{k-1})),$$

$$\tilde{L}_{n-1} = L_2(l_k), \ \tilde{U}_{n-1} = U_2(u_k).$$

Then (6.10) is equivalent to

$$A = \tilde{L}_{n-1}\tilde{L}_{n-2}\cdots\tilde{L}_1 D \tilde{U}_1\cdots\tilde{U}_{n-2}\tilde{U}_{n-1}.$$

The subclass of interest is defined as follows. The subclass STEP 1 is then defined as $A_1 = \{A : A = L_1 D U_1\}$. The name is chosen because of the restriction to one (namely, the first) stretch of elementary bidiagonal (lower and upper) factors in the factorization (6.10).

For this class, all the multiplicative principal-minor inequalities have been completely characterized in terms of a description of the generators for all such bounded ratios. As with the general case, all ratios are bounded by one and all may be deduced as a result of subtraction-free expressions for the difference between the denominator and numerator (see [LF08]). All generators are of the form

$$\frac{(i,j)(i-1,j+1)}{(i,j+1)(i-1,j)},$$

where $1 \leq i < j \leq n$ and we adopt the convention that if $j+1 > n$, then we drop $j+1$ from the ratio above, or if $i-1 < 1$, then we drop $i-1$ from the ratio above.

6.4 SOME NON-PRINCIPAL MINOR INEQUALITIES

Certainly the possible principal minor inequalities that exist among all TN/TP matrices are important and deserve considerably more attention. However, TN/TP matrices enjoy much more structure among their non-principal minors as well, and understanding the inequalities among all (principal and non-principal) minors throughout the TN/TP matrices is of great value and interest. In fact, as more analysis is completed on general determinantal inequalities, it is becoming more apparent (see, for example, [BF08, FJ00, LF08, Ska04]) that this analysis implies many new and interesting inequalities among principal minors.

We extend some of the notation developed earlier regarding minors to include non-principal minors. Recall that for an n-by-n matrix A and $\alpha_r, \alpha_c \subseteq N$, the submatrix of A lying in rows indexed by α_r and columns indexed by α_c is denoted by $A[\alpha_r, \alpha_c]$. For brevity, we may also let (α_r, α_c) denote $\det A[\alpha_r, \alpha_c]$. More generally, let $\alpha = \{\alpha_1, \alpha_2, \ldots, \alpha_p\}$ denote a collection of multisets (repeats allowed) of the form $\alpha_i = \{\alpha_r^i; \alpha_c^i\}$, where for each i, α_r^i denotes a row index set and α_c^i denotes a column index set ($|\alpha_r^i| = |\alpha_c^i|$, $i = 1, 2, \ldots, p$). If, further, $\beta = \{\beta_1, \beta_2, \ldots, \beta_q\}$ is another collection of such index sets with $\beta_i = \{\beta_r^i; \beta_c^i\}$ for $i = 1, 2, \ldots, q$, then, as in the principal case, we define the concepts such as

$$\alpha(A) := \prod_{i=1}^{p} \det A[\alpha_r^i, \alpha_c^i] = \prod_{i=1}^{p} (\alpha_r^i, \alpha_c^i),$$

and $\alpha \leq \beta$, the ratio $\frac{\alpha}{\beta}$ (assuming the denominator is not zero), bounded ratios and generators with respect to InTN.

6.4.1 Non-Principal Minor Inequalities: Multiplicative Case

Consider the following illustrative example. Recall that

$$\frac{\alpha}{\beta} = \frac{\{1, 2\}\{3\}}{\{1, 3\}\{2\}} \leq 1$$

with respect to InTN. Moreover, this ratio was identified as a generator for all bounded ratios with respect to 3-by-3 InTN matrices, and hence cannot be factored further into bounded ratios consist of principal minors.

On the other hand, consider the ratio of non-principal minors

$$\frac{(\{1, 2\}, \{1, 2\})(\{2\}, \{3\})}{(\{1, 2\}, \{1, 3\})(\{2\}, \{2\})}.$$

It is straightforward to verify that this ratio is bounded by 1 with respect to TP (this ratio involves a 2-by-3 TP matrix only), and is actually a factor in the factorization

$$\frac{\{1, 2\}\{3\}}{\{1, 3\}\{2\}} = \frac{(\{1, 2\}, \{1, 2\})(\{2\}, \{3\})}{(\{1, 2\}, \{1, 3\})(\{2\}, \{2\})} \cdot \frac{(\{1, 2\}, \{1, 3\})(\{3\}, \{3\})}{(\{1, 3\}, \{1, 3\})(\{2\}, \{3\})}.$$

Since each factor on the right is bounded by one, the ratio on the left is then also bounded by one.

Thus, as in the principal case, a similar multiplicative structure exists for bounded ratios of arbitrary minors of TP matrices (we consider TP to avoid zero minors). And like the principal case (cf. Section 6.3.1), characterizing all the bounded ratios via a complete description of the (multiplicative) generators is a substantial problem, which is still rather unresolved. However, for $n \leq 3$, all such generators have been determined (see [FK00]). In fact, the work in [FK00] may be viewed as an extension of the analysis begun in [FGJ03].

The main observation from [FK00] is the following result describing all the generators for all bounded ratios with respect to TP for $n \leq 3$.

Theorem 6.4.1 *Suppose $\frac{\alpha}{\beta}$ is a ratio of minors on three indices. Then $\frac{\alpha}{\beta}$ is bounded with respect to the totally positive matrices if and only if $\frac{\alpha}{\beta}$ can be written as a product of positive powers of the following bounded ratios:*

$$\frac{(13)(\phi)}{(1)(3)}, \frac{(2)(123)}{(12)(23)}, \frac{(\{1\},\{2\})(\{2\},\{1\})}{(\{1\},\{1\})(\{2\},\{2\})}, \frac{(\{2\},\{2\})(\{3\},\{1\})}{(\{2\},\{1\})(\{3\},\{2\})}, \frac{(\{1\},\{3\})(\{2\},\{2\})}{(\{1\},\{2\})(\{2\},\{3\})},$$

$$\frac{(\{2\},\{3\})(\{3\},\{2\})}{(\{2\},\{2\})(\{3\},\{3\})}, \frac{(\{13\},\{23\})(\{23\},\{13\})}{(\{13\},\{13\})(\{23\},\{23\})}, \frac{(\{13\},\{13\})(\{23\},\{12\})}{(\{13\},\{12\})(\{23\},\{13\})},$$

$$\frac{(\{12\},\{23\})(\{13\},\{13\})}{(\{12\},\{13\})(\{13\},\{23\})}, \frac{(\{12\},\{13\})(\{13\},\{12\})}{(\{12\},\{12\})(\{13\},\{13\})}, \frac{(\{1\},\{1\})(\{13\},\{23\})}{(\{1\},\{2\})(\{13\},\{13\})},$$

$$\frac{(\{1\},\{1\})(\{23\},\{13\})}{(\{2\},\{3\})(\{13\},\{13\})}, \frac{(\{1\},\{2\})(\{23\},\{23\})}{(\{2\},\{2\})(\{13\},\{23\})}, \frac{(\{2\},\{1\})(\{23\},\{23\})}{(\{2\},\{2\})(\{23\},\{13\})},$$

$$\frac{(\{2\},\{3\})(\{12\},\{12\})}{(\{2\},\{2\})(\{12\},\{13\})}, \frac{(\{3\},\{2\})(\{12\},\{12\})}{(\{2\},\{2\})(\{13\},\{12\})}, \frac{(\{3\},\{3\})(\{12\},\{13\})}{(\{2\},\{3\})(\{13\},\{13\})},$$

$$\frac{(\{3\},\{3\})(\{13\},\{12\})}{(\{3\},\{2\})(\{13\},\{13\})}.$$

Moreover, all the ratios above were shown to be bounded by one, so we may conclude that all such bounded ratios are necessarily bounded by one.

In an effort to extend Theorem 6.3.11 to general minors, one insight is to connect a minor of a TP matrix A with a corresponding Plücker coordinate. Recall from Chapter 1: The *Plücker coordinate*

$$[p_1, p_2, \ldots, p_n](B(A)),$$

where $1 \leq p_1 < p_2 < \cdots < p_n \leq 2n$, is defined to be the minor of $B(A)$ lying in rows $\{p_1, p_2, \ldots, p_n\}$ and columns $\{1, 2, \ldots, n\}$. That is,

$$[p_1, p_2, \ldots, p_n](B(A)) = \det B(A)[\{p_1, p_2, \ldots, p_n\}, N].$$

Suppose α, β are two index sets of N of the same size and γ is defined as

$$\gamma = \alpha \cup \{2n + 1 - i | i \in \beta^c\},$$

where β^c is the complement of β relative to N. Then for any n-by-n matrix A, we have

$$[\gamma](B(A)) = \det A[\alpha, \beta].$$

For simplicity, we will shorten $[\gamma](B(A))$ to just $[\gamma]$. Then, we may consider ratios of Plücker coordinates with respect to TP such as

$$\frac{[\gamma_1][\gamma_2]\cdots[\gamma_p]}{[\delta_1][\delta_2]\cdots[\delta_p]},$$

where $\gamma_i, \delta_j \subseteq \{1, 2, \ldots, 2n\}$ (see [BF08]).

In [BF08] two ratios were identified and played a central role in the analysis that was developed within that work. The first type of ratio was called *elementary* and has the form

$$\frac{[i, j', \Delta][i', j, \Delta]}{[i, j, \Delta][i', j', \Delta]},$$

where $|\Delta| = n - 2$, $1 \le i < i' < j < j' \le 2n$, and i, i', j, j' are not in Δ. It is evident that elementary ratios are connected with the short Plücker relation discussed in Chapter 1. The second key ratio type was referred to as *basic* and can be written as

$$\frac{[i, j+1, \Delta][i+1, j, \Delta]}{[i, j, \Delta][i+1, j+1, \Delta]},$$

with $|\Delta| = n - 2$, $1 \le i + 1 < j \le 2n$, and $i, i+1, j, j+1$ are not in Δ. It was shown in [BF08] that every bounded elementary ratio is a product of basic ratios and (see below) that the basic ratios are the generators for all bounded ratios consisting of two index sets over two index sets. We also note that both basic and elementary ratios were also studied in [Liu08].

Theorem 6.4.2 *Let $\alpha/\beta = [\gamma_1][\gamma_2]/[\delta_1][\delta_2]$ be a ratio where $\gamma_1, \gamma_2, \delta_1, \delta_2$ are subsets of $\{1, 2, \ldots, 2n\}$. Then the following statements are equivalent:*

(1) α/β satisfies STO and

$$\max(|\gamma_1 \cap L|, |\gamma_2 \cap L|) \ge \max(|\delta_1 \cap L|, |\delta_2 \cap L|)$$

for every interval $L \subseteq \{1, 2, \ldots, 2n\}$;

(2) α/β can be written as a product of basic ratios;

(3) α/β is bounded by one with respect to TP.

We note here that some of the techniques used in the proof of Theorem 6.4.2 extend similar ones used in [FGJ03] to general minors of TP matrices. Furthermore, the equivalence of statements (1) and (3) was first proved in [Ska04]. In addition, in [RS06], the notion of multiplicative determinantal inequalities for TP matrices has been related to certain types of positivity conditions satisfied by specific types of polynomials.

As in the principal case, narrowing focus to specific subclasses is often beneficial, and some research along these lines has taken place. The two classes of note, which were already defined relative to principal minor inequalities, are the tridiagonal InTN matrices and STEP 1. For both of these classes all multiplicative minor inequalities have been characterized via a complete description of the generators. For STEP 1, all generators have the form:

$$\frac{(\{i,j\},\{i,j+1\})(\{i,j+1\},\{i,j)\})}{(\{i,j\},\{i,j\})((\{i,j+1\},\{i,j+1\})},$$

where $1 \le i < j \le n$ (see [LF08]). Observe the similarity between the form of the generators above and the basic ratios defined previously. For more information on the tridiagonal case, consult [Liu08].

6.4.2 Non-Principal Minor Inequalities: Additive Case

Since all minors in question are nonnegative, studying additive relations among the minors of a TN matrix is a natural line of research. Although the multiplicative case has received more attention than the additive situation, there have been a number of significant advances in the additive case that have suggested more study.

For example, in Chapter 2 we needed an additive determinantal inequality to aid in establishing the original elimination result by Anne Whitney. Recall that for any n-by-n TN matrix $A = [a_{ij}]$ and for $p = 1, 2, \ldots, \lfloor n/2 \rfloor$ and for $q = 1, 2, \ldots, \lfloor (n+1)/2 \rfloor$, we have

$$a_{11}A_{11} - a_{21}A_{21} + \cdots - a_{2p,1}A_{2p,1}$$
$$\le \det A \le a_{11}A_{11} - a_{21}A_{21} + \cdots + a_{2q-1,1}A_{2q-1,1},$$

where, for each i, j, $A_{ij} = \det A(i, j)$.

Another instance of a valid and important additive determinantal inequality is one that results from the Plücker relations discussed in Chapter 1, as each constructed coordinate is positive.

As a final example, we discuss the notion of a compressed matrix and a resulting determinantal inequality that, at its root, is an additive minor inequality for TN matrices. Consider nk-by-nk partitioned matrices $A = [A_{ij}]_{i,j=1}^{k}$, in which each block A_{ij} is n-by-n. It is known that if A is a Hermitian positive definite matrix, then the k-by-k *compressed matrix* (or *compression of A*)

$$\tilde{C}_k(A) = [\det A_{ij}]_{i,j=1}^{k}$$

is also positive definite, and

$$\det \tilde{C}_k(A) > \det A.$$

Comparing and identifying common properties between positive definite matrices and totally positive matrices is both important and useful, and so we ask: if A is a totally positive nk-by-nk matrix, then is the compressed matrix $\tilde{C}_k(A)$ a totally positive matrix?

We address this question now by beginning with informative and interesting $k = 2$ case, including a proof from [FHGJ06].

Theorem 6.4.3 *Let A be a $2n$-by-$2n$ totally positive matrix that is partitioned as follows:*

$$A = \begin{pmatrix} A_{11} & A_{12} \\ A_{21} & A_{22} \end{pmatrix},$$

where each A_{ij} is an n-by-n block. Then the compression $\tilde{C}_2(A)$ is also totally positive. Moreover, in this case

$$\det \tilde{C}_2(A) > \det A.$$

Proof. Since A is totally positive, A can be written as $A = LU$, where L and U are ΔTP matrices. We can partition L and U into n-by-n blocks and rewrite $A = LU$ as

$$A = \begin{pmatrix} A_{11} & A_{12} \\ A_{21} & A_{22} \end{pmatrix} = \begin{pmatrix} L_{11} & 0 \\ L_{21} & L_{22} \end{pmatrix} \begin{pmatrix} U_{11} & U_{12} \\ 0 & U_{22} \end{pmatrix}.$$

Observe that

$$\det A_{11} \det A_{22} = \det(L_{11}U_{11}) \det \left([L_{21} L_{22}] \begin{bmatrix} U_{12} \\ U_{22} \end{bmatrix} \right).$$

Applying the classical Cauchy-Binet identity to the far right term above yields

$$\det \left([L_{21} L_{22}] \begin{bmatrix} U_{12} \\ U_{22} \end{bmatrix} \right)$$
$$= \sum_\gamma \det L[\{n+1, \ldots, 2n\}, \gamma] \det U[\gamma, \{n+1, \ldots, 2n\}], \quad (6.11)$$

where the sum is taken over all ordered subsets γ of $\{1, 2, \ldots, 2n\}$ with cardinality n. If we separate the terms with $\gamma = \{1, 2, \ldots, n\}$ and $\gamma = \{n+1, n+2, \ldots, 2n\}$, then the sum on the right in (6.11) reduces to

$$\sum_\gamma \det L[\{n+1, \ldots, 2n\}, \gamma] \det U[\gamma, \{n+1, \ldots, 2n\}]$$

$$= \det L_{21} \det U_{12} + \det L_{22} \det U_{22} + (\text{positive terms}).$$

Since L and U are ΔTP, all summands are positive. Hence

$\det A_{11} \det A_{22} = \det(L_{11}U_{11})[\det L_{21} \det U_{12} + \det L_{22} \det U_{22}] + (\text{positive terms}),$

which is equivalent to

$$\det A_{11} \det A_{22} = \det A_{12} \det A_{21} + \det A + (\text{positive terms}).$$

Thus we have

$$\det \tilde{C}_2(A) = \det A_{11} \det A_{22} - \det A_{12} \det A_{21}$$
$$= \det A + (\text{positive terms}),$$

and so $\det \tilde{C}_2(A) > \det A > 0$, which completes the proof. $\square$

The following is a consequence of Theorem 6.4.3 and the classical fact that the TP matrices are dense in the TN matrices.

Corollary 6.4.4 *Let A be a 2n-by-2n TN matrix that is partitioned as follows:*

$$A = \begin{pmatrix} A_{11} & A_{12} \\ A_{21} & A_{22} \end{pmatrix},$$

where A_{ij} is n-by-n. Then $\tilde{C}_2(A)$ is also TN and

$$\det \tilde{C}_2(A) \geq \det A.$$

To extend this result for larger values of k, it seemed necessary to identify the determinant of the compressed matrix as a generalized matrix function (for more details see [FHGJ06]). In fact, it was shown that the compression of a 3n-by-3n TP matrix is again TP, while for $k \geq 4$, compressions of nk-by-nk TP matrices need not necessarily be TP. Furthermore, in [FHGJ06] there are many related results on compressions of TP matrices.

Chapter Seven

Row and Column Inclusion and the Distribution of Rank

7.0 INTRODUCTION

Recall that an m-by-n matrix A is said to be *rank deficient* if rank$A <$ $\min\{m, n\}$.

Just as with zero entries (see Section 1.6), the distribution of ranks among submatrices of a TN matrix is much less free than in a general matrix. Rank deficiency of submatrices in certain positions requires rank deficiency elsewhere. Whereas, in the case of general matrices, rank deficiency of a large submatrix can imply rank deficiency of smaller, included submatrices, in the TN case rank deficiency of small submatrices can imply that of much larger ones. To be sure, in the general m-by-n case, a p-by-q submatrix of rank r does imply that the rank of the full m-by-n matrix is no more than $m + n - (p + q) + r$, so that rank deficiency of submatrices can imply rank deficiency of larger matrices. But, in the TN case a small rank deficient submatrix can imply that a much larger matrix has no greater rank, just as a zero entry can imply that other entries and entire blocks are zero.

Another, related phenomenon will be of interest here. Recall that in a positive semidefinite matrix, even if a principal submatrix is singular, any column outside it, but lying in the same rows, is in its column space (see, for example, [Joh98]). Thus, adding such columns cannot increase the rank, relative to that of the principal submatrix. Though principality of submatrices is less important in the TN case, there is a related phenomenon in the TN case, which is the subject of the next section. This phenomenon will be used later to develop general statements about the possible ranks of submatrices.

In the following sections we also show that, under a mild regularity assumption, the rank of a TN matrix must be concentrated in a single, contiguous, nonsingular, square submatrix (the "contiguous rank property").

7.1 ROW AND COLUMN INCLUSION RESULTS FOR TN MATRICES

It is known that if $A = [a_{ij}]$ is positive semidefinite, then any column vector of the form

$$\begin{bmatrix} a_{i_1j} \\ a_{i_2j} \\ \vdots \\ a_{i_kj} \end{bmatrix}$$

must lie in the column space of the principal submatrix $A[\{i_1, i_2, \ldots, i_k\}]$. This classical fact may be seen in a variety of ways. For example, consider a Hermitian matrix of the form

$$\begin{bmatrix} B & c \\ c^* & d \end{bmatrix}.$$

If the column vector c is *not* in the range of B, null space/range orthogonality implies there is a vector x in the null space of B such that $x^*c < 0$. There is then an $\varepsilon > 0$ such that

$$\begin{bmatrix} x^* & \varepsilon \end{bmatrix} \begin{bmatrix} B & c \\ c^* & d \end{bmatrix} \begin{bmatrix} x \\ \varepsilon \end{bmatrix} < 0.$$

Since A^T is also positive semidefinite, there is an analogous statement about rows. Moreover, for positive semidefinite matrices (though *both* inclusions are valid), it is equivalent to say that for each j either the row lies in the row space *or* the column lies in the column space of the indicated principal submatrix. It is precisely this statement that may be substantially generalized. The above fact has been known to several researchers for some time; see [Joh98] for more history and motivation.

For a given m-by-n matrix A, we let Row(A) (Col(A)) denote the row (column) space of A. Then rank$(A) = \dim$ Row$(A) = \dim$ Col(A).

Let $\mathcal{F}_0$ ($\mathcal{F}$) denote the set of n-by-n $P_0-(P-)$ matrices which also satisfy Fischer's inequality (see Section 6.0). The class $\mathcal{F}_0$ contains many familiar classes of matrices, for example, positive semidefinite matrices, M-matrices, TN matrices (and their permutation similarities) and triangular matrices with nonnegative main diagonal.

Using the notation developed above, we may reformulate the column inclusion result for positive semidefinite matrices as follows: If A is an n-by-n positive semidefinite matrix, then for any $\alpha \subseteq N$ and for each j, $A[\alpha, \{j\}] \in$ Col$(A[\alpha])$.

The next result can be found in [Joh98, Thm. 6.1].

Theorem 7.1.1 *If $A \in \mathcal{F}_0$ and $\alpha \subseteq N$ is such that* rank$(A[\alpha]) \geq |\alpha| - 1$, *then for each j either* $A[\{j\}, \alpha] \in$ Row$(A[\alpha])$ *or* $A[\alpha, \{j\}] \in$ Col$(A[\alpha])$.

Some condition imposed upon rank$(A[\alpha])$ is necessary, as may be seen from the following example.

Example 7.1.2 Let

$$A = \begin{bmatrix} 0 & 1 & 0 \\ 0 & 0 & 1 \\ 0 & 0 & 0 \end{bmatrix}.$$

Then $A \in \mathcal{F}_0$. Let $\alpha = \{1, 3\}$. Then $A[\alpha] = 0$. However, $A[\{2\}, \alpha] \neq 0$ and $A[\alpha, \{2\}] \neq 0$.

Note the above matrix is a reducible TN matrix. To generalize further, we use the following concept.

Definition 7.1.3 An n-by-n matrix A satisfies the *principal rank property* (PRP) if every principal submatrix A' of A has in turn an invertible rank(A')-by-rank(A') *principal* submatrix.

The following is simply a recasting of the previous definition. A satisfies the PRP if for every α, there exists $\beta \subseteq \alpha$, with $|\beta| = \text{rank}(A[\alpha])$ so that $A[\beta]$ is invertible.

Observe that the PRP is inherited by principal submatrices. The next lemma gives a sufficient condition for a matrix to satisfy the PRP.

Lemma 7.1.4 *Let A be an n-by-n matrix. Suppose the algebraic multiplicity of 0 equals the geometric multiplicity of 0, for every principal submatrix of A. Then A satisfies the PRP.*

Proof. Let A' be any k-by-k principal submatrix of A. If A' is nonsingular, then there is nothing to do. Thus suppose rank$(A') = r$ $(r < k)$. Since the algebraic multiplicity of 0 equals the geometric multiplicity of 0 for A', it follows that the characteristic polynomial of A' is equal to

$$\det(xI - A') = x^{k-r}(x^r - s_1 x^{r-1} + \cdots + (-1)^r s_r),$$

in which $s_r \neq 0$. Since s_r is equal to the sum of all the r-by-r principal minors of A', it follows that there exists at least one nonsingular r-by-r principal submatrix of A'. This completes the proof.   □

Note that symmetric matrices are examples of matrices that satisfy the above lemma, and hence satisfy the PRP. The converse to the above general lemma is easily seen not to hold in general. The simplest example demonstrating this fact is

$$A = \begin{bmatrix} 1 & -1 \\ 1 & -1 \end{bmatrix}.$$

However, for P_0-matrices the condition given in the lemma is clearly also necessary, since for a P_0-matrix the coefficient s_r will be nonzero if and only if there is a nonsingular r-by-r principal submatrix of A. The following is found in [Joh98, Thm. 6.5].

Theorem 7.1.5 *If $A \in \mathcal{F}_0$ satisfies the PRP, and $\alpha \subseteq N$, then for each j either $A[\{j\}, \alpha] \in \text{Row}(A[\alpha])$ or $A[\alpha, \{j\}] \in \text{Col}(A[\alpha])$.*

Theorem 7.1.5 has two hypotheses: $\mathcal{F}_0$ and PRP. Neither of these can be omitted in general. For example,

$$A = \begin{bmatrix} 0 & 1 \\ 1 & 0 \end{bmatrix}$$

satisfies PRP but not the row or column inclusion conclusion of Theorem 7.1.5, and

$$A = \begin{bmatrix} 0 & 1 & 0 \\ 0 & 0 & 1 \\ 0 & 0 & 0 \end{bmatrix}$$

is in $\mathcal{F}_0$, does not satisfy PRP, and does not satisfy the conclusion of Theorem 7.1.5 for $\alpha = \{1, 3\}$.

Since any positive semidefinite matrix satisfies the PRP (by the previous lemma) and is in $\mathcal{F}_0$ (see [HJ85]) we have the following.

Corollary 7.1.6 *If A is an n-by-n positive semidefinite matrix, then for any $\alpha \subseteq N$ and for each j, $A[\alpha, \{j\}] \in \mathrm{Col}(A[\alpha])$.*

We now specialize to the class of TN matrices, a subclass of $\mathcal{F}_0$. We may extend our row or column inclusion results somewhat to *non*-principal submatrices for this class. Before we state our next result, we need the following notation. Let α, $\beta \subseteq N$ and let $i_1 < i_2 < \cdots < i_k$ be the elements of α arranged in increasing order, and let $j_1 < j_2 < \cdots < j_k$ be the elements of β arranged in increasing order. Define $i_0 = 1$, $j_0 = 1$, $i_{k+1} = n$ and $j_{k+1} = n$. For $t = 0, 1, \ldots, k$, let $A[\{i_1, \ldots, i_t, (i_t, i_{t+1}), i_{t+1}, \ldots, i_k\}, \beta]$ denote the submatrix of A obtained from $A[\alpha, \beta]$ by inserting any row s for which $i_t < s < i_{t+1}$. Similarly, let $A[\alpha, \{j_1, \ldots, j_t, (j_t, j_{t+1}), j_{t+1}, \ldots, j_k\}]$ denote the submatrix obtained from $A[\alpha, \beta]$ by inserting any column q for which $j_t < q < j_{t+1}$.

Theorem 7.1.7 *Let A be an n-by-n TN matrix and let α, $\beta \subseteq N$, with $|\alpha| = |\beta|$. Suppose that $\mathrm{rank}(A[\alpha, \beta]) = p$, and either $p \geq |\alpha| - 1$ or at least one of $d(\alpha) = 0$ or $d(\beta) = 0$. Then either*

$$\mathrm{rank}(A[\{i_1, \ldots, i_t, (i_t, i_{t+1}), i_{t+1}, \ldots, i_k\}, \beta]) = p$$

or

$$\mathrm{rank}(A[\alpha, \{j_1, \ldots, j_t, (j_t, j_{t+1}), j_{t+1}, \ldots, j_k\}]) = p$$

for $t = 0, 1, \ldots, k$.

Proof. Let α, $\beta \subseteq N$, with $|\alpha| = |\beta|$. First suppose $d(\alpha) = 0$. The case when $d(\beta) = 0$ will then follow by transposition. Let s, $q \in N$, with $s < i_1$ and $q < j_1$ (if such s, q exist). The case in which $s > i_k$ and $q > j_k$ is handled similarly. Assume for the purpose of contradiction that $\mathrm{rank}(A[\alpha \cup \{s\}, \beta]) >$

p and $\text{rank}(A[\alpha, \beta \cup \{q\}]) > p$. Consider the submatrix $A[\alpha \cup \{s\}, \beta \cup \{q\}]$, which is TN and has rank equal to $p + 2$. Thus there exists a nonsingular submatrix A' of $A[\alpha \cup \{s\}, \beta \cup \{q\}]$ of order $p+2$. Since the $\text{rank}(A[\alpha, \beta]) = p$, it follows that $A' = A[\gamma \cup \{s\}, \delta \cup \{q\}]$, in which $\gamma \subseteq \alpha$, $|\gamma| = p+1$ and $\delta \subseteq \beta$, $|\delta| = p + 1$. Furthermore, $A[\gamma, \delta]$ is principal in A', since $s < i_1$ and $q < j_1$. Observe that since $\text{rank}(A[\alpha, \beta]) = p$, $A[\gamma, \delta]$ is singular. Consequently,

$$0 < \det A' \leq \det A[\gamma, \delta] \cdot a_{sq} = 0.$$

This is a contradiction and hence the conclusion holds.

Finally, suppose $\text{rank}(A[\alpha, \beta]) \geq |\alpha| - 1$, and let $|\alpha| = k$. Define $i_0 = j_0 = 1$ and $i_{k+1} = j_{k+1} = n$. Let t be any fixed integer, $0 \leq t \leq k$. Assume there exist a row indexed by s, $i_t < s < i_{t+1}$, and a column indexed by q, $j_t < q < j_{t+1}$, in which $\text{rank}(A[\{i_1, \ldots, i_t, s, i_{t+1}, \ldots, i_k\}, \beta]) > p$ and $\text{rank}(A[\alpha, \{j_1, \ldots, j_t, q, j_{t+1}, \ldots, j_k\}]) > p$. In fact, both must equal $p + 1$. Then $\text{rank} A[\alpha \cup \{s\}, \beta \cup \{q\}] = p + 2$. Again consider the nonsingular submatrix A' of $A[\alpha \cup \{s\}, \beta \cup \{q\}]$, which (as before) must be of the form $A[\gamma \cup \{s\}, \delta \cup \{q\}]$. Applying arguments similar to those above, we arrive at a contradiction. This completes the proof. $\qquad \square$

We complete this discussion with an example that illustrates three possibilities. The first two give insight into the necessity of the hypotheses in Theorem 7.1.7. The final one illustrates the previous remark.

Example 7.1.8 Let

$$A = \begin{bmatrix} 0 & 1 & 0 & 0 & 0 \\ 0 & 1 & 0 & 1 & 0 \\ 0 & 0 & 0 & 0 & 0 \\ 0 & 1 & 0 & 1 & 1 \\ 0 & 0 & 0 & 0 & 0 \end{bmatrix}.$$

It is not difficult to determine that A is TN.
(i) Let $\alpha = \{2, 4\}$ and $\beta = \{2, 4\}$. Then $\text{rank}(A[\alpha, \beta]) = 1$. However, if we let $s = 1$ and $q = 5$, then $\text{rank}(A[\alpha \cup \{s\}, \beta]) = 2$ and $\text{rank}(A[\alpha, \beta \cup \{q\}]) = 2$. Thus, it is necessary to include a row and a column from the same "gap" in α and β.
(ii) Similar arguments can be applied in the case $\alpha = \{1, 3\}$ and $\beta = \{4, 5\}$, and with $s = q = 2$.
(iii) Let $\alpha = \{2, 3\}$ and $\beta = \{1, 3\}$. Suppose $q = 2$. Then

$$\text{rank}(A[\alpha, \beta \cup \{q\}]) > \text{rank}(A[\alpha, \beta]).$$

Recall that a submatrix of a given matrix is called *contiguous* if both its rows and columns are based on consecutive sets of indices; sometimes such a matrix is called a *block*. A submatrix that extends a contiguous one by keeping the same rows (columns) and adding all earlier or later columns (rows) is called a *swath*. Figure 7.1 depicts two of the swaths determined

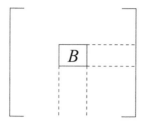

Figure 7.1 Swaths of a block B

by the swath B. A right (lower) swath determined by B includes B and the columns (rows) to the right (below). Left and upper swaths are defined similarly.

Theorem 7.1.9 *Suppose that $A = [a_{ij}]$ is an m-by-n TN matrix and that $A[\alpha, \beta]$ is a contiguous rank deficient (and not necessarily square) submatrix of A. Then, for each index i greater than those in α and each index j greater than those in β (and for each index i less than those in α and each index j less than those in β), either*

$$A[\{i\}, \beta] \in \text{Row}(A[\alpha, \beta])$$

or

$$A[\alpha, \{j\}] \in \text{Col}(A[\alpha, \beta]).$$

Proof. It suffices to prove one of the two claims, as the other is similar. We prove the first. Suppose the claim is not true. Let $\text{rank} A[\alpha, \beta] = k < \min\{|\alpha|, |\beta|\}$. Then

$$\text{rank} \begin{bmatrix} A[\alpha, \beta] & d \\ c^T & a_{ij} \end{bmatrix} = k + 2.$$

Hence the matrix on the left contains a $(k+2)$-by-$(k+2)$ submatrix B that is InTN. The matrix B must include the last row and column, because $\text{rank} A[\alpha, \beta] = k$, so that the submatrix A' of $A[\alpha, \beta]$ that lies in B must lie principally in B. Application of Fischer's determinantal inequality (6.3) to B implies that any principal submatrix of B is also invertible. In particular, A' is. But, as A' is $(k+1)$-by-$(k+1)$, its invertibility implies that $\text{rank} A[\alpha, \beta] > k$, a contradiction that completes the proof.   □

Corollary 7.1.10 *Suppose that A is an m-by-n TN matrix and that $A[\alpha, \beta]$ is a contiguous, rank deficient (and not necessarily square) submatrix of A. Then either the right or lower swath determined by $A[\alpha, \beta]$ has the same rank as $A[\alpha, \beta]$, and either the left or upper swath determined by $A[\alpha, \beta]$ has the same rank as $A[\alpha, \beta]$.*

Proof. Again, it suffices to prove one of the two conclusions, as the other is similar. Consider the first. Unless both the right and lower swaths have the same rank as $A[\alpha, \beta]$, one, say the right, has greater rank, and thus there is a vector outside $A[\alpha, \beta]$ that is not in $\mathrm{Col}(A[\alpha, \beta])$. Using this identified vector and Theorem 7.1.9 implies the desired conclusion. □

We note as a final remark that if $A[\alpha, \beta]$ in Corollary 7.1.10 is of full rank, then the same conclusions trivially hold.

7.2 SHADOWS AND THE EXTENSION OF RANK DEFICIENCY IN SUBMATRICES OF TN MATRICES

In the case of a zero entry in a TN matrix, a "shadow" of zero entries is cast to the northeast or southwest, unless the initial zero entry happens to lie on a zero line (see Section 1.6). For purposes of generalization, the zero entry may be viewed as a rank deficient 1-by-1 block, and a zero line as failure of a 1-regularity hypothesis. What, then, are the appropriately more general statements (both the regularity hypothesis and conclusion) when a larger, rank deficient submatrix is encountered?

Definition 7.2.1 Let $\alpha = \{i_1, i_2, \ldots, i_k\} \subseteq \{1, 2, \ldots, m\}$ and $\beta = \{j_1, j_2, \ldots, j_l\} \subseteq \{1, 2, \ldots, n\}$, with $i_1 < i_2 < \cdots < i_k$ and $j_1 < j_2 < \cdots < j_l$, and let A be an m-by-n matrix. Define $\bar{\alpha} = \{i_1, i_1 + 1, i_1 + 2, \ldots, i_{k-1}, i_k\}$ and $\bar{\beta} = \{j_1, j_1 + 1, j_1 + 2, \ldots, j_{l-1}, j_l\}$. Then the *convex hull of the submatrix* $A[\alpha, \beta]$, denoted by $\overline{A[\alpha, \beta]}$, is the submatrix $A[\bar{\alpha}, \bar{\beta}]$, that is, the submatrix with all intermediate rows and columns filled in.

Consider the following example. Let

$$A = \begin{bmatrix} 1 & 2 & 3 & 4 & 5 & 6 \\ 7 & 8 & 9 & 10 & 11 & 12 \\ 13 & 14 & 15 & 16 & 17 & 18 \\ 19 & 20 & 21 & 22 & 23 & 24 \end{bmatrix},$$

and suppose $\alpha = \{1, 3\}$ and $\beta = \{1, 3, 5\}$. Then

$$A[\alpha, \beta] = \begin{bmatrix} 1 & 3 & 5 \\ 13 & 15 & 17 \end{bmatrix},$$

and

$$\overline{A[\alpha, \beta]} = \begin{bmatrix} 1 & 2 & 3 & 4 & 5 \\ 7 & 8 & 9 & 10 & 11 \\ 13 & 14 & 15 & 16 & 17 \end{bmatrix}.$$

Definition 7.2.2 A *shadow* of a contiguous submatrix of a matrix A is either the convex hull of its upper and right swaths (northeast shadow) or the convex hull of its lower and left swaths (southwest shadow).

For example, if A is the 4-by-6 matrix above, then the southwest shadow of $A[\{1, 2\}, \{3, 4\}]$ is given by,

$$
\begin{bmatrix}
1 & 2 & \mathbf{3} & \mathbf{4} \\
7 & 8 & \mathbf{9} & \mathbf{10} \\
13 & 14 & \mathbf{15} & \mathbf{16} \\
19 & 20 & \mathbf{21} & \mathbf{22}
\end{bmatrix}.
$$

Our principal purpose here is to show that, under appropriate regularity conditions, in a TN matrix A, if B is a rank deficient noncontiguous submatrix of A, then the rank of its convex hull can be no greater than rank(B) (and, thus, will be equal to rank(B)) and that if B is a contiguous, rank deficient block, then the rank of one of its shadows will be no greater. To accomplish this, some further structure must be developed and thus may be of independent interest. We do note that others have studied rank deficiency in TN matrices and have considered the concept of shadows, along with other notions. For example, in [GK02] a number of interesting results were proved on the rank of a TN matrix, based on the ranks of "initial" and "trailing" rows (see, for example, [GK02, Lemma 2, p. 98]), which we actually state and use later in this section. More recently, rank deficient blocks and shadows have been observed in the context of TN matrices. See [dBP82], where it seems the term *shadow* was first used, and consult [Pin08] for some recent work on rank deficient submatrices of TN matrices, which overlaps part of the development here.

The next result is a reformulation of a similar lemma proved in [GK02, p. 98].

Lemma 7.2.3 *Suppose A is an m-by-n TN matrix that is the convex hull of rows $i_1 < i_2 < \cdots < i_p$, and that the rank of the matrix consisting of these rows is deficient. If rows $i_1, i_2, \ldots, i_{p-1}$ are linearly independent and rows $i_2, i_3, \ldots, i_p$ are linearly independent, then* rank$(A) = p - 1$.

Proof. For convenience we set $\alpha = \{i_2, i_3, \ldots, i_{p-1}\}$, and let r_i denote the ith row of A. Then there exist scalars (not all of which are trivial) $c_1, c_2, \ldots, c_{p-1}$ such that

$$
r_{i_p} = \sum_{k-1}^{p-1} c_k r_{i_k}.
$$

Furthermore, by hypothesis $c_1 \neq 0$. Let r_j be any other row of A (necessarily $i_1 < j < i_p$). Consider the p-by-p submatrix of A, $A[\alpha \cup \{j\} \cup \{i_p\}, \beta]$, where β is any collection of p columns. Then

$$
0 \leq \det A[\alpha \cup \{j\} \cup \{i_p\}, \beta] = (-1)^{p-1} c_1 \det A[\alpha \cup \{j\} \cup \{i_1\}, \beta],
$$

by the multilinearity of the determinant. On the other hand, there must exist a collection of $p - 1$ columns γ such that $0 < \det A[\alpha \cup \{i_1\}, \gamma]$ by hypothesis. Hence

$$
\det A[\alpha \cup \{i_p\}, \gamma] = (-1)^{p-2} c_1 \det A[\alpha \cup \{i_1\}, \gamma],
$$

from which it follows that $(-1)^{p-2}c_1 > 0$. From above we conclude that $\det A[\alpha \cup \{j\} \cup \{i_1\}, \beta] = 0$, for each such j. Hence r_j must be a linear combination of the rows $r_{i_1}, r_{i_2}, \ldots, r_{i_{p-1}}$. Thus we have shown that $\text{rank}(A) = p - 1$. □

The next lemma may be viewed as a substantial constraint on line insertion in a rank deficient TN matrix. However, it is convenient to state it slightly differently.

Lemma 7.2.4 *Suppose that A is an m-by-n TN matrix such that rows $1, 2, \ldots, k-1$ form a linearly independent set and rows $k+1, k+2, \ldots, m$ form a linearly independent set, while the rank of the matrix formed by rows $1, 2, \ldots, k-1, k+1, k+2, \ldots, m$ is $m-2$. Then, $\text{rank}(A) = m-2$, as well.*

Proof. The proof applies Lemma 7.2.3. Under the hypothesis, 0 is a non-trivial linear combination of rows $1, 2, \ldots, k-1, k+1, \ldots, m$ that is unique up to a factor of scale. In this linear combination, there must be a nonzero coefficient among the first $k-1$ and among the last $m-k$, else either the first $k-1$ or the last $m-k$ would constitute a linearly dependent set, contrary to hypothesis. Let r, $0 < r < k$, be the index of the first nonzero coefficient and s, $k-1 < s < m+1$, be the index of the last. Since deletion of any row with a zero coefficient in the nontrivial linear combination must leave a linearly independent set of rows, rows $r, r+1, \ldots, s$, except for k ($r < k < s$), satisfy the hypothesis of Lemma 7.2.3. Application of Lemma 7.2.3 implies that rows $r, r+1, \ldots, s$ (including k) are rank deficient by 2 and, thus, that A is rank deficient by 2, as was to be shown. □

Consider the following illustrative example. Using the notation in the lemma above, we let $m = 4$ and $k = 3$. Suppose A is a TN matrix of the form

$$\begin{bmatrix} 1 & 1 & 1 & 1 & 1 \\ 1 & 2 & 3 & 4 & 5 \\ * & * & * & * & * \\ 0 & 1 & 2 & 3 & 4 \end{bmatrix},$$

where the third row is to be determined. Observe that, as presented, A satisfies the hypotheses of Lemma 7.2.4 with $m = 4$ and $k = 3$. Then the only way to fill in the third row and preserve TN is if this row is a linear combination of rows 1 and 2 (the details are left to the reader).

We also note that in the case of TP matrices, a row may be inserted in any position so as to increase the rank, that is, to preserve TP (see Theorem 9.0.2).

We now turn to consideration of the consequences of a dispersed (not contiguous) rank deficient submatrix. Our purpose is to understand the circumstances under which its rank limits that of the smallest contiguous submatrix containing it. For this purpose, we need a hypothesis that extends the tight one used in Lemma 7.2.4.

Definition 7.2.5 Let $A[\alpha, \beta]$ be a dispersed submatrix of A, and suppose that $\alpha = \alpha_1 \cup \alpha_2 \cup \cdots \cup \alpha_p$ and $\beta = \beta_1 \cup \beta_2 \cup \cdots \cup \beta_q$, in which each α_i and β_j is a consecutive set of indices, while $\alpha_i \cup \alpha_{i+1}$ ($\beta_j \cup \beta_{j+1}$) is not, $i = 1, 2, \ldots, p-1$ ($j = 1, 2, \ldots, q-1$). We call $\alpha_1, \alpha_2, \ldots, \alpha_p$ ($\beta_1, \beta_2, \ldots, \beta_q$) the *contiguous components* of α (β). Now, we say that $A[\alpha, \beta]$ satisfies the *partitioned rank intersection (PRI) property* if, for each i, $i = 1, 2, \ldots, p-1$ (and $j = 1, 2, \ldots, q-1$), the span of the rows (and columns) of $A[\alpha_1 \cup \alpha_2 \cup \cdots \cup \alpha_i, \beta]$ (and $A[\alpha, \beta_1 \cup \beta_2 \cup \cdots \cup \beta_j]$) has a nontrivial intersection with the span of the rows of $A[\alpha_{i+1} \cup \cdots \cup \alpha_p, \beta]$ (and $A[\alpha, \beta_{j+1} \cup \cdots \cup \beta_q]$). By convention, we also say that any contiguous submatrix satisfies the PRI property.

Observe that the 4-by-5 matrix A above in the previous example satisfies PRI with $\alpha = \{1, 2, 4\}$.

Theorem 7.2.6 *Let A be an m-by-n TN matrix and let $A[\alpha, \beta]$ be a submatrix of A such that* $\mathrm{rank} A[\alpha, \beta] = r < \min\{|\alpha|, |\beta|\}$. *If $A[\alpha, \beta]$ satisfies PRI, then*

$$\mathrm{rank}(\overline{A[\alpha, \beta]}) = r,$$

as well.

Proof. The proof is based upon multiple applications of Lemma 7.2.4. First, we note that we may consider only the case in which the column index set has only one contiguous component, as when this case is proven, we may turn to the columns, under the assumption there is only one component to the row index set. Now we consider the gaps between row contiguous components one at a time. Without loss of generality, we may assume that our gap consists of just one row and that the rows falling in other gaps have been deleted. We are not yet in the case covered by Lemma 7.2.4, but we may obtain this case by (possibly) deleting some rows above and/or some rows below the gap. If, say, the rows above the gap are not linearly independent, then rows may be deleted from among them so as to leave a linearly independent set with the same span. Similarly, rows from below may be deleted, if necessary. The (nontrivial) overlap it spans will not have changed. If it has dimension greater than 1, then further rows may be deleted from above or below, until the hypothesis of Lemma 7.2.4 is achieved. Once it is, and Lemma 7.2.4 is applied, any deleted rows may be reintroduced without increasing rank. Continuing in this way for each row gap (and each vector in that gap), as well as for the columns, proves the claim of the theorem. $\square$

A straightforward way to see that a condition like PRI is necessary is to note that line insertion into a TP matrix is always a possibility (see Theorem 9.0.2), so as to preserve the property of being TP, and hence increasing the rank.

A nice illustration of the convex hull result above can be seen in the next example. Suppose A is a 4-by-4 TN matrix of the form

$$A = \begin{bmatrix} 1 & a & b & 1 \\ c & d & e & f \\ g & h & i & j \\ 1 & k & l & 1 \end{bmatrix}.$$

Assuming A has no zero lines (which does not affect the conclusion), which then will imply that A has no zero entries, we wish to conclude that A has rank equal to 1. Two applications of Sylvester's identity (1.4) applied to the principal submatrices $A[\{1,2,4\}]$ and $A[\{1,3,4\}]$ yield

$$A = \begin{bmatrix} 1 & a & b & 1 \\ c & ac & e & c \\ g & h & bg & g \\ 1 & a & b & 1 \end{bmatrix}.$$

Finally, it is easy to verify that $e = bc$ and $h = ag$, and thus A is a rank one TN matrix. Observe that A above satisfies PRI.

Though a rank deficient submatrix of a TN matrix gives no general implication about a specific submatrix that extends outside of it, there is a remarkably strong "either/or" implication about possibly much larger submatrices that lie in one of its shadows.

Definition 7.2.7 Given an m-by-n matrix A, the two *sections* determined by a submatrix $A[\alpha, \beta]$ are the submatrices $A[\alpha, \{1, 2, \ldots, n\}]$, the *latitudinal section*, and $A[\{1, 2, \ldots, m\}, \beta]$, the *longitudinal section*; the former is just the rows α and the latter is just the columns β. If $A[\alpha, \beta]$ is rank deficient, it is said to satisfy the *section rank property, SRP,* if each of its two sections have greater rank. Note that if $A[\alpha, \beta]$ is just a zero entry, this means that this zero entry does not lie on a zero line.

Keeping in mind the notions of sections and shadows, we introduce some further notation. If A is an m-by-n matrix and $A[\alpha, \beta]$ is a fixed contiguous submatrix of A with $\alpha = \{i_1, \ldots, i_p\}$ and $\beta = \{j_i, \ldots, j_l\}$, then the northeast shadow of A (that is, the convex hull of its upper and right swaths) will be denoted by $A[\alpha-, \beta+]$, where $\alpha- = \{1, 2, \ldots, i_1, \ldots, i_p\}$ and $\beta+ = \{j_1, \ldots, j_l, j_l + 1, \ldots, n\}$. Similarly, the southwest shadow of $A[\alpha, \beta]$ will be denoted by $A[\alpha+, \beta-]$.

Theorem 7.2.8 *Suppose that A is an m-by-n TN matrix and that $A[\alpha, \beta]$ is a contiguous, rank deficient submatrix satisfying the SRP. Then either its northeast shadow $(A[\alpha-, \beta+])$ or its southwest shadow $(A[\alpha+, \beta-])$ has the same rank as $A[\alpha, \beta]$. Moreover, which actually occurs is determined by which of the swath ranks exceeds $\mathrm{rank}(A[\alpha, \beta])$.*

Proof. The first statement in the theorem above will follow from two applications of the general row and column inclusion result (Theorem 7.1.9).

Since $A[\alpha, \beta]$ satisfies SRP, there exists a column in $A[\alpha, \beta-]$, not in positions β, that is not in the column space of $A[\alpha, \beta]$; or there is a column in $A[\alpha, \beta+]$, not in positions β, that is not in the column space of $A[\alpha, \beta]$. Suppose, without loss of generality, this column lies in $\beta+\backslash\beta$, and call it x. In a similar fashion we identify a row, y^T, not in the row space of $A[\alpha, \beta]$, that lies in, say, positions $\alpha - \backslash\alpha$. In this case, any row z^T in $A[\alpha+, \beta]$ must lie in the row space of $A[\alpha, \beta]$ by an application of Theorem 7.1.9. Thus rank$(A[\alpha+, \beta]) = $ rank$(A[\alpha, \beta])$.

Now, we repeat the above argument with $A[\alpha, \beta]$ replaced by $A[\alpha+, \beta]$. Let u be any column lying in rows $\alpha+$ and in $\beta-\backslash\beta$. Then, by Theorem 7.1.9, taking into account that y^T is not in the row space of $A[\alpha+, \beta]$, we conclude that u must lie in the column space of $A[\alpha+, \beta]$. Hence we have shown rank$(A[\alpha+, \beta-]) = $ rank$(A[\alpha, \beta])$; that is, the southwest shadow of $A[\alpha, \beta]$ has no larger rank.

Observe if $x \in \beta - \backslash\beta$ and $y \in \alpha + \backslash\alpha$, then we may deduce, via similar arguments, that rank$(A[\alpha-, \beta+]) = $ rank$(A[\alpha, \beta])$. Finally, we note that the case $x \in \beta-\backslash\beta$ and $y \in \alpha-\backslash\alpha$ or $x \in \beta+\backslash\beta$ and $y \in \alpha+\backslash\alpha$ cannot occur, as they violate general row and column inclusion (see Theorem 7.1.9). $\quad\square$

As an illustration, consider the following TN matrix. Let

$$A = \begin{bmatrix} 2 & 1 & 1 & 1 \\ 2 & 1 & 1 & 1 \\ 1 & 1 & 1 & 2 \\ 1 & 1 & 1 & 2 \end{bmatrix}.$$

Observe that A does not satisfy the SRP, as the longitudinal section of the $A[\{2,3\}]$ has rank 1, but the latitudinal section has rank 2. Furthermore, observe that both the northeast and southwest shadows of $A[\{2,3\}]$ also have rank equal to two, greater than the rank of $A[\{2,3\}]$.

Corollary 7.2.9 *Suppose A is an m-by-n TN matrix and that $A[\alpha, \beta]$ is a rank deficient submatrix satisfying PRI and such that $\overline{A[\alpha, \beta]}$ satisfies the SRP. Then, either the northeast shadow of $\overline{A[\alpha, \beta]}$ or the southwest shadow of $\overline{A[\alpha, \beta]}$ has the same rank as $A[\alpha, \beta]$.*

Proof. Since $A[\alpha, \beta]$ satisfies PRI, we know that rank$(\overline{A[\alpha, \beta]}) = $ rank$(A[\alpha, \beta])$ (by Theorem 7.2.6). Furthermore, since $\overline{A[\alpha, \beta]}$ is assumed to satisfy SRP, by Theorem 7.2.8 the desired conclusion follows. $\quad\square$

Corollary 7.2.10 *Suppose that A is an n-by-n InTN matrix and that $A[\alpha, \beta]$ is a rank deficient submatrix. If $A[\alpha, \beta]$ is contiguous, then one of its shadows has rank equal to that of $A[\alpha, \beta]$ (in particular, the one to the side of the diagonal of A on which more entries of $A[\alpha, \beta]$ lie). If $A[\alpha, \beta]$ is not contiguous, but satisfies PRI, then one of the shadows of $\overline{A[\alpha, \beta]}$ has the same rank as $A[\alpha, \beta]$ (in particular, the one to the side of the diagonal of A on which more entries of $\overline{A[\alpha, \beta]}$ lie). In neither case can both shadows have the same rank as $A[\alpha, \beta]$.*

7.3 THE CONTIGUOUS RANK PROPERTY

Our purpose here is to show that in a TN matrix the rank can be "explained" by a contiguous submatrix, given a necessary but simple regularity condition.

Definition 7.3.1 An m-by-n matrix A obeys the *contiguous rank property, CRP*, if it contains a contiguous rank(A)-by-rank(A) submatrix whose rank is equal to rank(A).

Example 7.3.2 The matrix

$$A = \begin{bmatrix} 2 & 1 & 1 \\ 0 & 0 & 0 \\ 1 & 1 & 1 \end{bmatrix}$$

is TN, and rank$A = 2$. But A does not enjoy the CRP, as each of its four 2-by-2 contiguous submatrices has rank 1. The difficulty is the zero row, which interrupts the submatrix

$$\begin{bmatrix} 2 & 1 \\ 1 & 1 \end{bmatrix}$$

of rank 2.

A regularity condition that rules out zero lines and additional degeneracies is the following.

Definition 7.3.3 An m-by-n matrix of rank at least r is called $(r-1)$-*regular* if any submatrix of $r-1$ consecutive lines (either $r-1$ full, consecutive rows or columns) has rank $r - 1$.

Of course, 1-regularity may be viewed as a strengthening of the "no zero-lines" hypothesis in Theorem 1.6.4 on zero/nonzero patterns of TN matrices. In some situations, this is the appropriate hypothesis.

Our main result of this section is then

Theorem 7.3.4 *Every $(r-1)$-regular, TN matrix of rank r satisfies the CRP.*

Proof. We first show that there exist r consecutive columns whose rank is r. Consider the first $r-1$ columns. They have rank $r-1$ by hypothesis. If the next column is not a linear combination of these, the first r columns are our desired consecutive columns. If the next column is a linear combination of the first $r-1$ columns, consider columns 2 through r, which then must have the same span as the first $r-1$ columns. Continue in this manner until a column, not in the span of the first $r-1$, is encountered (there must be such a column as the rank is r). This column, together with $r-1$ columns preceding it, will be the desired r consecutive columns of rank r.

We next show that within the above found r consecutive linearly independent columns, there are r consecutive rows that are linearly independent, thus determining the desired square, nonsingular r-by-r contiguous submatrix. To accomplish this, we make use of the shadow result (Theorem 7.2.8) to show that the above r columns will also satisfy $(r-1)$-regularity (on the rows); note that it is not immediate that this hypothesis on the full matrix conveys to our r columns. Then the argument of the first paragraph applied to the rows of our r columns will identify the desired nonsingular r-by-r block.

Suppose, for the sake of a contradiction, that the r consecutive columns of rank r, call them A', do not satisfy $r-1$ regularity with respect to the rows. Thus there exists a collection of $r-1$ consecutive rows in A' that is not of full rank. Starting from the bottom of A' and considering collections of $r-1$ consecutive rows, let B be the first such collection of $r-1$ consecutive rows of rank less than $r-1$. Suppose B occurs in rows $\{i+1,\dots,i+r-1\}$ and $i+r-1 \le n-2$. Since, in A, rows $\{i+1,\dots,i+r-1\}$ have rank $r-1$, there must be a column, x, outside A' in these rows, not in the span of B. Since there is a row below B in A' not in the row space of B, we conclude by row and column inclusion (Theorem 7.1.9) that this column must occur in the left swath of B. Hence, applying Theorem 7.1.9 again we have that the rank of the northeast shadow of B is the same as the rank of B (that is, the northeast shadow of B has rank less than $r-1$). It now follows that the submatrix of A' consisting of all columns and the bottom rows $\{i+2,i+3,\dots,n\}$ is $(r-1)$-regular. Observe that this submatrix has at least r rows, as $i+r-1 \le n-2$. If the submatrix B occurs in rows $\{i+1,\dots,i+r-1\}$ and $i+r-1 = n-1$, then we arrive at a contradiction, by following similar arguments, in that we may conclude the rank of A' is at most $r-1$. Observe that a similar argument may be applied from the top of A' to yield similar conclusions.

From the above analysis, we may assume that the only rank deficient collections of $r-1$ consecutive rows of A' occur in the first $r-1$ rows or the last $r-1$ rows. In this case, the submatrix obtained from A' by removing the first and last row will satisfy $(r-1)$-regularity. Now repeat the argument from the first paragraph to identify the desired nonsingular r-by-r block. $\square$

Example 7.3.5 Suppose A is a 3-by-3 TN matrix of the form

$$A = \begin{bmatrix} 2 & a & 1 \\ b & c & d \\ 1 & e & 1 \end{bmatrix}.$$

Assuming that the rank of A is 2 and that A is 1-regular, we know that none of a,b,c,d,e can be zero. Now, if A does not satisfy CRP, then each of its four 2-by-2 contiguous submatrices are of rank 1, from which we may conclude that $a = e$ and $d = b$. Thus A must be of the form

$$A = \begin{bmatrix} 2 & a & 1 \\ b & c & b \\ 1 & a & 1 \end{bmatrix}.$$

This leads to a contradiction as c must then be equal to both ab and $ab/2$, which can only occur if all $c = b = a = 0$.

Chapter Eight

Hadamard Products and Powers of TN Matrices

8.0 DEFINITIONS

The *Hadamard product* of two m-by-n matrices $A = [a_{ij}]$ and $B = [b_{ij}]$ is defined and denoted by
$$A \circ B = [a_{ij}b_{ij}].$$
Furthermore, if A is a m-by-n and entrywise nonnegative, then $A^{(t)}$ denotes the tth *Hadamard power* of A, and is defined to be
$$A^{(t)} = [a_{ij}^t],$$
for any $t \geq 0$.

Evidently, if A and B are entrywise positive (that is, TP$_1$), then $A \circ B$ is also TP$_1$. Furthermore, with a bit more effort it is not difficult to demonstrate that if A and B are TP$_2$ and 2-by-2, then
$$\det(A \circ B) > 0,$$
as both $\det A$ and $\det B$ are positive.

From the above simple inequality we are able to draw two nice consequences. The first is that TP$_2$ or TN$_2$ matrices are closed under Hadamard products. The second is that if A and B are both TP (or TN) and $\min\{m, n\} \leq 2$, then $A \circ B$ is TP (TN). Unfortunately, TP (TN) are not closed under Hadamard multiplication in general (see [Joh87, Mar70a]).

For example, consider the 3-by-3 TN matrix,
$$W = \begin{bmatrix} 1 & 1 & 0 \\ 1 & 1 & 1 \\ 1 & 1 & 1 \end{bmatrix}. \tag{8.1}$$
Then $W \circ W^T$ is TN$_2$, but $\det(W \circ W^T) = -1$, so $W \circ W^T$ is **not** TN. Similarly, TP is not closed under Hadamard multiplication, and it is straightforward to deduce that nonclosure under Hadamard multiplication extends to the m-by-n case, as both classes TP and TN are inherited by submatrices.

Nonclosure, under Hadamard powers, for TP$_k$ is also the case in general. However, if A is TP$_3$, then $A^{(t)}$ is TP$_3$ for all $t \geq 1$ (see Section 4 in this chapter).

The following example resolves the question of those dimensions in which there is Hadamard power closure. Let
$$A = \begin{bmatrix} 1 & 11 & 22 & 20 \\ 6 & 67 & 139 & 140 \\ 16 & 182 & 395 & 445 \\ 12 & 138 & 309 & 375 \end{bmatrix}.$$

Then A is TP but $\det(A \circ A) < 0$. Some classes of matrices, such as positive semidefinite matrices, are closed under Hadamard multiplication (see [HJ85, p. 458]), but this class does not form a semigroup. Of course, the opposite situation occurs for TN matrices.

The remainder of this chapter is devoted to a number of concepts connected with Hadamard products and Hadamard powers of TP_k matrices. We treat Hadamard products first, and then consider the interesting situation of Hadamard powers.

8.1 CONDITIONS UNDER WHICH THE HADAMARD PRODUCT IS TP/TN

The well-known fact that TP matrices are not closed under Hadamard multiplication has certainly not deterred researchers from this interesting topic. In fact, it can be argued that this has fueled considerable investigations on when such closure issues, with particular attention being paid to specific subclasses of TP/TN matrices.

One such instance is the case when both A and B are tridiagonal and TN. This was first proved in [Mar70b]. For completeness, we will carefully write out a proof following this original idea. Then we will outline a second proof which makes use of diagonal scaling.

Proposition 8.1.1 *Suppose $A = [a_{ij}]$ and $B = [b_{ij}]$ are both n-by-n tridiagonal and IrTN. Then $A \circ B$ is also IrTN and tridiagonal.*

Proof. Since it is known that TN_2 matrices are closed under Hadamard product, the result holds for $n = 2$. Assume, by induction, that the statement is valid for all such matrices of order at most $n - 1$. To verify the claim, it is sufficient to show that $\det(A \circ B) \geq 0$. The nonnegativity of all other minors of $A \circ B$ is implied by induction and the fact that any principal minor is a product of contiguous principal minors. From here the nonnegativity of all the principal minors is sufficient to deduce that a tridiagonal matrix is TN.

To this end, observe that

$$\det(A \circ B) = a_{11}b_{11} \det \left(A(1) \circ \begin{bmatrix} b_{22} - \frac{a_{12}a_{21}}{a_{11}a_{22}}\frac{b_{12}b_{21}}{b_{11}} & b_{23} & 0 & \\ b_{32} & b_{33} & \ddots & \\ 0 & & \ddots & \ddots \\ & & & & b_{nn} \end{bmatrix} \right).$$

Upon a closer examination, it is clear that

$$E := \begin{bmatrix} b_{22} - \frac{a_{12}a_{21}}{a_{11}a_{22}}\frac{b_{12}b_{21}}{b_{11}} & b_{23} & 0 & \\ b_{32} & b_{33} & \ddots & \\ 0 & & \ddots & \ddots \\ & & & & b_{nn} \end{bmatrix}$$

can be rewritten as

$$
\begin{bmatrix}
b_{22} - \frac{b_{12}b_{21}}{b_{11}} & b_{23} & 0 & \\
b_{32} & b_{33} & \ddots & \\
0 & & \ddots & \ddots & \\
& & & & b_{nn}
\end{bmatrix} + tE_{11},
$$

where $t \geq 0$, since $\frac{a_{12}a_{21}}{a_{11}a_{22}} \leq 1$. Thus it follows that E, as defined above, is TN. So by induction, $\det(A(1) \circ E) \geq 0$, which implies that $\det(A \circ B) \geq 0$.
□

The proof presented above basically follows the ideas as they were laid out in [Mar70b].

We note that Proposition 8.1.1 also follows from the well-known result that positive semidefinite matrices are closed under Hadamard multiplication (see, for example, [HJ85]). If A and B are tridiagonal and IrTN, then there exist positive diagonal matrices D_A and D_B such that both $A' = D_A A D_A^{-1}$ and $B' = D_B B D_B^{-1}$ are positive semidefinite. Hence $A' \circ B'$ is also positive semidefinite, from which we may apply Lemma 0.1.1 and conclude that $A \circ B$ is IrTN.

As noted in Chapter 0, Gantmacher and Krein verified that a generalized Vandermonde matrix, $V := [x_i^{\alpha_j}]$, where $0 < x_1 < x_2 < \cdots < x_n$ and $\alpha_1 < \alpha_2 < \cdots < \alpha_n$ is TP. Thus it is evident that if V_1 and V_2 are two generalized Vandermonde TP matrices (of the same size), then so is $V_1 \circ V_2$.

Another important example along these lines is the Hadamard product of two TN Routh-Hurwitz matrices (recall Example 0.1.7). It was proved in [GW96a]) that the Hadamard product of two TN Routh-Hurwitz matrices is in turn a totally nonnegative matrix. We note that in [Asn70] it was first verified that a polynomial is stable if and only if the associated Routh-Hurwitz matrix is TN. So combining this with the former fact, we may deduce an important result on the entrywise product of two stable polynomials.

There are other examples of subclasses of TN matrices that enjoy closure under Hadamard multiplication, including Green's matrices [GK02, p. 91]; finite moment matrices of probability measures that are either symmetric about the origin or possesses support on the nonnegative real line (see [Hei94]). As a final example, it was verified in [Wag92] that triangular TN infinite Toeplitz matrices are closed under Hadamard multiplication, but this is definitely not true in the finite case (see the 5-by-5 example presented in [GW96b]).

8.2 THE HADAMARD CORE

Under Hadamard multiplication, it is clear that the all ones matrix, J, plays the role of the identity. Thus there is a TN matrix, namely J, with the property that the Hadamard product of J with any other TN (TP) matrix remains TN (TP). Building on this, it is natural to ask for a description of all TP/TN matrices with the property that the Hadamard product with *any* TN matrix is again TN. Before we venture too far, observe that if A is an m-by-n matrix that satisfies $A \circ B$ is TN for every m-by-n TN matrix B, then certainly A must also be TN, as $A \circ J = A$ will have to be TN.

Thus we define the *Hadamard core* of the m-by-n TN matrices as follows:

$$\mathrm{TN}^{(C)} = \{A : \ B \in \mathrm{TN} \Rightarrow A \circ B \in \mathrm{TN}\}.$$

(Note that we drop the dependence on m and n, as the dimensions will be clear from the context.)

Often we may just refer to the Hadamard core to mean the Hadamard core for TN matrices. We now present various elementary properties for matrices in the Hadamard core (the reader is urged to consult [CFJ01], where the Hadamard core originated).

From the discussion above, it is clear that $\mathrm{TN}^{(C)}$ is nonempty and that $\mathrm{TN}^{(C)} \subseteq \mathrm{TN}$. Furthermore, since TN_2 matrices are closed under Hadamard multiplication, it follows that $\mathrm{TN}^{(C)} = \mathrm{TN}$ whenever $\min\{m, n\} \leq 2$. However, for $m, n \geq 3$, $\mathrm{TN}^{(C)}$ is strictly contained in TN (e.g., the matrix W in (8.1) is not in $\mathrm{TN}^{(C)}$). Hence $\mathrm{TN}^{(C)}$ becomes an interesting subclass of the m-by-n TN matrices with $m, n \geq 3$.

We begin our analysis on the Hadamard core by describing some general properties of matrices in this core and about the subclass $\mathrm{TN}^{(C)}$ (see [CFJ01] for the original definition of the Hadamard core and related properties).

The first property demonstrates that the Hadamard core is closed under Hadamard multiplication–a natural and fundamental property.

Proposition 8.2.1 *Suppose A and B are two m-by-n matrices in the Hadamard core. Then $A \circ B$, the Hadamard product of A and B, is in the Hadamard core.*

Proof. Let C be any m-by-n TN matrix. Then $B \circ C$ is TN since B is in the Hadamard core. Hence $A \circ (B \circ C)$ is TN. But $A \circ (B \circ C) = (A \circ B) \circ C$. Thus $A \circ B$ is in the Hadamard core, since C was arbitrary. $\square$

The next basic fact mirrors the inheritance property enjoyed by submatrices of TN matrices.

Proposition 8.2.2 *An m-by-n TN matrix A is in the Hadamard core if and only if every submatrix of A is in the corresponding Hadamard core.*

Proof. The sufficiency is trivial since A is a submatrix of itself. To prove necessity, suppose there exists a submatrix, say $A[\alpha, \beta]$, that is not in the Hadamard core. Then there exists a TN matrix B such that $A[\alpha, \beta] \circ B$ is not TN. Embed B into an m-by-n TN matrix $C = [c_{ij}]$ such that $C[\alpha, \beta] = B$, and $c_{ij} = 0$ otherwise. Since $A[\alpha, \beta] \circ B$ is a submatrix of $A \circ C$ and $A[\alpha, \beta] \circ B$ is not TN, we have that $A \circ C$ is not TN. $\square$

The next interesting property follows directly from the multilinearity of the determinant and is concerned with the collection of columns that can be inserted into a matrix so as to preserve the property of being in the Hadamard core ([CFJ01]). We say that a column m-vector v is *inserted in column k* $(k = 1, 2, \ldots, n, n + 1)$ of an m-by-n matrix $A = [b_1, b_2, \ldots, b_n]$, with columns $b_1, b_2, \ldots, b_n$, if we obtain the new m-by-$(n + 1)$ matrix of the form $[b_1, \ldots, b_{k-1}, v, b_k, \ldots b_n]$.

Proposition 8.2.3 *The set of columns (or rows) that can be appended to an m-by-n TN matrix in the Hadamard core so that the resulting matrix remains in the Hadamard core is a nonempty convex set.*

Proof. Suppose A is an m-by-n TN matrix in the Hadamard core. Let S denote the set of columns that can be appended to A so that the new matrix remains in the Hadamard core. Since $0 \in S$, we have that $S \neq \phi$. Let $x, y \in S$. Then the augmented matrices $[A, x]$ and $[A, y]$ are both in the Hadamard core. Suppose $t \in [0, 1]$ and consider the matrix $[A, tx + (1 - t)y]$. Let $[B, z]$ be any m-by-$(n + 1)$ TN matrix. Then

$$[A, tx + (1 - t)y] \circ [B, z] = [A \circ B, t(x \circ z) + (1 - t)(y \circ z)].$$

We only consider submatrices of $[A, tx + (1 - t)y] \circ [B, z]$ that involve column $n + 1$. Let $[A', tx' + (1 - t)y'] \circ [B', z']$ denote any such square submatrix of $[A, tx + (1 - t)y] \circ [B, z]$. Then

$$\begin{aligned}
\det &([A', tx' + (1 - t)y'] \circ [B', z']) \\
&= \det ([A' \circ B', t(x' \circ z')]) \\
&= t\det ([A' \circ B', x' \circ z']) + (1 - t)\det ([A' \circ B', y' \circ z']) \\
&= t\det ([A', x'] \circ [B', z']) + (1 - t)\det ([A', y'] \circ [B', z']) \geq 0,
\end{aligned}$$

as $[A, x]$ and $[A, y]$ are both in the Hadamard core. This completes the proof. $\square$

Since any entrywise nonnegative rank one matrix is diagonally equivalent to J, it is not difficult to deduce that any rank one TN matrix is in the Hadamard core. In fact, the core itself is clearly closed under diagonal equivalence in general.

Proposition 8.2.4 *Suppose A is an m-by-n TN matrix in the Hadamard core. Then DAE is in the Hadamard core for any $D, E \in \mathcal{D}_n$. In particular, any rank one TN matrix is in the Hadamard core.*

Proof. Suppose A is in $\text{TN}^{(C)}$ and that $D, E \in \mathcal{D}_n$. Then for any TN matrix B, we have

$$D(A \circ B)E = (DAE) \circ B \in \text{TN},$$

as $A \circ B$ is TN.

To verify the second statement, observe that if A is a rank one TN matrix, then it is easy to show that $A = DJE$, where $D, E \in \mathcal{D}_n$. Since J is always a member of the Hadamard core, we have that DJE is in the Hadamard core. Thus A is in the Hadamard core. $\square$

Note that W in (8.1) implies that not all rank two TN matrices are in the Hadamard core, and in fact by directly summing the matrix W with an identity matrix it follows that there exist TN matrices of all ranks greater than one that are not in the Hadamard core.

In Section 8.1 we showed that the Hadamard product of any two tridiagonal TN matrices is again a TN matrix. In fact, we can extend this to conclude that all tridiagonal TN matrices are in $\text{TN}^{(C)}$ ([CFJ01]).

Theorem 8.2.5 *Let T be an n-by-n TN tridiagonal matrix. Then T is in the Hadamard core.*

Proof. The proof is by induction on n; the base case $(n = 2)$ follows as TN_2 matrices are closed under the Hadamard product. Suppose the result holds for all such tridiagonal matrices of order less than n. Let B be an arbitrary tridiagonal TN matrix and let A be a TN matrix. Without loss of generality, we may assume that both A and B are IrTN, as the reducible case follows from the irreducible one. As in the proof of Proposition 8.1.1, it is sufficient to prove that $\det(A \circ B) \geq 0$. In this case, the proof is essentially the same as in Proposition 8.1.1, since

$$
\det(A \circ B) = a_{11}b_{11} \det \left(A(1) \circ \begin{bmatrix} b_{22} - \frac{a_{12}a_{21}}{a_{11}a_{22}} \frac{b_{12}b_{21}}{b_{11}} & b_{23} & 0 & \\ b_{32} & b_{33} & \ddots & \\ 0 & \ddots & \ddots & \\ & & & b_{nn} \end{bmatrix} \right).
$$

By induction, the latter determinant is nonnegative.   □

We note here that the proof presented in [CFJ01] (the original reference) made use of the fact that tridiagonal TN matrices are signature similar to M-matrices, which in turn are diagonally similar to diagonally dominant matrices.

We obtain a result from [Mar70b] as a special case.

Corollary 8.2.6 *The Hadamard product of any two n-by-n tridiagonal TN matrices is again TN.*

We now proceed onto a complete description of $\text{TN}^{(C)}$ for $\min\{m, n\} = 3$, which differs significantly from the case $m = 2$. As we have seen, not all 3-by-3 TN matrices are in $\text{TN}^{(C)}$; recall that

$$
W = \begin{bmatrix} 1 & 1 & 0 \\ 1 & 1 & 1 \\ 1 & 1 & 1 \end{bmatrix}
$$

is not a member of $\text{TN}^{(C)}$. In the sequel, W plays a key role in describing $\text{TN}^{(C)}$.

Since TN_2 matrices are closed under Hadamard multiplication, it is not surprising that describing 3-by-3 TN matrices in $\text{TN}^{(C)}$ with at least one zero entry is not that complicated. For example, if $A = [a_{ij}]$ is TN and 3-by-3, with $a_{ij} = 0$ for some pair i, j with $|i - j| \leq 1$, then A is reducible, and hence A is necessarily in $TN^{(C)}$. The next result provides some key insight into the potential necessary and sufficient conditions for general membership in $TN^{(C)}$, and we encourage the reader to consult [CFJ01].

Lemma 8.2.7 *Let $A = [a_{ij}]$ be a 3-by-3 TN matrix with $a_{13} = 0$ (or $a_{31} = 0$). Then A is in $TN^{(C)}$ if and only if $A \circ W$ and $A \circ W^T$ are both TN.*

Proof. Necessity is obvious as both W and its transpose are TN. For sufficiency assume that both $A \circ W$ and $A \circ W^T$ are TN. Since $a_{13} = 0$, it is clear that $A \circ W = A$, so the only real assumption being made here is that $A \circ W^T$ is TN. Observe that $A \circ W^T = T$ is a tridiagonal TN matrix. Thus, by Theorem 8.2.5, for any 3-by-3 TN matrix B, we have $\det(T \circ B) \geq 0$. Moreover, since the $(3,1)$ entry enters positively into the determinant (keeping in mind that all 2-by-2 minors in question are nonnegative), it follows that

$$\det(A \circ B) \geq \det(T \circ B) \geq 0,$$

which completes the proof.   □

Toward a characterization of the Hadamard core for 3-by-3 TN matrices, we turn our focus to the SEB factorization of TN matrices. The only minor of concern at this stage is the determinant, as all 2-by-2 minors will nonnegative. So it is natural to ask, how negative can $\det(A \circ B)$ be for arbitrary 3-by-3 TN matrices?

Assume A, B are two arbitrary 3-by-3 InTN matrices with SEB factorizations

$$A = L_3(a)L_2(b)L_3(c)U_3(d)U_2(e)U_3(f)$$

and

$$B = L_3(u)L_2(v)L_3(w)U_3(x)U_2(y)U_3(z),$$

for given nonnegative numbers $a, b, c, d, e, f, u, v, w, x, y, z$. Thus, we are assuming, by diagonal scaling, that $\det A = \det B = 1$. It is a computation to verify that

$$\det(A \circ B) = 1 + xuyv(1 - befc) + zwvy(1 - beda) \qquad (8.2)$$
$$+ \text{(sum of 42 nonnegative terms)}.$$

Therefore, if A is a 3-by-3 InTN matrix that satisfies $befc, beda \leq 1$, then certainly $\det(A \circ B) > 0$, for every InTN matrix B. Thus, by continuity we have $\det(A \circ B) \geq 0$, for all TN matrices B, and so A is in $TN^{(C)}$. The key issue now is to better understand the conditions $befc, beda \leq 1$. Observe that for the matrix A given above, we have

$$\det(A \circ W) = \det A - a_{13} \det A[\{2, 3\}, \{1, 2\}] = 1 - befc$$

and

$$\det(A \circ W^T) = \det A - a_{31} \det A[\{1, 2\}, \{2, 3\}] = 1 - beda.$$

Thus the required conditions are equivalent to

$$\det(A \circ W), \det(A \circ W^T) \geq 0.$$

As a consequence of the above arguments, we have established the following characterization of $\mathrm{TN}^{(C)}$ (see [CFJ01]).

Theorem 8.2.8 *Let A be a 3-by-3 InTN matrix. Then A is in the Hadamard core if and only if $A \circ W$ and $A \circ W^T$ are both TN, where*

$$W = \begin{bmatrix} 1 & 1 & 0 \\ 1 & 1 & 1 \\ 1 & 1 & 1 \end{bmatrix}.$$

The singular case will follow in a similar manner. However, since $\det(A) = 0$, the assumptions $\det(A \circ W) \geq 0$ and $\det(A \circ W^T) \geq 0$, impose some very restrictive constraints on the entries of A (see [CFJ01] for more details).

As a consequence of the above arguments, we have a complete characterization of the 3-by-3 TN matrices that belong to $\mathrm{TN}^{(C)}$.

Corollary 8.2.9 *Let A be a 3-by-3 TN matrix. Then A is in the Hadamard core if and only if $A \circ W$ and $A \circ W^T$ are both TN.*

We take a moment to observe that if $A \in \mathrm{TN}^{(C)}$ and 3-by-3, then the following determinantal inequality holds,

$$\det(A \circ B) \geq \det AB,$$

for any 3-by-3 TN matrix B. In the case A is singular, there is nothing to prove, but if both A and B are InTN, then from (8.2) we infer that

$$\det(A \circ B) \geq 1 = \det A \cdot \det B = \det AB.$$

We will revisit this inequality in the next section on Oppenheim's inequality.

We now present some useful facts and consequences of Corollary 8.2.9, which may also be found in [CFJ01].

Corollary 8.2.10 *Let $A = [a_{ij}]$ be a 3-by-3 TN matrix. Then A is in the Hadamard core if and only if*

$$a_{11}a_{22}a_{33} + a_{31}a_{12}a_{23} \geq a_{11}a_{23}a_{32} + a_{21}a_{12}a_{33},$$
$$a_{11}a_{22}a_{33} + a_{21}a_{13}a_{32} \geq a_{11}a_{23}a_{32} + a_{21}a_{12}a_{33}.$$

The next example, which originally appeared in [CFJ01], illustrates the conditions above regarding membership in $\mathrm{TN}^{(C)}$.

Example 8.2.11 (Pólya matrix) Let $q \in (0, 1)$. Define the n-by-n Polya matrix Q whose (i, j)th entry is equal to q^{-2ij}. Then it is well known (see [Whi52]) that Q is TP for all n. (In fact, Q is diagonally equivalent to a TP Vandermonde matrix.) Suppose Q represents the 3-by-3 Pólya matrix. By Corollary 8.2.10 and the fact that Q is symmetric, Q is in TN$^{(C)}$ if and only if $q^{-28} + q^{-22} \geq q^{-26} + q^{-26}$, which is equivalent to $q^{-28}(1 - q^2 - q^2(1 - q^4)) \geq 0$. This inequality holds if and only if $1 - q^2 \geq q^2(1 - q^4) = q^2(1 - q^2)(1 + q^2)$. Thus q must satisfy $q^4 + q^2 - 1 \leq 0$. It is easy to check that the inequality holds for $q^2 \in (0, 1/\mu)$, where $\mu = \frac{1 + \sqrt{5}}{2}$ (the golden mean). Hence Q is in TN$^{(C)}$ for all $q \in (0, \sqrt{1/\mu})$.

Corollary 8.2.12 *Let $A = [a_{ij}]$ be a 3-by-3 TN matrix. Suppose $B = [b_{ij}]$ is the unsigned classical adjoint matrix of A. Then A is in the Hadamard core if and only if $a_{11}b_{11} - a_{12}b_{12} \geq 0$, and $a_{11}b_{11} - a_{21}b_{21} \geq 0$; or, equivalently,*

$$a_{11} \det A[\{2, 3\}] - a_{12} \det A[\{2, 3\}, \{1, 3\}] \geq 0$$

and

$$a_{11} \det A[\{2, 3\}] - a_{21} \det A[\{1, 3\}, \{2, 3\}] \geq 0.$$

Even though Corollary 8.2.12 is simply a recapitulation of Corollary 8.2.10, the conditions rewritten in the above form aid in the proof of the next fact. Recall that if A is a nonsingular TN matrix, then $SA^{-1}S$ is a TN matrix in which $S = \text{diag}(1, -1, 1, -1, \ldots, \pm 1)$ (see, e.g., [GK02, p. 109] or Section 1.3).

Theorem 8.2.13 *Suppose A is a 3-by-3 nonsingular TN matrix in the Hadamard core. Then $SA^{-1}S$ is in the Hadamard core.*

Proof. Observe that $SA^{-1}S$ is TN and, furthermore $SA^{-1}S = \frac{1}{\det A}B$, where $B = [b_{ij}]$ is the unsigned classical adjoint of A. Hence $SA^{-1}S$ is in TN$^{(C)}$ if and only if B is a member of TN$^{(C)}$. Observe that the inequalities in Corollary 8.2.12 are symmetric in the corresponding entries of A and B. Thus B is in TN$^{(C)}$. This completes the proof.   □

Corollary 8.2.14 *Let A be a 3-by-3 TN matrix whose inverse is tridiagonal. Then A is in the Hadamard core.*

In [GK60] it was shown that the set of all inverse tridiagonal TN matrices is closed under Hadamard multiplication. (In the symmetric case, which can be assumed without loss of generality, an inverse tridiagonal matrix is often called a Green's matrix, as was the case in [GK60] and [GW96b], and see Chapter 0.) The above result strengthens this fact in the 3-by-3 case. However, it is not true in general that inverse tridiagonal TN matrices are contained in TN$^{(C)}$. For $n \geq 4$, TN$^{(C)}$ does not enjoy the "inverse closure" property as in Theorem 8.2.13. Consider the following example from [CFJ01].

Example 8.2.15 Let $A = \begin{bmatrix} 1 & a & ab & abc \\ a & 1 & b & bc \\ ab & b & 1 & c \\ abc & bc & c & 1 \end{bmatrix}$, where $a, b, c > 0$ are

chosen so that A is positive definite. Then it is easy to check that A is TN, and the inverse of A is tridiagonal. Consider the upper right 3-by-3 submatrix of A, namely $M = \begin{bmatrix} a & ab & abc \\ 1 & b & bc \\ b & 1 & c \end{bmatrix}$, which is TN. By Proposition

8.2.2, if A is in TN$^{(C)}$, then M is in TN$^{(C)}$. However, $\det(M \circ W) = abc(b^2 - 1) < 0$, since $b < 1$. Thus A is not in TN$^{(C)}$.

We can extend this analysis to the rectangular 3-by-n case in the following manner. For $3 \le k \le n$, let $W^{(k)} = (w_{ij}^{(k)})$ be the 3-by-n TN matrix consisting of entries

$$w_{ij}^{(k)} = \begin{cases} 0 & \text{if } i = 1, j \ge k, \\ 1 & \text{otherwise.} \end{cases}$$

For $1 \le k \le n - 2$, let $U^{(k)} = (u_{ij}^{(k)})$ be the 3-by-n TN matrix consisting of entries

$$u_{ij}^{(k)} = \begin{cases} 0 & \text{if } i = 3, 1 \le j \le k, \\ 1 & \text{otherwise.} \end{cases}$$

For example, if $n = 5$ and $k = 3$, then

$$W^{(3)} = \begin{bmatrix} 1 & 1 & 0 & 0 & 0 \\ 1 & 1 & 1 & 1 & 1 \\ 1 & 1 & 1 & 1 & 1 \end{bmatrix},$$

and

$$U^{(3)} = \begin{bmatrix} 1 & 1 & 1 & 1 & 1 \\ 1 & 1 & 1 & 1 & 1 \\ 0 & 0 & 0 & 1 & 1 \end{bmatrix}.$$

See the work [CFJ01] for an original reference.

Theorem 8.2.16 *Let A be a 3-by-n ($n \ge 3$) TN matrix. Then A is in the Hadamard core if and only if $A \circ W^{(k)}$ is TN for $3 \le k \le n$ and $A \circ U^{(j)}$ is TN for $1 \le j \le n - 2$.*

Proof. The necessity is obvious, since $W^{(k)}$ and $U^{(j)}$ are both TN. Observe that it is enough to show that every 3-by-3 submatrix of A is in TN$^{(C)}$ by Proposition 8.2.2. Let B be any 3-by-3 submatrix of A. Consider the matrices $A \circ W^{(k)}$ and $A \circ U^{(j)}$ for $3 \le k \le n$ and $1 \le j \le n - 2$. By hypothesis $A \circ W^{(k)}$ and $A \circ U^{(j)}$ are TN. Hence by considering appropriate submatrices it follows that $B \circ W$ and $B \circ W^T$ are both TN. Therefore B is in TN$^{(C)}$ by Corollary 8.2.9. Thus A is in TN$^{(C)}$.   □

Of course by transposition, we may obtain a similar characterization of TN$^{(C)}$ in the n-by-3 case.

At present no characterization of the Hadamard core for 4-by-4 TN matrices is known. A potential reason for the complications that arise in the 4-by-4 case is that all known proofs regarding membership in $\text{TN}^{(C)}$ for 3-by-3 TN matrices are rather computational in nature. We expect there is more to learn about $\text{TN}^{(C)}$ in the 3-by-3 case.

In any event the question is is there a finite collection of (test) matrices that are needed to determine membership in $\text{TN}^{(C)}$? If so, must they have some special structure? For example, in the 3-by-3 case all the entries of the test matrices are either zero or one.

8.3 OPPENHEIM'S INEQUALITY

As mentioned in the previous sections, the Hadamard product plays a substantial role within matrix analysis and in its applications (see, for example, [HJ91, chap. 5]).

Some classes of matrices, such as the positive semidefinite matrices, are closed under Hadamard multiplication, and with such closure, investigations into inequalities involving the Hadamard product, usual product, determinants, eigenvalues, are interesting to consider. For example, Oppenheim's inequality states that

$$\det(A \circ B) \geq \det B \prod_{i=1}^{n} a_{ii}$$

for any two n-by-n positive semidefinite matrices A and B (see [HJ85, p. 480]). Combining Oppenheim's inequality with Hadamard's determinantal inequality ($\det A \leq \prod_i a_{ii}$) gives

$$\det(A \circ B) \geq \det(AB).$$

Hence the Hadamard product dominates the conventional product in determinant throughout the class of positive semidefinite matrices.

For the case in which A and B are TN, it is certainly not true in general that $\det(A \circ B) \geq 0$, so there is no hope that Oppenheim's inequality holds for arbitrary TN matrices. However, in [Mar70b] it was shown that Oppenheim's inequality holds for the special class of tridiagonal TN matrices. This suggests that an Oppenheim type inequality may be valid in the context of the Hadamard core.

Indeed this is the case, as we see in the next result ([CFJ01]), which generalizes Markham's inequality for the case of tridiagonal TN matrices.

Theorem 8.3.1 *Let A be an n-by-n TN matrix in the Hadamard core, and suppose B is any n-by-n TN matrix. Then*

$$\det(A \circ B) \geq \det B \prod_{i=1}^{n} a_{ii}.$$

Proof. If B is singular, then there is nothing to show, since $\det(A \circ B) \geq 0$, as A is in the Hadamard core. Assume B is nonsingular. If $n = 1$, then the inequality is trivial. Suppose, by induction, that Oppenheim's inequality holds for all $(n-1)$-by-$(n-1)$ TN matrices A and B with A in the Hadamard core. Suppose A and B are n-by-n TN matrices and assume that A is in the Hadamard core. Let A_{11} (B_{11}) denote the principal submatrix obtained from A (B) by deleting row and column 1. Then by induction

$$\det(A_{11} \circ B_{11}) \geq \det B_{11} \prod_{i=2}^{n} a_{ii}.$$

Since B is nonsingular, by Fischer's inequality B_{11} is nonsingular. Consider the matrix $\tilde{B} = B - xE_{11}$, where $x = \frac{\det B}{\det B_{11}}$ and E_{11} is the $(1,1)$ standard basis matrix. Then $\det \tilde{B} = 0$, and $\tilde{B}$ is TN (see Lemma 9.5.2). Therefore $A \circ \tilde{B}$ is TN and $\det(A \circ \tilde{B}) \geq 0$. Observe that $\det(A \circ \tilde{B}) = \det(A \circ B) - xa_{11}\det(A_{11} \circ B_{11}) \geq 0$. Thus

$$\det(A \circ B) \geq xa_{11}\det(A_{11} \circ B_{11})$$

$$\geq xa_{11}\det B_{11} \prod_{i=2}^{n} a_{ii}$$

$$= \det B \prod_{i=1}^{n} a_{ii},$$

as desired.   □

As in the positive semidefinite case, as any TN matrix also satisfies Hadamard's determinantal inequality, we have

Corollary 8.3.2 *Let A be an n-by-n TN matrix in the Hadamard core, and suppose B is any n-by-n TN matrix. Then*

$$\det(A \circ B) \geq \det(AB).$$

We close this section with some further remarks about Oppenheim's inequality. In the case in which $A = [a_{ij}]$ and $B = [b_{ij}]$ are n-by-n positive semidefinite matrices, it is clear from Oppenheim's inequality that

$$\det(A \circ B) \geq \max\left\{ \det B \prod_{i=1}^{n} a_{ii},\ \det A \prod_{i=1}^{n} b_{ii} \right\}.$$

However, in the case in which A is in the Hadamard core and B is an n-by-n TN matrix, it is not true in general that $\det(A \circ B) \geq \det A \prod_{i=1}^{n} b_{ii}$. Consider the following example ([CFJ01]).

Example 8.3.3 Let A be any 3-by-3 TP matrix in $\mathrm{TN}^{(C)}$, and let $B = W$, the 3-by-3 TN matrix equal to $\begin{bmatrix} 1 & 1 & 0 \\ 1 & 1 & 1 \\ 1 & 1 & 1 \end{bmatrix}$. Then since the $(1,3)$ entry of

A enters positively into $\det A$ it follows that

$$\det(A \circ B) < \det A = \det A \prod_{i=1}^{3} b_{ii}.$$

If, however, both A and B are in $\text{TN}^{(C)}$, then we have the next result, which is a direct consequence of Theorem 8.3.1.

Corollary 8.3.4 *Let* $A = [a_{ij}]$ *and* $B = [b_{ij}]$ *be two n-by-n matrices in* $TN^{(C)}$. *Then*

$$\det(A \circ B) \geq \max \left\{ \det B \prod_{i=1}^{n} a_{ii}, \ \det A \prod_{i=1}^{n} b_{ii} \right\}.$$

The next example sheds some light on the necessity that A be in $\text{TN}^{(C)}$ in order for Oppenheim's inequality to hold. In particular, we show that if A and B are TN and $A \circ B$ is TN, then Oppenheim's inequality need not hold ([CFJ01]).

Example 8.3.5 Let $A = \begin{bmatrix} 1 & .84 & .7 \\ .84 & 1 & .84 \\ 0 & .84 & 1 \end{bmatrix}$ and $B = A^T$. Then A (and hence B) is TN, and $\det A = \det B = .08272$. Now

$$A \circ B = \begin{bmatrix} 1 & .7056 & 0 \\ .7056 & 1 & .7056 \\ 0 & .7056 & 1 \end{bmatrix},$$

and it is not difficult to verify that $A \circ B$ is TN with $\det(A \circ B) \approx .00426$. However, in this case

$$\det(A \circ B) \approx .00426 < .08272 = \begin{cases} \det A \prod_{i=1}^{3} b_{ii}, \\ \det B \prod_{i=1}^{3} a_{ii}. \end{cases}$$

8.4 HADAMARD POWERS OF TP$_2$

If $A \geq 0$ is an entrywise nonnegative m-by-n matrix, then for any $t \geq 0$,

$$A^{(t)} = [a_{ij}^t],$$

denotes the tth Hadamard power of A. It is easily checked that if A is TN$_2$ (TP$_2$), then $A^{(t)}$ is TN$_2$ (TP$_2$).

In previous sections we observed that TN$_3$ and TP$_3$ are not in general closed under Hadamard multiplication; however, it is the case that if A is TN$_3$, then $A^{(t)}$ is TN$_3$ for all $t \geq 1$. For completeness' sake, we use an example to indicate the need for the power t to be at least one in the above claim. Let

$$A = \begin{bmatrix} 1 & .5 & 0 \\ 1 & 1 & .5 \\ 0 & 1 & 1 \end{bmatrix}.$$

Then A is TN, but $\det(A^{(t)}) = 1 - 2^{1-t} < 0$, for all t with $0 < t < 1$. To verify the claim above that $A^{(t)}$ is TN$_3$ for all $t \geq 1$, we analyze the function $\det A^{(t)}$, for $t \geq 0$, which turns out to be an exponential polynomial in the entries of the 3-by-3 TP matrix A, assuming A is of the form:

$$A = \begin{bmatrix} 1 & 1 & 1 \\ 1 & a & b \\ 1 & c & d \end{bmatrix},$$

where $1 < a < b \leq c < d$. It then follows that $\det A^{(t)}$ is an exponential polynomial, and when the terms in the determinant expansion are ordered in descending order in terms of the bases $1 < a < b \leq c < d$, there are three changes in sign. From this it follows that such a polynomial will have at most three nonnegative roots. The remainder of the proof, which is only sketched here, requires that $A^{(2)}$ be TP whenever A is TP. This follows from an easy application of Sylvester's determinantal formula (see Section 1.2).

Let $A = \begin{bmatrix} 1 & 1 & 1 \\ 1 & a & b \\ 1 & c & d \end{bmatrix}$ be TP. Then, using Sylvester's expansion for the determinant, we have that

$$\frac{(ad - bb)(a - 1)}{a} > \frac{(b - a)(c - a)}{a}.$$

Since A is TP, the following inequalities hold:

(1) $d > 1 \Rightarrow a^2 d > a^2$;

(2) $d > b \Rightarrow ad > ab$;

(3) $b > 1 \Rightarrow acb > ac$.

These inequalities and $\det A > 0$ are then used to establish that $A \circ A$ is TP.

Returning to Hadamard powers for larger order matrices, we may observe by example that TN (TP) matrices are not, in general, Hadamard power closed.

Example 8.4.1 Let

$$A = \begin{bmatrix} 1 & 11 & 22 & 20 \\ 6 & 67 & 139 & 140 \\ 16 & 182 & 395 & 445 \\ 12 & 138 & 309 & 375 \end{bmatrix}.$$

Then A is TP but $\det(A \circ A) < 0$. Thus $A^{(2)}$ is not TP.

Hadamard powers of TN matrices have been considered by others, motivated in part by applications. One of the most noteworthy works (in the context of TN matrices) is [GT98], where it was shown that all Hadamard powers (at least one) of a TN Routh-Hurwitz matrix (see Example 0.1.7) remain TN.

Recall as well from Section 8.2, that if A is in $\mathrm{TN}^{(C)}$, then $A^{(p)}$ is in $\mathrm{TN}^{(C)}$ for all positive integers p.

We now consider continuous Hadamard powers of TP_2 matrices and show that eventually such powers become TP (see [FJ07b] for more work along these lines). We say that an entrywise nonnegative matrix A is *eventually TN (TP)*, or is *eventually of positive determinant* (assuming that A is square), if there exists a $T \geq 0$ such that $A^{(t)}$ is TN (TP) or has a positive determinant for all $t \geq T$. We also say that a nonnegative matrix A *eternally* has a given property if $A^{(t)}$ has the desired property for all $t \geq 1$.

It is clear that if A is an n-by-n matrix that is diagonally equivalent (via positive diagonal matrices) to a matrix that is eventually of positive determinant, then A must be eventually of positive determinant.

We call an n-by-n nonnegative matrix $A = [a_{ij}]$ *normalized dominant* if $a_{ii} = 1$, $i = 1, 2, \ldots, n$ and $a_{ij} \leq 1$, for $1 \leq i \neq j \leq n$.

Lemma 8.4.2 *If A is an n-by-n normalized dominant matrix, then A is eventually of positive determinant.*

The crux of the argument relies on the fact that since the off-diagonal entries of $A^{(t)}$ tend to zero as t increases, it is then clear that $\det(A^{(t)})$ will approach one as t increases.

To better understand the matrices that are eventually of positive determinant, we let TN_2^+ denote the subset of TN_2 in which all *contiguous* 2-by-2 principal minors are positive. Observe that if A is in TN_2^+, then A is reducible if and only if at least one entry of the form $a_{i,i+1}$ or $a_{i+1,i}$ is zero. The next result first appeared in [FJ07b], which we include with a proof for completeness.

Lemma 8.4.3 *If A is n-by-n and $A \in TN_2^+$, then A is diagonally equivalent (via positive diagonal matrices) to a normalized dominant matrix.*

Proof. Suppose $A \in \mathrm{TN}_2^+$ and assume that A is irreducible. Then, the diagonal, and the super- and subdiagonal entries of A must all be positive. Hence there exist positive diagonal matrices $D, E \in \mathcal{D}_n$ such that $B = DAE$, and such that B has ones on its main diagonal. Furthermore, we can find a positive diagonal matrix F such that $C = FBF^{-1}$ has ones on its main diagonal and has symmetric tridiagonal part (i.e., $c_{i,i+1} = c_{i+1,i}$, for each $i = 1, 2, \ldots, n - 1$). Since C is in TN_2^+ and is normalized, it follows that $0 < c_{i,i+1}, c_{i+1,i} < 1$, for each $i = 1, 2, \ldots, n - 1$. We establish that each of the off-diagonal entries are strictly less than one by sequentially moving along each diagonal of the form $c_{i,i+k}$ for $k = 2, 3, \ldots, n - 1$. Let $k \geq 2$; to demonstrate that $c_{i,i+2} < 1$, consider the 2-by-2 minor of C based on rows $\{i, i + 1\}$ and columns $\{i + 1, i + 2\}$. Assuming that all the entries of C of the form $c_{i,i+s}$ for $s = 2, 3, \ldots, k - 1$ have been shown to be strictly less than one, it follows that $c_{i,i+k} < 1$ by using the nonnegativity of the 2-by-2 minor of C based on rows $\{i, i + k - 1\}$ and columns $\{i + k - 1, i + k\}$. By induction all the entries above the main diagonal of C are strictly less than one. Similar arguments apply to the entries below the main diagonal. Hence C is a normalized dominant matrix.

Suppose $A \in \mathrm{TN}_2^+$ and A is reducible. We verify this case by induction on n. If $n = 2$, then A is triangular and the conclusion follows easily. Suppose that any $A \in \mathrm{TN}_2^+$ of size less than n is diagonally equivalent to a normalized dominant matrix. Assume that A is an n-by-n TN_2^+ matrix and is reducible. Then for some i, with $2 \leq i \leq n$, we have $a_{i,i+1} = 0$ (if the only zero entry occurs on the subdiagonal consider transposition). Since $A \in \mathrm{TN}_2^+$, all the main diagonal entries of A must be positive, from which it follows that $a_{kl} = 0$ for all $1 \leq k \leq i$ and $i + 1 \leq l \leq n$. In particular, A must be in the form

$$A = \begin{bmatrix} A_1 & 0 \\ A_2 & A_3 \end{bmatrix},$$

where A_1 is i-by-i, A_3 is $(n-i)$-by-$(n-i)$, and both A_1 and A_3 are TN_2^+. By induction, both A_1 and A_3 are diagonally equivalent to a normalized dominant matrix. To complete the argument, consider a diagonal similarity of A via a diagonal matrix of the form $D = \begin{bmatrix} I & 0 \\ 0 & \varepsilon I \end{bmatrix}$, with $\varepsilon > 0$ and small. $\qquad \square$

A simple consequence of the above results is the following ([FJ07b]).

Corollary 8.4.4 *If A is n-by-n and $A \in \mathrm{TN}_2^+$, then A is eventually of positive determinant. In particular, if A is n-by-n and $A \in \mathrm{TP}_2$, then A is eventually of positive determinant.*

We may now offer a characterization of eventually TP matrices under Hadamard powers, which does appear in [FJ07b].

Theorem 8.4.5 *Suppose A is an m-by-n matrix. Then the following statements are equivalent:*

(1) A is eventually TP;

(2) A is TP_2;

(3) A is entrywise positive, and all 2-by-2 contiguous minors of A are positive.

Proof. (1) $\Longrightarrow$ (2): Suppose A is eventually TP. Then clearly A must have positive entries. Since the 2-by-2 minor in rows $\{i, j\}$ and columns $\{p, q\}$ in $A^{(t)}$ is positive for some t, we have $a_{ip}^t a_{jq}^t > a_{iq}^t a_{pj}^t$. By taking tth roots, this minor is positive in A. Since this is true of any 2-by-2 minor, $A \in \mathrm{TP}_2$.

(2) $\Longrightarrow$ (3): Trivial.

(3) $\Longrightarrow$ (1): If (3) holds, then, by Corollary 3.1.6, each submatrix of A is TP_2, and so by Corollary 8.4.4, this submatrix is eventually of positive determinant. Hence A is eventually TP. $\qquad \square$

Corollary 8.4.6 *If A is TP, then A is eventually TP.*

We observe that some of the work in [KV06] is very much related to eventuality of TP_2 matrices.

If we relax the conditions in Theorem 8.4.5 that A be just in TN_2, then A need not be eventually TN or TP. Consider the example,

$$A = \begin{bmatrix} 10 & 3 & 2 & 1 \\ 2 & 1 & 1 & 1 \\ 1 & 1 & 1 & 2 \\ 1 & 2 & 3 & 7 \end{bmatrix}.$$

Then A is TN_2 and has positive entries, but no Hadamard power of A is TN. This is easily seen since for any Hadamard power the principal minor lying in rows and columns $\{2, 3, 4\}$ is always negative. Notice that not all 2-by-2 minors of A based on consecutive indices are positive. Thus we need an additional condition on a matrix A in TN_2 to ensure that A is eventually TN, which is determined in what follows.

A matrix in TN_2 is called *2-regular* if each two consecutive lines (rows or columns) constitute a linearly independent set (see also Section 7.3). Note that a 2-regular TN_2 matrix has no zero lines. For the purposes of the next result, we introduce a new term that is a variant of the "shadow" terminology defined in Chapter 7. Suppose that A in TN_2 is 2-regular. Then we say that A satisfies a *2-shadow condition* if when the rank$A[\{i, i+1\}, \{j, j+1\}] = 1$, then either the submatrix of A of the form $A[\{1, 2, \ldots, i+1\}, \{j, j+1, \ldots, n\}]$ or $A[\{i, i+1, \ldots, n\}, \{1, 2, \ldots, j+1\}]$ has rank one. In other words, such an A satisfies the 2-shadow property if whenever a 2-by-2 contiguous submatrix B of A is singular, then at least one of the contiguous submatrices of A such that B lies in the northeast corner or the southwest corner (i.e., one of the shadows of $A[\{i, i+1\}, \{j, j+1\}] = 1$) has rank one (see also Section 7.2).

If we let $A = \begin{bmatrix} 1 & 1 & 1 \\ 1 & 1 & 2 \\ 1 & 2 & 5 \end{bmatrix}$, then A is in TN_2 and is 2-regular, but A does not satisfy the 2-shadow condition.

Suppose A had two consecutive rows (or columns) that were multiples of each other. Then A is eventually TN if and only if the matrix obtained from A by deleting one of these rows is eventually TN. Thus, with regard to assessing eventually TN, there is no loss of generality in assuming that A is 2-regular.

The next result, characterizing the TN_2 matrices that are eventually TN, first appeared in [FJ07b], where details of the proof can be found.

Theorem 8.4.7 *Suppose that A is an m-by-n 2-regular matrix. Then A is eventually TN if and only if $A \in TN_2$ and A satisfies the 2-shadow condition.*

We may now also conclude that

Corollary 8.4.8 *If A is TN, then A is eventually TN.*

Chapter Nine

Extensions and Completions

9.0 LINE INSERTION

The property that a matrix be TP is sufficiently strong that, at first glance, construction seems even more difficult than recognition. Of course, the elementary bidiagonal factorization provides an easy way simply to write down an example, but with this factorization it is very difficult to "design" many entries of the resulting matrix. Here, we review a variety of construction, extension and completion ideas for both TP and TN matrices.

Another way to produce a TP matrix is a technique that follows the simple diagram

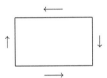

This "exterior bordering" technique represents a strategy for extending any existing TP matrix by adding a line (a row) above or below or (a column) to the right or left. Since a positive number is also a TP matrix, the procedure may be begun, easily, with any positive number. To obtain, for example, an $(m+1)$-by-n TP matrix from an m-by-n one by adding an additional row at the bottom, choose an arbitrary positive entry for the $(m+1, 1)$ position. Then choose an entry for the $(m+1, 2)$ position that is sufficiently large as to make the lower left 2-by-2 minor positive, then an entry for the $(m+1, 3)$ position that makes the lower left 3-by-3 minor positive, and so on until the additional row is completed. Note that, by making the lower left k-by-k minor positive, the lower right $(k-1)$-by-$(k-1)$ minor of it will necessarily be positive by Sylvester's determinantal identity (1.5) and other noncontiguous minors that include the new entry will be positive by the sufficiency of the contiguous ones (see Corollary 3.1.5). Thus, to produce the new TP matrix with one more row, all numbers greater than the threshold that makes each k-by-k lower left minor zero will suffice. This gives well-defined and calculable bounds. Addition of a row at the top is similar, but by working right-to-left (instead of left-to-right). Adding a column on the left (resp. right) may be accomplished by working bottom-to-top (resp. top-to-bottom) in a similar way, all as indicated in the diagram above. Note that in each case, one works toward either the northwes or the southeast corner. Extension of a TN matrix to a larger TN one may be accomplished more simply, either by adding zeros or by repeating the adjacent line. We note that making an entry larger than necessary to retain TN may be problematic, as, because of Sylvester's identity, it may preclude a later choice to make all minors nonnegative, as the following simple example demonstrates:

$$\begin{bmatrix} ? & x & 1 \\ 2 & 0 & 2 \\ 1 & 0 & 1 \end{bmatrix}.$$

Observe that if x is chosen positive, then no matter what nonnegative value is chosen for the $(1,1)$ entry the resulting matrix will not be TN.

We then have

Theorem 9.0.1 *Let $0 < m \le m'$ and $0 < n < n'$. Any m-by-n TP (resp. TN) matrix occurs as a submatrix of an m'-by-n' one in any desired contiguous positions.*

If we ask, instead, to increase the size of a given TN (or TP) matrix by inserting a line *between* two given adjacent lines, this is easy in the TN case by making the new line 0 or repeating one of the adjacent lines. The TP case of line insertion is more interesting and requires a careful argument. This was carried out in [JS00] to produce the following result.

Theorem 9.0.2 *Let $0 < m \le m'$ and $0 < n < n'$. Given any m-by-n TP (resp. TN) matrix A, there is an m'-by-n' TP (resp. TN) matrix B such that A is a submatrix of B in any desired m rows and n columns.*

9.1 COMPLETIONS AND PARTIAL TN MATRICES

We say that a rectangular array is a *partial matrix* if some of its entries are specified, while the remaining, unspecified, entries are free to be chosen. A partial matrix is *partial TP (TN)* if each of its fully specified submatrices is TP (TN). A *completion* of a partial matrix is a choice of values for the unspecified entries, resulting in a conventional matrix that agrees with the original partial matrix in all its specified positions. Of course, for a completion $\hat{A}$ of a partial matrix A to be TP (TN), A must have been partial TP (TN). It is a question of interest which partial TP (TN) matrices have a TP (TN) completion. It is not the case that every partial TP (TN) matrix has a TP (TN) completion.

Example 9.1.1 Let

$$A = \begin{bmatrix} 1 & 1 & .4 & ? \\ .4 & 1 & 1 & .4 \\ .2 & .8 & 1 & 1 \\ ? & .2 & .4 & 1 \end{bmatrix}.$$

It is then easily checked that A is partial TP, but A has no TN completion. To see this, suppose that

$$\hat{A} = \begin{bmatrix} 1 & 1 & .4 & x \\ .4 & 1 & 1 & .4 \\ .2 & .8 & 1 & 1 \\ y & .2 & .4 & 1 \end{bmatrix}$$

is a completion of A. Then

$$\det(\hat{A}) = -.0016 - .008x - .328y - .2xy,$$

which is always negative for $x, y \geq 0$.

It is natural to ask which patterns for the specified entries of a partial TP (TN) matrix guarantee that a partial TP (TN) matrix has a TP (TN) completion. From the discussion of bordering/line insertion, we already know some: those partial TP (TN) matrices whose unspecified entries constitute a collection of lines. Another simple example is the following.

Example 9.1.2 Let

$$A = \begin{bmatrix} a_{11} & a_{12} & ? \\ a_{21} & a_{22} & a_{23} \\ ? & a_{32} & a_{33} \end{bmatrix}$$

be a 3-by-3 partial TN matrix in which the entries a_{ij} are specified and $a_{22} > 0$. Then the completion

$$\hat{A} = \begin{bmatrix} a_{11} & a_{12} & \frac{a_{12}a_{23}}{a_{22}} \\ a_{21} & a_{22} & a_{23} \\ \frac{a_{21}a_{32}}{a_{22}} & a_{32} & a_{33} \end{bmatrix}$$

is TN (easily checked). If A is partial TN with $a_{22} = 0$, then a_{21} or a_{12} must be zero and a_{23} or a_{32} must be zero, in which case the completion,

$$\hat{A} = \begin{bmatrix} a_{11} & a_{12} & 0 \\ a_{21} & a_{22} & a_{23} \\ 0 & a_{32} & a_{33} \end{bmatrix}$$

is TN. We will see that this strategy for a TN completion generalizes nicely. If A is partial TP, the completion $\hat{A}$ will not be TP (only TN, as the upper left and lower right 2-by-2 minors are 0), but one nearby (with slightly smaller (1,3) and (3,1) entries) is TP.

In the case of other completion problems (such as the positive definite one [GJdSW84]), square, combinatorially symmetric (the i, j entry is specified if and only if the j, i one is) patterns play a fundamental role.

We digress here to mention a result in the case of positive (semi-) definite completion theory for comparison purposes. Since the positive semidefinite work occurred earlier, it also provided a backdrop for the TN work that we mention. Since positive semidefinite matrices are necessarily symmetric

(Hermitian), the specified entries of a partial matrix may always be identified with an undirected graph, and since the property of being positive semidefinite is inherited by principal submatrices, a partial Hermitian matrix must be partial positive semidefinite (all fully specified principal submatrices are positive semidefinite) to have a chance at a positive semidefinite completion.

This raises the question of which graphs (for the specified entries, the definition will follow) ensure that a partial positive semidefinite matrix has a positive semidefinite completion. The answer is exactly the chordal graphs [GJdSW84]. An undirected graph is *chordal* if it has no induced cycles of length 4 or more (that is, any long cycle has a "chord"), but the chordal graphs may also be characterized as sequential "clique sums" (identification along a clique) of cliques. (Recall that a clique is also known as a complete graph, or a graph that includes all possible edges.) The MLBC graphs, to be defined, are special chordal graphs in which, not only is there labeling of vertices, but the sequential clique summing is "path-like" and the summing is always along a clique consisting of just one vertex.

In the TN case, the question of which square, combinatorially symmetric patterns with specified diagonal insure that a partial TN matrix has a TN completion has been settled. A key component is the following special completion problem that generalizes Example 9.1.2. Consider the partial matrix

$$A = \begin{bmatrix} A_{11} & a_{12} & ? \\ a_{21}^T & a_{22} & a_{23}^T \\ ? & a_{32} & A_{33} \end{bmatrix},$$

in which A_{11} is n_1-by-n_1, A_{33} is n_3-by-n_3, and a_{22} is a scalar. If A is partial TN (that is, $\begin{bmatrix} A_{11} & a_{12} \\ a_{21}^T & a_{22} \end{bmatrix}$ and $\begin{bmatrix} a_{22} & a_{23}^T \\ a_{32} & A_{33} \end{bmatrix}$ are TN) and $a_{22} > 0$, then the completion

$$\hat{A} = \begin{bmatrix} A_{11} & a_{12} & \frac{1}{a_{22}} a_{12} a_{23}^T \\ a_{21}^T & a_{22} & a_{23}^T \\ \frac{1}{a_{22}} a_{32} a_{21}^T & a_{32} & A_{33} \end{bmatrix}$$

is TN (see [JKL98]). (Note that a_{12} is n_1-by-1, a_{23}^T is 1-by-n_3, a_{21}^T is 1-by-n_1, a_{32} is n_3-by-1, so that $\frac{a_{12} a_{23}^T}{a_{22}}$ is n_1-by-n_3 and $\frac{a_{32} a_{21}^T}{a_{22}}$ is n_3-by-n_1.) This is so by observing that

$$\begin{bmatrix} a_{12} & \frac{1}{a_{22}} a_{12} a_{23}^T \\ a_{22} & a_{23}^T \end{bmatrix} \text{ and } \begin{bmatrix} a_{21}^T & a_{22} \\ \frac{1}{a_{22}} a_{32} a_{21}^T & a_{32} \end{bmatrix}$$

are nonnegative and of rank one, so that, by Sylvester's determinantal identity, $\det(\hat{A}) \geq 0$ and, by induction on $n = n_1 + 1 + n_3$, $\hat{A}$ is TN. In case A is partial TN and $a_{22} = 0$, there is also a TN completion (see [DJK08]).

9.2 CHORDAL CASE—MLBC GRAPHS

The pattern of a combinatorially symmetric n-by-n partial matrix with specified main diagonal may be described with a simple undirected graph on
n vertices (which in our case can always be represented as $\{1, 2, \ldots, n\}$).
There is an undirected edge (often interpreted as an unordered pair from
$\{1, 2, \ldots, n\}$) between vertices i and j if and only if the (i, j) entry, with
$i \neq j$, is specified. A *monotonically labeled block-clique graph* (an MLBC
graph) is a labeled, undirected graph, in which each maximal clique (complete, induced subgraph) is on consecutively labeled vertices, and each two
cliques that intersect (vertex-wise) intersect in exactly one vertex which is
the highest labeled in one of the cliques and lowest labeled in the other. The
maximal clique containing vertex 1 and each other maximal clique, except
the one containing vertex n, has a successor clique, so that the maximal
cliques may be viewed as in a line and numbered $1, 2, \ldots, k$, if convenient.
For example, Figure 9.1 shows a MLBC graph on 11 vertices with four max-

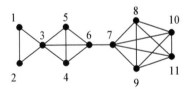

Figure 9.1 MLBC graph on 11 vertices

imal cliques: $\{1, 2, 3\}$, $\{3, 4, 5, 6\}$, $\{6, 7\}$, and $\{7, 8, 9, 10, 11\}$, connected (or
summed) along vertices 3, 6, and 7. As mentioned above, it is clear that an
MLBC graph is just a special case of a chordal graph.

In general, the pattern of the specified entries of an n-by-n partial matrix $A = [a_{ij}]$ with specified main diagonal entries may be described by a
directed graph G on vertices $1, 2, \ldots, n$, in which there is a directed edge
(or arc) (i, j) from i to j if and only if a_{ij} is specified. In case the partial
matrix is combinatorially symmetric, the graph of the specified entries may
be taken to be undirected (as seen above). A key issue when considering TN
completion problems and the graph of the specified entries than the labeling
of the vertices. This is because TN is not closed under arbitrary permuta-
tions, so the same graph with a different labeling yields a (possibly) different
completion problem, and often a much different result.

A partial matrix with the 11-vertex MLBC graph in Figure 9.1 as the graph of its specified entries would then have the form

$$
\begin{bmatrix}
x & x & x & ? & ? & ? & ? & ? & ? & ? & ? \\
x & x & x & ? & ? & ? & ? & ? & ? & ? & ? \\
x & x & x & x & x & x & ? & ? & ? & ? & ? \\
? & ? & x & x & x & x & ? & ? & ? & ? & ? \\
? & ? & x & x & x & x & ? & ? & ? & ? & ? \\
? & ? & x & x & x & x & x & ? & ? & ? & ? \\
? & ? & ? & ? & ? & x & x & x & x & x & x \\
? & ? & ? & ? & ? & ? & x & x & x & x & x \\
? & ? & ? & ? & ? & ? & x & x & x & x & x \\
? & ? & ? & ? & ? & ? & x & x & x & x & x \\
? & ? & ? & ? & ? & ? & x & x & x & x & x
\end{bmatrix},
$$

where x represents a specified entry and ? represents and unspecified entry, as usual.

Note that the partial matrix

$$
A = \begin{bmatrix}
A_{11} & a_{12} & ? \\
a_{21}^T & a_{22} & a_{23}^T \\
? & a_{32} & A_{33}
\end{bmatrix}
$$

discussed earlier has the MLBC graph on $n_1 + n_3 + 1$ vertices with just two maximal cliques of $n_1 + 1$ and $n_3 + 1$ vertices, respectively. As indicated above, if A is partial TN, then A has a TN completion (a special completion was provided above). Now, if a partial TN matrix has a general MLBC graph for its specified entries, with $k > 2$ maximal cliques, it is not difficult to see inductively that it also has a TN completion. First, complete the square partial submatrix corresponding to the first two maximal cliques (say starting from the one that contains the vertex labeled 1). The resulting partial matrix is still partial TN but now has $k - 1$ maximal cliques, and reduction to the case of two maximal cliques (done above) shows completability. Interestingly, such patterns are the only ones for which a partial TN matrix is necessarily completable, among square, combinatorially symmetric patterns with a specified main diagonal (see [JKL98] and [DJK08]).

Theorem 9.2.1 *All square, combinatorially symmetric partial TN matrices with specified main diagonal and with G as the labeled undirected graph of its specified entries have a TN completion if and only if G is MLBC.*

The proof of sufficiency of the condition was outlined above. Necessity of the condition is proven by an intricate collection of noncompletable, partial TN examples that may be found in [JKL98], where sufficiency was first proven. The case in which some diagonal entries are zero was given in [DJK08].

Because the property TP depends so much upon the entire matrix, there does not appear to be a simple way to move easily between TN and TP completability theory ([JN09] notwithstanding). For example, a complete result in the partial TP case, analogous to Theorem 9.2.1, is not known at present. However, it has been shown that the MLBC condition is sufficient for TP completability of partial TP matrices. The proof, however, is much more intricate [JN09].

Theorem 9.2.2 *Let G be an MLBC graph on n vertices. Then every n-by-n partial TP matrix, with specified main diagonal, the graph of whose specified entries is G, has a TP completion.*

When the specified entries of a partial matrix do not align as a MLBC graph (including noncombinatorial symmetry, nonsquareness and the possibility that some diagonal entries are not specified), some partial TN (TP) matrices may and some may not have a TN (TP) completion. The difficulty, at present, is if there is a way to tell. According to real algebraic geometry ([BCR98]), the TN or TP matrices form a semi-algebraic set in real matrix space, that is, membership may be described by a finite list of polynomial equalities and/or inequalities in the entries. The definition by nonnegativity or positivity of minors suffices for this. It follows from the Tarski-Seidenburg principle that a projection of a semi-algebraic set onto a subspace is also semi-algebraic. Using this, it follows that for any fixed pattern of the specified entries, there is a finite list of conditions, polynomial inequalities in the specified entries, that are necessary and sufficient for the existence of a TN (TP) completion. In general, however, it is quite difficult to generate a good list of conditions, though it is convenient to realize that such a list exists. A list might be found using Gröbner bases in computational algebra, but such procedures quickly become computationally prohibitive. The MLBC result, Theorem 9.2.1, may be viewed as a case in which the conditions that the partial matrix be partial TN (these are polynomial, as each fully specified minor is a polynomial in its entries and must be nonnegative).

There are situations in which nice conditions are known and we close this chapter by mentioning a selection of such results along with other completion results.

9.3 TN COMPLETIONS: ADJACENT EDGE CONDITIONS

The simplest connected, undirected graph, corresponding to a pattern with specified diagonal, that is not an MLBC graph shown in Figure 9.2, corresponding

Figure 9.2 Labeled path on 3 vertices—Non-MLBC

to the partial matrix

$$A = \begin{bmatrix} a_{11} & x & a_{13} \\ y & a_{22} & a_{23} \\ a_{31} & a_{32} & a_{33} \end{bmatrix}$$

with x and y unspecified. For A to be partial TN (TP), we must have

$$a_{ij} \geq (>) 0 \text{ for all } i, j, \text{ and}$$
$$a_{11}a_{33} - a_{13}a_{31}, a_{22}a_{33} - a_{23}a_{32} \geq (>), 0.$$

In the TP case, for example, we may assume, without loss of generality by diagonal scaling, that $a_{11} = a_{22} = a_{33} = 1$. For $\det(A[\{1,2\}, \{2,3\}])$ to be positive, we must then have

$$x > \frac{a_{13}}{a_{23}},$$

and for $\det(A[\{1,3\}, \{1,2\}]) > 0$,

$$x < \frac{a_{32}}{a_{31}}.$$

In order that both inequalities hold, we must have

$$\frac{a_{13}}{a_{23}} < \frac{a_{32}}{a_{31}}$$

or

$$a_{13}a_{31} < a_{23}a_{32}.$$

Thus, for the adjacent edges $\{1,3\}$ and $\{3,2\}$, the edge product with the "outer" indices is required to be smaller than the one with the inner indices. Thus, this was called the *adjacent edge conditions* in [DJK08]. On the other hand, if the adjacent edge condition holds, the completion

$$\hat{A} = \begin{bmatrix} a_{11} & \frac{a_{13}}{a_{23}} & a_{13} \\ \frac{a_{31}}{a_{32}} & a_{22} & a_{23} \\ a_{31} & a_{32} & a_{33} \end{bmatrix}$$

is easily checked to be TP$_2$ and is TP, as

$$\det \hat{A} = \left(1 - \frac{a_{13}a_{31}}{a_{23}a_{32}}\right)(1 - a_{23}a_{32}) > 0,$$

by Sylvester's determinantal identity. The TN case with positive main diagonal is similar. We summarize the 3-by-3 case as follows.

Theorem 9.3.1 *If A is a partial TN (TP) matrix with ones on the main diagonal and the graph of its specified entries is the graph from Figure 9.2, then A has a TN (TP) completion if and only if the adjacent edge condition*

$$a_{13}a_{31} \leq (<) a_{23}a_{32}$$

holds.

Of course, there is a similar condition

$$a_{13}a_{31} \leq (<) a_{12}a_{21},$$

for the graph in Figure 9.3.

The key point is that the edge with indices closer together has larger product. For an n-by-n partial matrix, such adjacent edge conditions are necessary for completability because the properties TN and TP are inherited by submatrices. This means that the adjacent edge conditions are necessary in general (see also [DJK08]).

Figure 9.3 Another labeled path on 3 vertices—Non-MLBC

Theorem 9.3.2 *Let A be an n-by-n partial TN (TP) matrix with ones on its main diagonal and G as the graph of its specified entries. Then a necessary condition for A to have a TN (TP) completion is that whenever undirected edges $\{i,j\}$ and $\{j,k\}$ occur with either $j < i < k$ or $j > i > k$ in G, we must have the adjacent edge conditions*

$$a_{ij}a_{ji} \geq (>) a_{jk}a_{kj}.$$

Implicitly in these results, it is assumed, even in the TN case, that the main diagonal entries are all positive. If, however, a main diagonal entry is 0, then the other forced zeros may be exploited to simplify certain completions problems. In [DJK08] (and earlier in [DJ97]) graphs are discussed for which the adjacent edge conditions, in addition to partial TN (TP), are sufficient for TN (TP) completion.

Typically, more than the adjacent edge conditions are needed. The case of a 4-cycle, as the graph of the specified entries, is an illustrative, small example. Consider the case in which the vertices of the cycle are numbered consecutively (as in Figure 9.4). We first note that, in a general n-by-n TN

Figure 9.4 A 4-cycle with vertices labeled consecutively

matrix $A = [a_{ij}]$ with ones on the main diagonal, we have such inequalities as

$$a_{i,i+1}a_{i+1,i+2} \cdots a_{i+k,i+k+1} \geq a_{i,i+k+1}.$$

This simply follows from just TN$_2$ by working outward from the main diagonal. Specializing and using the transposed inequality, it follows that for the partial TN matrix

$$A = \begin{bmatrix} 1 & a_{12} & ? & a_{14} \\ a_{21} & 1 & a_{23} & ? \\ ? & a_{32} & 1 & a_{34} \\ a_{41} & ? & a_{43} & 1 \end{bmatrix}$$

(the graph of whose specified entries is the one in Figure 9.4) to have a TN completion, we must have

$$a_{12}a_{23}a_{34} \geq a_{14}$$

and

$$a_{21}a_{32}a_{43} \geq a_{41}.$$

Remarkably, these conditions are sufficient for TN completability. If a_{12} or a_{21} is zero, the problem quickly reduces to a simpler one. If both are positive, the very particular completion

$$A = \begin{bmatrix} 1 & a_{12} & a_{12}a_{23} & a_{14} \\ a_{21} & 1 & a_{23} & \frac{a_{14}}{a_{12}} \\ \frac{a_{41}}{a_{43}} & a_{32} & 1 & a_{34} \\ a_{41} & a_{43}a_{32} & a_{43} & 1 \end{bmatrix}$$

is TN (see [JK10]). Also, interestingly, no more symmetric completion works, even if the data are symmetric!

This 4-cycle case may be used to initiate an induction in which a general n-cycle result is proven with the aid of the MLBC result (Theorem 9.2.1). For more details, the reader is urged to consult [JK10], which includes the next result.

Theorem 9.3.3 *Let G be the sequentially labeled cycle on n vertices. A partial TN matrix $A = [a_{ij}]$, with ones on its main diagonal and graph G for its specified entries, has a TN completion if and only if the two cycle conditions*

$$a_{12}a_{23} \cdots a_{n-1,n} \geq a_{1n}$$

and

$$a_{21}a_{32} \cdots a_{n,n-1} \geq a_{n1}$$

hold

As indicated above, the proof in the n-by-n case relies on the 4-by-4 case outlined. Suppose A is an n-by-n partial matrix with specified entries as presented by

$$A = \begin{bmatrix} 1 & a_{12} & ? & \cdots & a_{1n} \\ a_{21} & 1 & a_{23} & ? & ? \\ ? & \ddots & \ddots & \ddots & \vdots \\ \vdots & ? & \ddots & \ddots & a_{n-1,n} \\ a_{n1} & ? & \cdots & a_{n,n-1} & 1 \end{bmatrix},$$

with the assumed necessary conditions $a_{12}a_{23} \cdots a_{n-1,n} \geq a_{1n}$ and $a_{21}a_{32} \cdots a_{n,n-1} \geq a_{n1}$.

In the $(2, n)$ position, we put a_{1n}/a_{12} and in the $(n, 2)$ position, we insert a_{n1}/a_{21}. So at this point our partial matrix has the form:

$$A = \begin{bmatrix} 1 & a_{12} & ? & \cdots & a_{1n} \\ a_{21} & 1 & a_{23} & ? & a_{1n}/a_{12} \\ ? & \ddots & \ddots & \ddots & \vdots \\ \vdots & ? & \ddots & \ddots & a_{n-1,n} \\ a_{n1} & a_{n1}/a_{21} & \cdots & a_{n,n-1} & 1 \end{bmatrix}.$$

We may now complete the lower right $(n-1)$-by-$(n-1)$ block by induction. Having completed this block, we may view the remainder of the completion, namely the $(n-3)$ unspecified entries in the first column and the $(n-3)$ unspecified entries in the first row, as completing an MLBC graph consisting of two cliques.

We also note that a similar result holds in the TP case as well, with analogous necessary conditions. The proof in the TP case, found in [JK10], is similar to the TN case above, but requires a far more delicate bordering technique that we omit here. The result, however, can be stated as (see [JK10]),

Theorem 9.3.4 *Let G be the sequentially labeled cycle on n vertices. A partial TP matrix $A = [a_{ij}]$, with ones on its main diagonal and graph G for its specified entries, has a TP completion if and only if the two cycle conditions*

$$a_{12}a_{23} \cdots a_{n-1,n} > a_{1n}$$

and

$$a_{21}a_{32} \cdots a_{n,n-1} > a_{n1}$$

hold.

9.4 TN COMPLETIONS: SINGLE ENTRY CASE

We continue this chapter by relaxing the conditions of combinatorial symmetry of the partial matrix and squareness of the matrix entirely. A natural question, then, is for which single entries does every partial TP matrix with that unspecified entry have a TP completion? Suppose the pattern is m-by-n. The $(1,1)$ and (m,n) entries play a special role; they are unbounded above as they enter positively into every minor in which they lie. Thus, if either of these is the single entry, then a TP completion is always possible. In addition, we can make this more precise in the following manner, which will be useful in what follows.

Lemma 9.4.1 *Let*

$$A = \begin{bmatrix} x & a_{12} & \cdots & a_{1n} \\ a_{21} & a_{22} & \cdots & a_{2n} \\ \vdots & \vdots & & \vdots \\ a_{n1} & a_{n2} & \cdots & a_{nn} \end{bmatrix},$$

$n \geq 2$, be a partial TP matrix in which x is the only unspecified entry. Then, if x is chosen so that $\det A \geq 0$, we must then have that $\det A[\{1, \ldots, k\}] > 0$, $k = 1, \ldots, n-1$. (In particular, $x > 0$.)

Proof. The proof follows inductively using Sylvester's identity (1.5) for determinants applied to the leading principal minors of A in decreasing order of size. For example,

$$0 \leq \det A \det A[\{2, 3, \ldots, n-1\}]$$
$$= \det A[\{1, 2, \ldots, n-1\}] \det A[\{2, 3, \ldots, n\}]$$
$$-\det A[\{1, 2, \ldots, n-1\}, \{2, \ldots, n\}] \det A[\{2, \ldots, n\}, \{1, 2, \ldots, n-1\}],$$

and since A is partial TP, we have that $\det A[\{1, 2, \ldots, n-1\}] > 0$. If A is replaced by $A[\{1, 2, \ldots, n-1\}]$ in the above identity, then it follows that $\det A[\{1, 2, \ldots, n-2\}] > 0$, etc.   □

On the other hand, consider the partial matrix

$$\begin{bmatrix} 100 & 100 & 40 & x \\ 40 & 100 & 100 & 40 \\ 20 & 80 & 100 & 100 \\ 3 & 20 & 40 & 100 \end{bmatrix}.$$

To make its determinant positive, it has to be that $x < -1144/14$, so that no TP completion is possible, as $x > 0$. Thus, the $(1, 4)$ entry cannot be such a single entry. This example may be bordered (as described at the beginning of this chapter) to give such an example for any $m \geq 4$ and $n \geq 4$. The example

$$\begin{bmatrix} 1000 & 10 & 10 & 10 \\ 20 & 9 & x & 10 \\ 2 & 1 & 9 & 10 \\ 1 & 2 & 40 & 1000 \end{bmatrix}$$

shows that the $(2, 3)$ position also cannot be such an entry. To see this, observe that in order for $\det A[\{1, 2\}, \{3, 4\}]$ to be positive we must have $x < 1$. But, then, $\det A[\{2, 3, 4\}, \{1, 2, 3\}]$ is positive only if $x > 199/3$. Hence there can be no TP completion. Of course, the $(4, 1)$ and $(3, 2)$ are similarly excluded. These, in some sense, provide the dividing line, as was shown in [FJS00].

Before we come to the main result, we state and prove a very interesting observation that was not only needed for the proofs in [FJS00], but is also a fact that seems to arise in a number of places in the theory of total positivity (e.g., eigenvectors of TP matrices).

Lemma 9.4.2 *Let $A = [a_1, a_2, \ldots, a_n]$ be an $(n-1)$-by-n TP matrix, with a_i representing the ith column of A. Then, for $k = 1, 2, \ldots, n$,*

$$a_k = \sum_{\substack{i=1 \\ 1 \neq k}}^{n} y_i a_i, \tag{9.1}$$

has a unique solution y for which

$$\operatorname{sgn}(y_i) = \begin{cases} \operatorname{sgn}(-1)^i & \text{if } k \text{ is odd}, \\ \operatorname{sgn}(-1)^{i-1} & \text{if } k \text{ is even}. \end{cases}$$

Proof. If $k = 1$, (9.1) has solution

$$y = [a_2, a_3, \ldots, a_n]^{-1} a_1$$

$$= \frac{1}{\det[a_2, a_3, \ldots, a_n]} \begin{bmatrix} \det[a_1, a_3, a_4, \ldots, a_n] \\ \det[a_2, a_1, a_4, \ldots, a_n] \\ \vdots \\ \det[a_2, a_3, \ldots, a_{n-1}, a_1] \end{bmatrix},$$

and $\text{sgn}(y_i) = \text{sgn}(-1)^i$. If $k > 1$, then (9.1) has solution

$$y = [a_1, a_2, \ldots, a_{k-1}, a_{k+1}, \ldots, a_n]^{-1} a_k$$

$$= \frac{1}{\det[a_1, \ldots, a_{k-1}, a_{k+1}, \ldots, a_n]} \begin{bmatrix} \det[a_k, a_2, \ldots, a_{k-1}, a_{k+1}, \ldots, a_n] \\ \det[a_1, a_k, \ldots, a_{k-1}, a_{k+1}, \ldots, a_n] \\ \vdots \\ \det[a_1, \ldots, a_{k-2}, a_k, a_{k+1}, \ldots, a_n] \\ \det[a_1, \ldots, a_{k-1}, a_k, a_{k+2}, \ldots, a_n] \\ \det[a_1, \ldots, a_{k-1}, a_{k+1}, a_k, \ldots, a_n] \\ \vdots \\ \det[a_1, \ldots, a_{k-1}, a_{k+1}, \ldots, a_{n-1}, a_k] \end{bmatrix}$$

and we see that if k is odd, $\text{sgn}(y_i) = \text{sgn}(-1)^i$, while if k is even, $\text{sgn}(y_i) = \text{sgn}(-1)^{i-1}$. $\qquad \square$

Getting back to single entry completion problems, the case when $m = 2$ is quite elegant, and as such, we include it here for reference.

Lemma 9.4.3 *Let A be a 2-by-n partial TP matrix with exactly one unspecified entry. Then A is completable to a TP matrix.*

Proof. By reversing the indices we may assume, without loss of generality, that

$$A = \begin{bmatrix} a_1 & a_2 & \cdots & a_{k-1} & x & a_{k+1} & \cdots & a_n \\ b_1 & b_2 & \cdots & b_{k-1} & b_k & b_{k+1} & \cdots & b_n \end{bmatrix}.$$

Since A is partial TP, it follows that $\frac{a_i}{b_i} > \frac{a_j}{b_j}$ for $i < j$ and distinct from k. Then, if we choose x such that $\frac{a_{k-1}}{b_{k-1}} > \frac{x}{b_k} > \frac{a_{k+1}}{b_{k+1}}$, the resulting completion of A is TP. $\qquad \square$

We now state and include a proof of just one case of the main result of this section (the remainder of the argument, along with many other related facts, can be found in [FJS00]).

Theorem 9.4.4 *Let A be an m-by-n partial TP matrix with only one unspecified entry in the (s, t) position. If $\min\{m, n\} \leq 3$, then A has a TP completion. If $\min\{m, n\} \geq 4$, then any such A has a TP completion if and only if $s + t \leq 4$ or $s + t \geq m + n - 2$.*

Proof. As stated above we only consider one case, and we assume for simplicity's sake that $m = n \geq 4$. Suppose $(s, t) = (1, 2)$. Then let
$$F = A[\{2, \ldots, n-1\}, \{2, \ldots, n\}] = [f_1, f_2, \ldots, f_{n-1}].$$
By Lemma 9.4.2, $f_1 = \sum_{i=2}^{n-1} y_i f_i$ in which $\operatorname{sgn}(y_i) = \operatorname{sgn}(-1)^i$.

Let $B_{n-1}(x_B) = [b_1, b_2, \ldots, b_{n-1}]$, with x_B chosen so that $b_1 = \sum_{i=2}^{n-1} y_i b_i$. Thus, $\det B_{n-1}(x_B) = 0$ and, by Lemma 9.4.1, we then have $\det B_k(x_B) > 0, k = 1, 2, \ldots, n-2$; in particular, $x_B > 0$. So

$$
\begin{aligned}
\det A_{n-1}(x_B) &= \det[b_0, b_1, b_2, \ldots, b_{n-2}] \\
&= \det[b_0, \sum_{i=2}^{n-1} y_i b_i, \quad b_2, \ldots, b_{n-2}] \\
&= \det[b_0, y_{n-1} b_{n-1}, b_2, \ldots, b_{n-2}] \\
&= (-1)^{n-3} y_{n-1} \det[b_0, b_2, \ldots, b_{n-2}, b_{n-1}] \\
&= |y_{n-1}| \det[b_0, b_2, \ldots, b_{n-2}, b_{n-1}] \\
&> 0.
\end{aligned}
$$

Applying Sylvester's identity (1.5) for determinants, we obtain $\det A > 0$ and we can continue to apply this identity to obtain:
$$\det A_k(x_B) > 0, \ k = 2, \ldots, n-2.$$
We can then increase x_B (so as to make $\det B_{n-1}(x) > 0$) and obtain a TP completion of A.   □

To close this section, we include some further interesting and related discussion. For $3 \leq m, n \leq 4$, consider partial m-by-n TP matrices with exactly two unspecified entries. We note that completing a 2-by-n partial TP matrix with two unspecified entries follows easily from Lemma 9.4.3. For the case when $\min\{m, n\} = 3$, all such completable patterns have been characterized in [FJS00]. For example, all partial 3-by-3 TP matrices with exactly two unspecified entries are completable except for the following four patterns:

$$
\begin{bmatrix} x & ? & x \\ ? & x & x \\ x & x & x \end{bmatrix}, \quad
\begin{bmatrix} x & x & x \\ x & x & ? \\ x & ? & x \end{bmatrix}, \quad
\begin{bmatrix} x & ? & x \\ x & x & ? \\ x & x & x \end{bmatrix}, \text{ and } \quad
\begin{bmatrix} x & x & x \\ ? & x & x \\ x & ? & x \end{bmatrix}.
$$

Finally, recall from earlier in this section we demonstrated that the $(1, 4)$ and $(2, 3)$ entries do not correspond to positions for which every partial TP matrix with that unspecified entry have a TP completion. However, curiously, it is a fact that if A is a partial TP matrix with both the $(1,4)$ and $(3,2)$ entries unspecified, then A may be completed to a TP matrix (see [FJS00]).

9.5 TN PERTURBATIONS: THE CASE OF RETRACTIONS

We have already seen that if we increase the $(1,1)$ or the (m, n) entry of an m-by-n TN (TP) matrix, then the resulting matrix is TN (TP). In this section we investigate further which entries of a TN (TP) matrix may be perturbed (that is, increased or decreased) so that the result is a TN (TP) matrix. Suppose A is an n-by-n matrix. Then $\det(A - tE_{11}) = \det A - t\det A(\{1\})$. Therefore, if $\det A(\{1\}) \neq 0$, then $\det(A - tE_{11}) = 0$, when $t = \frac{\det A}{\det A(\{1\})}$. We are now in a position to make the following definitions on retractions and retractable sets, and we direct the reader to [FJ07a] where some recent work on this topic was considered for various positivity classes of matrices, including TN.

Definition 9.5.1 Let $\mathcal{C}$ denote a given subclass of the n-by-n P_0-matrices, and suppose A is an n-by-n matrix. Then we define

(i) $A^R = \{A - tE_{11} : t \in [0, \frac{\det A}{\det A(\{1\})}]\}$ — "the set of retractions of A",

(ii) $\mathcal{C}^{(R)} = \{A \in \mathcal{C}: A^R \subseteq \mathcal{C}\}$ — "the retractable subset of $\mathcal{C}$",

(iii) $\mathcal{C}_R = \bigcup_{A \in \mathcal{C}} A^R$ — "the set of all retractions of matrices in $\mathcal{C}$".

If, in (i), $\det A(\{1\}) = 0$, then the interval for t is defined to be the single point zero. The notions of retraction (of a matrix) and retractable sets are very important for studying certain determinantal inequalities, such as Oppenheim's inequality for entrywise (or Hadamard) products, see, for example, Section 8.3 within this work. It is known (see [FJ07a]), and is not difficult to prove, that if $\mathcal{C} = PSD$, the set of all positive semidefinite matrices, then $PSD^{(R)} = PSD_R = PSD$. Also if $\mathcal{C} = M$, the set of all M-matrices, then $M^{(R)} = M_R = M$. A somewhat more subtle result is that if $\mathcal{C} = TN$, the set of all TN matrices, then $TN^{(R)} = TN_R = TN$. This fact will follow immediately from the next lemma which can also be found in [FJS00].

Lemma 9.5.2 Let A be an n-by-n TN matrix with $\det A(\{1\}) \neq 0$. Then $A - xE_{11}$ is TN for all $x \in [0, \frac{\det A}{\det A(\{1\})}]$.

Proof. First, observe that for every value $x \in [0, \frac{\det A}{\det A(\{1\})}]$,

$$\det(A - xE_{11}) = \det A - x \det A(\{1\}) \geq 0,$$

as $\det A \geq 0$. Recall that A admits a UL-factorization (follows from the LU-factorization result and reversal) into TN matrices (see Chapter 2). Partition A as follows,

$$A = \begin{bmatrix} a_{11} & a_{12}^T \\ a_{21} & A_{22} \end{bmatrix},$$

where a_{11} is 1-by-1 and $A_{22} = A(\{1\})$. Partition L and U conformally with A. Then

$$A = \begin{bmatrix} a_{11} & a_{12}^T \\ a_{21} & A_{22} \end{bmatrix} = UL = \begin{bmatrix} u_{11} & u_{12}^T \\ 0 & U_{22} \end{bmatrix} \begin{bmatrix} l_{11} & 0 \\ l_{21} & L_{22} \end{bmatrix}$$

$$= \begin{bmatrix} u_{11}l_{11} + u_{12}^T l_{21} & u_{12}^T L_{22} \\ U_{22}l_{21} & U_{22}L_{22} \end{bmatrix}.$$

Consider the matrix $A - xE_{11}$, with $x \in [0, \frac{\det A}{\det A(\{1\})}]$. Then

$$A - xE_{11} = \begin{bmatrix} u_{11}l_{11} + u_{12}^T l_{21} - x & u_{12}^T L_{22} \\ U_{22}l_{21} & U_{22}L_{22} \end{bmatrix}$$

$$= \begin{bmatrix} u_{11} - \frac{x}{l_{11}} & u_{12}^T \\ 0 & U_{22} \end{bmatrix} \begin{bmatrix} l_{11} & 0 \\ l_{21} & L_{22} \end{bmatrix} = U'L,$$

if $l_{11} \neq 0$. Note that if $l_{11} = 0$, then L, and hence A, is singular. In this case $x = 0$ is the only allowed value for x. But, then, the desired result is trivial. Thus we assume that $l_{11} > 0$. To show that $A - xE_{11}$ is TN, it is enough to verify that $u_{11} - x/l_{11} \geq 0$. Since if this were the case, U' would be TN, and, as L is TN by assumption, we have that their product, $A - xE_{11}$, is TN. Since $l_{11} > 0$ and $\det A(\{1\}) > 0$, it follows that L and U_{22} are nonsingular. Hence $0 \leq \det(A - xE_{11}) = (u_{11} - x/l_{11})\det U_{22}\det L$, from which it follows that $u_{11} - x/l_{11} \geq 0$.   □

Note that a similar result holds for decreasing the (n, n) entry by considering the matrix $\tilde{T}A\tilde{T}$, which reverses both the row and column indices of A.

Corollary 9.5.3 *Let TN denote the class of all n-by-n TN matrices. Then $TN^{(R)} = TN_R = TN$.*

For the remainder of this section we restrict ourselves to the set of TP matrices. The first result for this class is a slight strengthening of Lemma 9.5.2 (see [FJS00]). Recall that an n-by-n triangular matrix A is said to a *triangular TP matrix*, ΔTP, if all minors of A are positive, except for those that are zero by virtue of the zero pattern of A. (Recall the definitions of TN_k and TP_k from Chapter 0.)

Theorem 9.5.4 *Let A be an n-by-n TP matrix. Then $A - tE_{11}$ is a TP_{n-1} matrix, for all $t \in [0, \frac{\det A}{\det A(\{1\})}]$.*

Proof. Following the proof of Lemma 9.5.2 we can write

$$A - tE_{11} = \begin{bmatrix} u_{11} - \frac{t}{l_{11}} & u_{12}^T \\ 0 & U_{22} \end{bmatrix} \begin{bmatrix} l_{11} & 0 \\ l_{21} & L_{22} \end{bmatrix} = U'L,$$

where $U' = U - (\frac{t}{l_{11}})E_{11}$, and both U, L are triangular TP matrices (see [Cry76] or Chapter 2). If $u_{11} - t/l_{11} > 0$, then U' and L are triangular TP matrices and hence $A - tE_{11}$ is TP. So consider the case $u_{11} - t/l_{11} = 0$, or equivalently, $\det(A - tE_{11}) = 0$, or $t = \frac{\det A}{\det A(\{1\})}$. Let $B = A - tE_{11}$. Observe that

$B[\{1, 2, \ldots, n\}, \{2, 3, \ldots, n\}]$ and $B[\{2, 3, \ldots, n\}, \{1, 2, \ldots, n\}]$ are TP matrices since A is TP. Thus the only contiguous minors left to verify are the leading contiguous minors of B. Consider the submatrix $B[\{1, 2, \ldots, k\}]$ for $1 \leq k < n$. Then $\det B[\{1, 2, \ldots, k\}] = \det A[\{1, 2, \ldots, k\}] - t \det A[\{2, \ldots, k\}]$. This minor is positive if and only if

$$\det A[\{1, 2, \ldots, k\}] > t \det A[\{2, \ldots, k\}],$$

which is equivalent to $\det A[\{1, 2, \ldots, k\}] \det A(\{1\}) > \det A \det A[\{2, \ldots, k\}]$, an example of a Koteljanskiĭ inequality (recall this from Chapter 6). The only issue left to settle is whether equality holds for the above Koteljanskiĭ inequality. We claim here that for a TP matrix every Koteljanskiĭ inequality is strict. Suppose the contrary, that is, assume there exist two index sets α and β such that $\det A[\alpha \cup \beta] \det A[\alpha \cap \beta] = \det A[\alpha] \det A[\beta]$. For simplicity, we may assume that $\alpha \cup \beta = N$; otherwise replace A by $A[\alpha \cup \beta]$ in the following. By Jacobi's identity (1.2), we have

$$\det A^{-1}[(\alpha \cup \beta)^c] \det A^{-1}[(\alpha \cap \beta)^c] = \det A^{-1}[\alpha^c] \det A^{-1}[\beta^c].$$

Let $C = SA^{-1}S$, for $S = \mathrm{diag}(1, -1, \ldots, \pm 1)$. Then C is TP and the above equation implies $\det C = \det C[\alpha^c] \det C[\beta^c]$. By a result in [Car67], C is reducible, which is nonsense since C is TP. Thus $A - tE_{11}$ is TP_{n-1}, by Fekete's criterion (see Corollary 3.1.5). This completes the proof. □

It can be deduced from the proof above that $A - tE_{11}$ is TP for all $t \in [0, \frac{\det A}{\det A(\{1\})})$, and is TP_{n-1} when $t = \frac{\det A}{\det A(\{1\})}$. An obvious next question is what other entries can be increased/decreased to the point of singularity so that the matrix is TP_{n-1}? As it turns out retracting (or decreasing) the $(2,2)$ entry of a TP matrix results in a TP_{n-1} matrix as well (see [FJS00]).

Theorem 9.5.5 *Let A be an n-by-n TP matrix. Then $A - tE_{22}$ is TP_{n-1} for all $t \in [0, \frac{\det A}{\det A(\{2\})}]$.*

Proof. Using the fact that all Koteljanskiĭ inequalities are strict, it follows that all the leading principal minors of $A - tE_{22}$ are positive. Consider the submatrix $B = (A - tE_{22})[\{1, 2, \ldots, n\}, \{2, 3, \ldots, n\}]$. To show that B is TP, we need only consider the contiguous minors of B that involve the first and second row and first column, as all other minors are positive by assumption. Let C denote such a submatrix of B. To compute $\det C$, expand the determinant along the second row of C. Then

$$\det C = (-1)^{1+2}(-t)\det C(\{2\}, \{1\}) + \det A[\alpha, \beta],$$

where $\det A[\alpha, \beta]$ is some minor of A. Thus $\det C$ is a positive linear combination of minors of A, and hence is positive. Therefore B is TP. Similar arguments show that $(A - tE_{22})[\{2, \ldots, n\}, \{1, 2, \ldots, n\}]$ is TP. This completes the proof. □

A similar fact holds for the retraction of the $(n-1, n-1)$ entry of a TP matrix. The next result follows directly from Theorems 9.5.4 and 9.5.5.

Corollary 9.5.6 *Let* $n \le 4$. *If* A *is an* n-*by*-n *TP matrix and* $1 \le i \le n$, *then* $A - tE_{ii}$ *is* TP_{n-1} *for all* $t \in [0, \frac{\det A}{\det A(\{i\})}]$.

According to the next example, we cannot decrease any other interior main diagonal entry (in general) of a TP matrix and stay TP_{n-1} (see [FJS00]).

Example 9.5.7 Consider the following matrix

$$A = \begin{bmatrix} 100 & 10 & 7/5 & 2 & 1 \\ 22 & 5 & 2 & 3 & 2 \\ 3 & 1 & 1.01 & 2 & 3 \\ 1 & 1 & 2 & 5 & 12 \\ 1/2 & 2 & 5 & 15 & 50 \end{bmatrix}.$$

Then A is a TP matrix with $\frac{\det A}{\det A(\{3\})} \approx .03$. However,

$$\frac{\det A[\{1, 2, 3\}, \{3, 4, 5\}]}{\det A[\{1, 2\}, \{4, 5\}]} = .01.$$

Thus for $t \in (.01, .03]$, we have $\det(A - tE_{33})[\{1, 2, 3\}, \{3, 4, 5\}] < 0$, and hence $A - tE_{33}$ is not TP_4.

Up to this point we have only considered retracting on a single diagonal entry. An obvious next step is to consider increasing or decreasing off-diagonal entries in a TP matrix. We begin a study of perturbing off-diagonal entries by considering the (1,2) entry (see also [FJS00]).

Theorem 9.5.8 *Let* A *be an* n-*by*-n *TP matrix. Then* $A + tE_{12}$ *is* TP_{n-1} *for all* $t \in [0, \frac{\det A}{\det A(\{1\}, \{2\})}]$.

Proof. Since the (1,2) entry of A enters negatively into $\det A$ we increase a_{12} to $a_{12} + \frac{\det A}{\det A(\{1\}, \{2\})}$ so that $\det(A + tE_{12}) = 0$. Observe that the submatrix $(A + tE_{12})[\{2, \ldots, n\}, \{1, 2, \ldots, n\}]$ is equal to $A[\{2, \ldots, n\}, \{1, 2, \ldots, n\}]$ and hence is TP. Moreover, $(A + tE_{12})[\{1, 2, \ldots, n\}, \{2, \ldots, n\}]$ is TP since we have increased the "(1,1)" entry of a TP matrix. The only remaining minors to verify are the leading principal minors of $A + tE_{12}$. The nonnegativity of these leading minors follows from Sylvester's identity (1.5). Hence $\det(A + tE_{12})[\{1, 2, \ldots, n-1\}] > 0$. Replacing $A + tE_{12}$ by $(A + tE_{12})[\{1, 2, \ldots, n-1\}]$ in the above identity yields $\det(A + tE_{12})[\{1, 2, \ldots, n-2\}] > 0$ and so on. This completes the proof. $\square$

Using transposition and reversal, the conclusion of Theorem 9.5.8 holds when the (1,2) entry of a TP matrix A is replaced by the (2,1), $(n-1, n)$ and $(n, n-1)$ entries of A. Unfortunately, this is all that can be said positively concerning increasing or decreasing off-diagonal entries. Consider the following example which can also be found in [FJS00].

Example 9.5.9 Let

$$A = \begin{bmatrix} 13 & 33 & 31 & 10 \\ 132 & 383 & 371 & 120 \\ 13 & 38 & 37 & 12 \\ 1 & 3 & 3 & 1 \end{bmatrix}.$$

Then

$$\frac{\det A}{\det A(\{1\}, \{3\})} = 1 \text{ and } \frac{\det A[\{1,2\}, \{3,4\}]}{a_{24}} = \frac{1}{12}.$$

Thus for $t \in (1/12, 1]$, $\det(A - tE_{13})[\{1,2\}, \{3,4\}] < 0$. Hence $A - tE_{13}$ is not TP_3.

Similar examples can also be constructed in the case of the $(2,3)$ and $(1,4)$ entries, and consequently in all remaining off-diagonal entries of an n-by-n TP matrix.

Chapter Ten

Other Related Topics on TN Matrices

10.0 INTRODUCTION AND TOPICS

Part of the motivation for this work grew out of the fact that total positivity and, in particular, TP matrices arise in many parts of mathematics and have a broad history, and, therefore, a summary of this topic seems natural. This is what we have tried to accomplish here. There are many facts and properties about TP and TN matrices that neither fit well in another chapter nor are as lengthy as a separate chapter. Yet, these topics are worthy of being mentioned and discussed. Thus, in some sense this final chapter represents a catalog of additional topics dealing with other aspects of TN matrices. These topics include powers of TP/TN matrices, a generalization of direct summation in conjunction with TN matrices, and TP/TN polynomial matrices.

10.1 POWERS AND ROOTS OF TP/TN MATRICES

It is an important feature of square TN and TP matrices that each class is closed under positive integer (conventional) powers (Cauchy-Binet). Moreover, sufficiently high powers of IITN matrices are necessarily TP (by definition). Indeed, even the InTN and TP matrices are closed under negative integer powers, except for the "checkerboard" signature similarity (see Chapter 1).

In addition to integer powers, continuous real powers of TP matrices and InTN matrices are well defined in a natural way, just as they are for positive definite matrices. If A is InTN and $A = UDW$ with D a positive diagonal matrix (and $WU = I$), then $A^t := UD^tW$ is uniquely defined for all $t \in \mathbb{R}$. This definition also gives continuous dependence on t that agrees with conventional integer powers at all integer values of t. What, then, about A^t when A is TP and $t > 1$ or $0 < t < 1$ (e.g., $t = 1/k$, for k a positive integer)? If A is n-by-n and $n > 2$, in neither case is A^t necessarily TP. (For $n \leq 2$, it is a nice exercise that A^t is TP for $t > 0$.) This is not surprising for $0 < t < 1$, though it lies in contrast with the positive definite case, but for $t > 1$, it is not so clear what to expect. Consider the TP matrix

$$A = \begin{bmatrix} 2 & 1 & .1 \\ 1 & 2 & 1 \\ .1 & 1 & 2 \end{bmatrix}.$$

Then A is TP and its unique square root with all positive eigenvalues is (approximately) given by

$$B = A^{1/2} = \begin{bmatrix} 1.362 & .3762 & -.0152 \\ .3762 & 1.3103 & .3762 \\ -.0152 & .3762 & 1.3632 \end{bmatrix}.$$

It is clear, via the negative entries, that B is not TP.

In general, under what circumstances roots are TP, or continuous powers larger than 1 are TP, is not known. Because of the threshold conditions of Theorem 5.3.7, continuous powers will *eventually* be TP when the eigenvalue ratios become large enough.

Theorem 10.1.1 *For each TP (resp. InTN) matrix A, there is a positive number T such that for all $t > T$, A^t is TP (resp. InTN).*

We note the similarity of this result for continuous conventional powers with the result (Corollary 8.4.6) for continuous Hadamard powers. It follows that

Corollary 10.1.2 *For each TP (resp. InTN) matrix A there is a positive number T such that for all $t > T$, A^t and $A^{(t)}$ are both TP (resp. InTN).*

The subset of TP consisting of those matrices for which the above constant is at most 1 has not yet been fully studied. It is continuous power closed in both senses.

We close this section by mentioning a couple of related results on this topic. As we discussed above, the class IITN is not closed under extraction of square roots, or any other roots for that matter. We may deduce this in another interesting fashion by choosing a tridiagonal example. Suppose A is IITN and tridiagonal, and assume there is a kth root, $k > 1$; call it B. If the matrix B is even in IITN, it must be irreducible, else A would be reducible. Since B would have to be irreducible, its entries on the main diagonal, and sub- and superdiagonals must be positive. It follows that the kth power, $k \geq 2$, of B will have more positive entries than a tridiagonal matrix. Since A is tridiagonal, this contradicts $B^k = A$. We know of no simple characterization of IITN matrices with square roots in IITN.

What about TP matrices? This class rules out the tridiagonal example above. Could it be an $(n-1)$st power of a tridiagonal matrix in InTN? This, in fact, can never happen, unless the TP matrix, call it A, is symmetrizable via similarity from $\mathcal{D}_n$. This is because the supposed tridiagonal $(n-1)$st root, call it B, must be irreducible, and any irreducible nonnegative tridiagonal matrix is diagonally similar to a symmetric one. If $D \in \mathcal{D}_n$ is such that $D^{-1}BD$ is symmetric, then $(D^{-1}BD)^{n-1} = D^{-1}B^{n-1}D = D^{-1}AD$ would be symmetric. However, even symmetric TP matrices need not have (symmetric) tridiagonal InTN $(n-1)$st root, and it is not known which do.

10.2 SUBDIRECT SUMS OF TN MATRICES

Let $0 \leq k \leq m, n$ and suppose that

$$A = \begin{bmatrix} A_{11} & A_{12} \\ A_{21} & A_{22} \end{bmatrix} \in M_m(\mathbb{C}) \text{ and } B = \begin{bmatrix} B_{22} & B_{23} \\ B_{32} & B_{33} \end{bmatrix} \in M_n(\mathbb{C}),$$

in which $A_{22}, B_{22} \in M_k(\mathbb{C})$. Then we call

$$C = \begin{bmatrix} A_{11} & A_{12} & 0 \\ A_{21} & A_{22} + B_{22} & B_{23} \\ 0 & B_{32} & B_{33} \end{bmatrix}$$

the k-subdirect sum of A and B, which we denote by $A \oplus_k B$. When the value of k is irrelevant or clear, we may just refer to a subdirect sum, and when $k = 0$, a 0-subdirect sum is a familiar direct sum, and $\oplus_0$ will be abbreviated to $\oplus$. Of course, the k-subdirect sum of two matrices is generally not commutative. See [FJ99] for more information and background (including references) on subdirect sums of other positivity classes of matrices.

For a given class of matrices Π four natural questions may be asked:

(I) if A and B are in Π, must a 1-subdirect sum C be in Π;

(II) if

$$C = \begin{bmatrix} C_{11} & C_{12} & 0 \\ C_{21} & C_{22} & C_{23} \\ 0 & C_{32} & C_{33} \end{bmatrix}$$

is in Π, may C be written as $C = A \oplus_1 B$, such that A and B are both in Π, when C_{22} is 1-by-1; and

(III) and (IV) the corresponding questions with 1 replaced by $k > 1$.

We begin with a simple determinantal identity, that is useful in consideration of 1-subdirect sums. We provide a proof here for completeness (see [FJ99]).

Lemma 10.2.1 *Let*

$$A = \begin{bmatrix} A_{11} & a_{12} \\ a_{21} & a_{22} \end{bmatrix} \in M_m(\mathbb{C}) \text{ and } B = \begin{bmatrix} b_{22} & b_{23} \\ b_{32} & B_{33} \end{bmatrix} \in M_n(\mathbb{C}),$$

in which a_{22} and b_{22} are both 1-by-1. Then

$$\det(A \oplus_1 B) = \det A_{11} \det B + \det A \det B_{33}. \tag{10.1}$$

Proof. It is routine to verify that if A_{11} and B_{33} are both singular, then $A \oplus_1 B$ is necessarily singular; hence (10.1) holds in this special case. Suppose, then, without loss of generality, that A_{11} is nonsingular. Using Schur complements (see Chapter 1), it follows that

$$\det \begin{bmatrix} A_{11} & a_{12} & 0 \\ a_{21} & a_{22} + b_{22} & b_{23} \\ 0 & b_{32} & B_{33} \end{bmatrix} = \det A_{11} \det \begin{bmatrix} a_{22} + b_{22} - a_{21} A_{11}^{-1} a_{12} & b_{23} \\ b_{32} & B_{33} \end{bmatrix}.$$

Expanding the latter determinant by the first row gives

$$\det A_{11} \det \begin{bmatrix} b_{22} & b_{23} \\ b_{32} & B_{33} \end{bmatrix} + \det A_{11}(a_{22} - a_{21}A_{11}^{-1}a_{12}) \det B_{33}$$

$$= \det A_{11} \det B + \det A \det B_{33}.$$

□

In [FJ99] we considered questions (I)–(IV) for various positivity classes: positive (semi-) definite; M-matrices; symmetric M-matrices; P-matrices; P_0-matrices; doubly nonnegative matrices (entrywise nonnegative positive semidefinite matrices); and completely positive matrices (matrices of the form BB^T with B entrywise nonnegative). One of the results proved in [FJ99] is the next result, which addresses specifically questions (I) and (II) for each of the above positivity classes.

Theorem 10.2.2 *Let* Π *denote any one of the above positivity classes. Suppose*

$$C = \begin{bmatrix} C_{11} & c_{12} & 0 \\ c_{21} & c_{22} & c_{23} \\ 0 & c_{32} & C_{33} \end{bmatrix}$$

is an n-by-n matrix with C_{11} *and* C_{33} *square, and with* c_{22} *1-by-1. Then* C *is in* Π *if and only if* C *may be written as* $C = A \oplus_1 B$, *in which both* A *and* B *are in* Π.

For more information on questions (III) and (IV) for these positivity classes, see [FJ99]. We now turn our attention to the class of TN matrices. The next result answers questions (I) and (II) for the class TN.

Theorem 10.2.3 *Let*

$$C = \begin{bmatrix} C_{11} & c_{12} & 0 \\ c_{21} & c_{22} & c_{23} \\ 0 & c_{32} & C_{33} \end{bmatrix} \in M_n(\mathbb{R})$$

be an n-by-n matrix with C_{11} *and* C_{33} *square, and with* c_{22} *1-by-1. Then* C *is TN if and only if* C *may be written as* $C = A \oplus_1 B$, *in which both* A *and* B *are TN.*

Proof. First, suppose both A and B are TN, and let

$$A = \begin{bmatrix} A_{11} & a_{12} \\ a_{21} & a_{22} \end{bmatrix} \in M_m(\mathbb{R}) \quad \text{and} \quad B = \begin{bmatrix} b_{22} & b_{23} \\ b_{32} & B_{33} \end{bmatrix} \in M_r(\mathbb{R}),$$

where a_{22} and b_{22} are both 1-by-1. Note that any principal submatrix of $A \oplus_1 B$ either involves the overlapping entry and hence is nonnegative by (10.1), or does not involve the overlapping entry and is nonnegative since it is a direct sum of TN matrices. Let $C = A \oplus_1 B$, and let $C[\alpha, \beta]$ be any square submatrix of C. Let $\alpha_1 = \alpha \cap \{1, 2, \ldots, m-1\}$, $\alpha_2 = \alpha \cap \{m+1, m+2, \ldots, n\}$, $\beta_1 = \beta \cap \{1, 2, \ldots, m-1\}$, and $\beta_2 = \beta \cap \{m+1, m+2, \ldots, n\}$.

Furthermore, we can assume $\alpha_1, \alpha_2, \beta_1$, and β_2 are all nonempty; otherwise it is straightforward to verify that $\det C[\alpha, \beta] \geq 0$. Suppose $m \notin \alpha \cap \beta$. Then either $m \notin \alpha$ and $m \notin \beta$ or, without loss of generality, $m \in \alpha$ and $m \notin \beta$. First assume $m \notin \alpha$ and $m \notin \beta$. Then

$$C[\alpha, \beta] = \begin{bmatrix} A_{11}[\alpha_1, \beta_1] & 0 \\ 0 & B_{33}[\alpha_2, \beta_2] \end{bmatrix}.$$

If $|\alpha_1| = |\beta_1|$, then $C[\alpha, \beta]$ is a direct sum of TN matrices, and hence is TN. Otherwise $|\alpha_1| \neq |\beta_1|$, and without loss of generality, assume $|\alpha_1| > |\beta_1|$ (the case $|\alpha_1| < |\beta_1|$ follows by symmetry). Therefore the size of the larger zero block is $|\alpha_1|$-by-$|\beta_2|$. Furthermore,

$$|\alpha_1| + |\beta_2| \geq |\beta_1| + |\beta_2| + 1 = |\beta| + 1 = |\alpha| + 1,$$

hence $\det C[\alpha, \beta] = 0$. Now assume $m \in \alpha$ and $m \notin \beta$. Then

$$C[\alpha, \beta] = \begin{bmatrix} A_{11}[\alpha_1, \beta_1] & 0 \\ a_{21}[\beta_1] & b_{23}[\beta_2] \\ 0 & B_{33}[\alpha_2, \beta_2] \end{bmatrix}.$$

If $|\alpha_1| = |\beta_1|$, then $|\alpha_2| + 1 = |\beta_2|$. Hence $C[\alpha, \beta]$ is block triangular and it follows that $\det C[\alpha, \beta] \geq 0$. Otherwise, $|\alpha_1| \neq |\beta_1|$. If $|\alpha_1| > |\beta_1|$, then $\det C[\alpha, \beta] = 0$, and if $|\alpha_1| < |\beta_1|$, then $C[\alpha, \beta]$ is either block triangular or singular. Thus suppose $m \in \alpha \cap \beta$. Again there are two cases to consider. Suppose $|\alpha_1| = |\beta_1|$. Then $|\alpha_2| = |\beta_2|$, and $\det C[\alpha, \beta] \geq 0$ follows from (10.1). Otherwise suppose $|\alpha_1| \neq |\beta_1|$, and without loss of generality, assume $|\alpha_1| > |\beta_1|$ (the case $|\alpha_1| < |\beta_1|$ follows by symmetry). The order of the larger zero block is $|\alpha_1|$-by-$|\beta_2|$, and

$$|\alpha_1| + |\beta_2| \geq |\beta_1| + |\beta_2| + 1 = |\beta| = |\alpha|.$$

If $|\alpha_1| + |\beta_2| > |\beta|$, then, as before, $\det C[\alpha, \beta] = 0$. So assume $|\alpha_1| + |\beta_2| = |\beta|$, in which case, it follows that $C[\alpha, \beta]$ is a block triangular matrix with the diagonal blocks being TN, hence $\det C[\alpha, \beta] \geq 0$.

Conversely, suppose C is TN. Since

$$\begin{bmatrix} C_{11} & c_{12} \\ c_{21} & c_{22} \end{bmatrix} \quad \text{and} \quad \begin{bmatrix} c_{22} & c_{23} \\ c_{32} & C_{33} \end{bmatrix}$$

are TN, choose $\hat{a_{22}} \geq 0$ and $\hat{b_{22}} \geq 0$ as small as possible so that

$$\hat{A} = \begin{bmatrix} C_{11} & c_{12} \\ c_{21} & \hat{a_{22}} \end{bmatrix} \quad \text{and} \quad \hat{B} = \begin{bmatrix} \hat{b_{22}} & c_{23} \\ c_{32} & C_{33} \end{bmatrix}$$

are both TN. Assume $\hat{A} \in M_m(\mathbb{R})$ and $\hat{B} \in M_r(\mathbb{R})$. In what follows we are only concerned with square submatrices of $\hat{A}$ and $\hat{B}$. Let

$\Gamma = \{\hat{A}[\alpha, \beta] \ : \ m \in \alpha, \ \beta \subseteq \{1, 2, \ldots, m\}$ and $\hat{A}[\alpha, \beta]$ is singular$\}$, and let $\Lambda = \{\hat{B}[\gamma, \delta] \ : \ 1 \in \gamma, \ \delta \subseteq \{1, 2 \ldots, r\}$ and $\hat{B}[\gamma, \delta]$ is singular$\}$. Observe that Γ (similarly Λ) is nonempty, since if Γ were empty, then for every $m \in \alpha, \ \beta \subseteq \{1, 2, \ldots, m\}$, $\det\hat{A}[\alpha, \beta] > 0$. Then by continuity we may decrease $\hat{a}_{22}$, while $\hat{A}$ remains TN, which contradicts the minimality of $\hat{a}_{22}$. Therefore Γ is nonempty and, similarly, so is Λ.

Suppose for the moment that $\hat{a}_{22} + \hat{b}_{22} \leq c_{22}$. Then increase $\hat{a}_{22}$ to a_{22} and increase $\hat{b}_{22}$ to b_{22} so that $a_{22} + b_{22} = c_{22}$, and let

$$A = \begin{bmatrix} C_{11} & c_{12} \\ c_{21} & a_{22} \end{bmatrix} \quad \text{and} \quad B = \begin{bmatrix} b_{22} & c_{23} \\ c_{32} & C_{33} \end{bmatrix}.$$

By the observation preceding Theorem 10.2.3, A and B are TN, and $C = A \oplus_1 B$. Thus, if we can show $\hat{a}_{22} + \hat{b}_{22} \leq c_{22}$, then the proof is complete. So suppose $\hat{a}_{22} + \hat{b}_{22} > c_{22}$. Then one of two possibilities can occur. Either $\hat{A}[\alpha - \{m\}, \beta - \{m\}]$ is singular for every $\hat{A}[\alpha, \beta] \in \Gamma$, or there exists $\hat{A}[\alpha_0, \beta_0] \in \Gamma$ such that $\hat{A}[\alpha_0 - \{m\}, \beta_0 - \{m\}]$ is nonsingular. Suppose that $\hat{A}[\alpha - \{m\}, \beta - \{m\}]$ is singular for every $\hat{A}[\alpha, \beta] \in \Gamma$. In this case, each such $\det\hat{A}[\alpha, \beta]$ does not depend on $\hat{a}_{22}$, and hence $\hat{a}_{22}$ may be decreased without affecting $\det\hat{A}[\alpha, \beta]$. Also, as previously noted, if $m \in \alpha', \ \beta' \subseteq \{1, 2, \ldots, m\}$ and $\det\hat{A}[\alpha', \beta'] > 0$, then $\hat{a}_{22}$ may be decreased in this case. However, this contradicts the minimality of $\hat{a}_{22}$. Thus there exists $\hat{A}[\alpha_0, \beta_0] \in \Gamma$ such that $\hat{A}[\alpha_0 - \{m\}, \beta_0 - \{m\}]$ is nonsingular. Similar arguments also show that there exists $\hat{B}[\gamma_0, \delta_0] \in \Lambda$ such that $\hat{B}[\gamma_0 - \{1\}, \delta_0 - \{1\}]$ is nonsingular. Furthermore, if $\hat{a}_{22}$ is decreased, then $\det\hat{A}[\alpha_0, \beta_0] < 0$. Since $\hat{a}_{22} + \hat{b}_{22} > c_{22}$, decrease $\hat{a}_{22}$ to a'_{22} such that $a'_{22} + \hat{b}_{22} = c_{22}$. Then

$$C' = \begin{bmatrix} \hat{A}[\alpha_0 - \{m\}, \beta_0 - \{m\}] & c_{12}[\alpha_0 - \{m\}] \\ c_{21}[\beta_0 - \{m\}] & a'_{22} \end{bmatrix} \oplus_1 \hat{B}[\gamma_0, \delta_0]$$

is a submatrix of C. However, $0 \leq \det C' < 0$, by (10.1). This is a contradiction, hence $\hat{a}_{22} + \hat{b}_{22} \leq c_{22}$, which completes the proof. $\qquad \square$

The class TN is not closed under addition, hence it follows that the answer to question (III) is negative. However, even more goes wrong.

Example 10.2.4

$$A = \begin{bmatrix} 2 & 1 & 0 \\ 1 & 1 & 0 \\ 0 & 2 & 1 \end{bmatrix} \quad \text{and } B = \begin{bmatrix} 1 & 2 & 0 \\ 0 & 1 & 0 \\ 0 & 0 & 1 \end{bmatrix}.$$

Then A and B are both TN matrices. However,

$$A \oplus_2 B = \begin{bmatrix} 2 & 1 & 0 & 0 \\ 1 & 2 & 2 & 0 \\ 0 & 2 & 2 & 0 \\ 0 & 0 & 0 & 1 \end{bmatrix}$$

is not a TN matrix since $\det(A \oplus_2 B) = -2$. Note that the sum in the overlapping positions is a totally nonnegative matrix.

To address question (IV), as in [FJ99], we will make strong use of the fact that TN matrices can be factored into TN matrices L and U (see Chapter 2).

Lemma 10.2.5 *Let*

$$
C = \begin{bmatrix} C_{11} & C_{12} & 0 \\ C_{21} & C_{22} & C_{23} \\ 0 & C_{32} & C_{33} \end{bmatrix},
$$

in which C_{22} and C_{33} are square and C_{11} is m-by-m $(m \geq 1)$. Suppose C is TN. Then

$$
C = LU = \begin{bmatrix} L_{11} & 0 & 0 \\ L_{21} & L_{22} & 0 \\ 0 & L_{32} & L_{33} \end{bmatrix} \cdot \begin{bmatrix} U_{11} & U_{12} & 0 \\ 0 & U_{22} & U_{23} \\ 0 & 0 & U_{33} \end{bmatrix},
$$

in which L and U (partitioned conformally with C) are both TN.

We are now in a position to prove an affirmative answer to question (IV) for the class TN ([FJ99]).

Theorem 10.2.6 *Let*

$$
C = \begin{bmatrix} C_{11} & C_{12} & 0 \\ C_{21} & C_{22} & C_{23} \\ 0 & C_{32} & C_{33} \end{bmatrix},
$$

in which C_{11} and C_{33} are square, and with C_{22} k-by-k. If C is TN, then C can be written as $C = A \oplus_k B$ so that A and B are TN.

Proof. By Lemma 10.2.5, $C = LU$, in which both L and U are TN, and $L_{31} = 0$ and $U_{13} = 0$. Then it is easy to check that

$$
C = LU = \begin{bmatrix} L_{11}U_{11} & L_{11}U_{12} & 0 \\ L_{21}U_{11} & L_{22}U_{22} + L_{21}U_{12} & L_{22}U_{23} \\ 0 & L_{32}U_{22} & L_{33}U_{33} + L_{32}U_{23} \end{bmatrix}.
$$

Hence C can be written as

$$
C = \begin{bmatrix} L_{11}U_{11} & L_{11}U_{12} & 0 \\ L_{21}U_{11} & L_{21}U_{12} & 0 \\ 0 & 0 & 0 \end{bmatrix} + \begin{bmatrix} 0 & 0 & 0 \\ 0 & L_{22}U_{22} & L_{22}U_{23} \\ 0 & L_{32}U_{22} & L_{33}U_{33} + L_{32}U_{23} \end{bmatrix}.
$$

Notice that if

$$
A = \begin{bmatrix} L_{11}U_{11} & L_{11}U_{12} \\ L_{21}U_{11} & L_{21}U_{12} \end{bmatrix} = \begin{bmatrix} L_{11} & 0 \\ L_{21} & L_{22} \end{bmatrix} \cdot \begin{bmatrix} U_{11} & U_{12} \\ 0 & 0 \end{bmatrix},
$$

and

$$B = \begin{bmatrix} L_{22}U_{22} & L_{22}U_{23} \\ L_{32}U_{22} & L_{33}U_{33} + L_{32}U_{23} \end{bmatrix} = \begin{bmatrix} L_{22} & 0 \\ L_{32} & L_{33} \end{bmatrix} \cdot \begin{bmatrix} U_{22} & U_{23} \\ 0 & U_{33} \end{bmatrix},$$

then $C = A \oplus_k B$. Each of the four matrices on the right is easily seen to be TN because L and U are, and it follows from the multiplicative closure of the class TN that both A and B are TN, which completes the proof of the theorem. $\qquad\square$

A final result of interest, which has also been mentioned in the context of tridiagonal matrices in Chapter 0, is that for nonnegative tridiagonal matrices the property of being TN or P_0 are equivalent.

Corollary 10.2.7 *Suppose that A is an n-by-n entrywise nonnegative tridiagonal matrix. Then A is TN if and only if A is a P_0-matrix (that is, has nonnegative principal minors).*

Proof. Since being TN is a stronger property than being P_0, we need only assume that A is P_0. The proof is by induction on n, with the case $n = 2$ being obvious. Since A is tridiagonal and P_0, it follows that $A = B \oplus_1 C$, with both B, C being P_0 and tridiagonal. By induction, both B and C are then TN as well. Hence, by Theorem 10.2.3 $A = B \oplus_1 C$ is TN. $\qquad\square$

10.3 TP/TN POLYNOMIAL MATRICES

A matrix function may be viewed as one in which each entry is a function of the same collection of variables. Here, we consider square matrix functions $A(t)$ of a single variable t, in which each entry is a polynomial in t, a so-called *polynomial matrix*. In other words, $A(t) = [a_{ij}(t)]$ in which $a_{ij}(t)$ is a polynomial in t, $i, j = 1, 2, \ldots, n$, and the coefficients of all our polynomials will be real.

Under what circumstances should a polynomial matrix be called TP, TN or InTN? The definition here is that $A(t)$ is TP (TN, InTN) if for each $t \in \mathbb{R}$, $A(t)$ is TP (TN, InTN). Consider the following example: Let

$$A(t) = \begin{bmatrix} 3 + t^2 & 2 - 2t + t^2 \\ 1 & 1 + t^2 \end{bmatrix}.$$

Then is it easily checked that $A(t)$ is a TP polynomial matrix.

To what extent do TP polynomial matrices have properties that mimic TP matrices (or TN or InTN)? In [JS00] it is shown that TP line insertion (Theorem 9.0.2) is still possible for polynomial matrices, that is, if $A(t)$ is an m-by-n polynomial matrix, a polynomial line may be inserted anywhere to produce an $(m + 1)$-by-n or m-by-$(n + 1)$ TP polynomial matrix $\hat{A}(t)$. This is trivial when only TN is required of both $A(t)$ and $\hat{A}(t)$.

We also note here that if $A(t)$ is InTN, then $A(t)$ has an elementary bidiagonal factorization in which each bidiagonal factor is a polynomial matrix (with only one nontrivial polynomial entry) that takes on only nonnegative values and the diagonal factor has polynomial diagonal entries, taking on only positive values.

Theorem 10.3.1 *Let $A(t)$ be a polynomial matrix. Then $A(t)$ is an InTN polynomial matrix if and only if $A(t)$ has a bidiagonal factorization in which each bidiagonal factor is a polynomial matrix that takes on only nonnegative values and the diagonal factor has polynomial diagonal entries, taking on only positive values.*

In addition, $A(t)$ is a TP polynomial matrix if and only if $A(t)$ has a bidiagonal factorization in which all factors are polynomial matrices taking on only positive values.

10.4 PERRON COMPLEMENTS OF TN MATRICES

Perron complements were conceived in connection with a divide and conquer algorithm for computing the stationary distribution vector for a Markov chain (see [Mey89]). For an n-by-n nonnegative and irreducible matrix A, the *Perron complement* of $A[\beta]$ in A, is given by

$$\mathcal{P}(A/A[\beta]) = A[\alpha] + A[\alpha,\beta](\rho(A)I - A[\beta])^{-1}A[\beta,\alpha], \qquad (10.2)$$

in which $\beta \subset N$, $\alpha = N \setminus \beta$, and $\rho(\cdot)$ denotes the spectral radius of a matrix. Recall that since A is irreducible and nonnegative, $\rho(A) > \rho(A[\beta])$, so that the expression on the right hand side of (10.2) is well defined. It is known that for A nonnegative, $\mathcal{P}(A/A[\beta])$ is nonnegative and $\rho(\mathcal{P}(A/A[\beta])) = \rho(A)$ (see [Mey89] and also [JX93]).

Recall from Chapter 1 that the Schur complement was defined as:

$$A/A[\gamma] = A[\gamma^c] - A[\gamma^c,\gamma](A[\gamma])^{-1}A[\gamma,\gamma^c].$$

Thus, it is evident that Perron and Schur complements are similar in definition. In [FN01] Perron complements of TN matrices were studied and a generalized version of a Perron complement was defined and studied in connection with TN matrices.

For any $\beta \subset N$ and for any $t \geq \rho(A)$, let the *extended Perron complement at t* be the matrix

$$\mathcal{P}_t(A/A[\beta]) = A[\alpha] + A[\alpha,\beta](tI - A[\beta])^{-1}A[\beta,\alpha], \qquad (10.3)$$

which is also well defined since $t \geq \rho(A) > \rho(A[\beta])$.

Before we address some results on Perron complements of TN matrices, we state an important result, proved in [And87], on Schur complements of TN matrices and the compound ordering (defined in Chapter 5). A key ingredient needed is Sylvester's determinantal identity (see Chapter 1),

$$\det(A/A[\{k,\ldots,n\}])[\gamma,\delta] = \frac{\det A[\gamma \cup \{k,\ldots,n\}, \delta \cup \{k,\ldots,n\}]}{\det A[\{k,\ldots,n\}]}, \qquad (10.4)$$

where $2 \leq k \leq n$, and $\gamma, \delta \subseteq \{1,2,\ldots,k-1\}$ with $|\gamma| = |\delta|$.

Then we have,

Lemma 10.4.1 *If A is an n-by-n TN matrix, and $\beta = \{1, 2, \dots, k\}$ or $\beta = \{k, k+1, \dots, n\}$, then*

$$A[\alpha] \overset{(t)}{\geq} (A/A[\beta]) \overset{(t)}{\geq} 0,$$

where $\alpha = N \setminus \beta$ and provided that $A[\beta]$ is invertible.

When is the Perron complement of a TN matrix TN? To address this, we verify a quotient formula for the Perron complement that is reminiscent of Haynsworth's quotient formula for Schur complements.

Recall from Proposition 1.5.1 that Schur complements of TN matrices are again TN whenever the index set in question is contiguous. For Perron complements, we have an analogous result ([FN01]), which we include here with proof.

Lemma 10.4.2 *Let A be an n-by-n irreducible TN matrix, and let $\beta = \{1\}$ or $\beta = \{n\}$ and define $\alpha = N \setminus \beta$. Then for any $t \in [\rho(A), \infty)$, the matrix*

$$\mathcal{P}_t(A/A[\beta]) = A[\alpha] + A[\alpha, \beta](tI - A[\beta])^{-1}A[\beta, \alpha]$$

is TN. In particular, the Perron complement $\mathcal{P}(A/A[\beta])$ is TN for $\beta = \{1\}$ or $\beta = \{n\}$.

Proof. Assume $\beta = \{n\}$. (The arguments for the case $\beta = \{1\}$ are similar.) Let A be partitioned as follows

$$A = \begin{bmatrix} B & c \\ d^T & e \end{bmatrix},$$

where B is $(n-1)$-by-$(n-1)$ and e is a scalar. Then $\mathcal{P}_t(A/e) = B + \frac{cd^T}{(t-e)}$, where $t - e > 0$. Consider the matrix

$$X = \begin{bmatrix} B & -c \\ d^T & t-e \end{bmatrix}.$$

Observe that $X/(t - e) = \mathcal{P}_t(A/e)$. Thus any minor of $\mathcal{P}_t(A/e)$ is just a minor of a related Schur complement. Using the formula (10.4) we have

$$\det(X/(t-e))[\gamma, \delta] = \frac{\det X[\gamma \cup \{n\}, \delta \cup \{n\}]}{(t-e)},$$

where $\gamma, \delta \subset \{1, 2, \dots, n-1\}$. Observe that

$$\det X[\gamma \cup \{n\}, \delta \cup \{n\}] = \det \begin{bmatrix} B[\gamma, \delta] & -c[\gamma] \\ d^T[\delta] & t-e \end{bmatrix}$$

$$= t \det B[\gamma, \delta] + \det \begin{bmatrix} B[\gamma, \delta] & -c[\gamma] \\ d^T[\delta] & -e \end{bmatrix}$$

$$= t \det B[\gamma, \delta] - \det \begin{bmatrix} B[\gamma, \delta] & c[\gamma] \\ d^T[\delta] & e \end{bmatrix}$$

$$\geq t \det B[\gamma, \delta] - e \det B[\gamma, \delta]$$

$$= (t-e) \det B[\gamma, \delta]$$

$$\geq 0.$$

The first inequality follows since the matrix on the left is TN and TN matrices satisfy Fischer's inequality (see Chapter 6). This completes the proof.
□

Unfortunately, TN matrices are not closed under arbitrary Perron complementation, even when β is a singleton as the next example demonstrates. Note that the same example appears in [FN01], as such examples are not easy to produce.

Example 10.4.3 Recall from Chapter 0 that a polynomial $f(x)$ is said to be *stable* if all the zeros of $f(x)$ have nonpositive real parts. Further, if $f(x)$ is a stable polynomial, then the Routh-Hurwitz matrix formed from f is TN.
   Consider the polynomial

$$f(x) = x^{10} + 6.2481x^9 + 17.0677x^8 + 26.7097x^7 + 26.3497x^6 + 16.9778x^5$$
$$+7.1517x^4 + 1.9122x^3 + .3025x^2 + .0244x + .0007.$$

It can be shown that f is a stable polynomial. Hence its associated Routh–Hurwitz array is TN call it H. Let $P \equiv \mathcal{P}(H/H[\{7\}])$ (which is 9-by-9). Then P is not TN, as $\det P[\{8,9\},\{5,6\}] < 0$, for example.

Recall Haynsworth's quotient formula (see, for example, [FN01]) which can be stated as follows. For $\phi \neq \alpha \subset \beta \subset N$,

$$(A/A[\beta]) = (A/A[\alpha])/(A[\beta]/A[\alpha]).$$

Rather than denote Schur and Perron complements with respect to principal submatrices, we will denote them only by their index sets. For example, if A is n-by-n and $\beta \subset N$, then we denote $\mathcal{P}_t(A/A[\beta])$ by $\mathcal{P}_t(A/\beta)$ and $A/A[\beta]$ by A/β. Along these lines, we have (as in [FN01])

Theorem 10.4.4 Let A be any n-by-n irreducible nonnegative matrix, and fix any nonempty set $\beta \subset N$. Then for any $\phi \neq \gamma_1, \gamma_2 \subset \beta$ with $\gamma_1 \cup \gamma_2 = \beta$ and $\gamma_1 \cap \gamma_2 = \phi$, we have

$$\mathcal{P}_t(A/\beta) = \mathcal{P}_t(\mathcal{P}_t(A/\gamma_1)/\gamma_2)$$

for any $t \in [\rho(A), \infty)$.

Proof. Observe that for any index set $\beta \subset N$, the following identity holds:
$$\mathcal{P}_t(A/\beta) = tI - ((tI - A)/\beta).$$
   Hence we have
$$\begin{aligned}
\mathcal{P}_t(\mathcal{P}_t(A/\gamma_1)/\gamma_2) &= \mathcal{P}_t((tI - ((tI - A)/\gamma_1))/\gamma_2) \\
&= tI - ([tI - (tI - ((tI - A)/\gamma_1))]/\gamma_2) \\
&= tI - ((tI - A)/\gamma_1)/\gamma_2) \\
&= tI - ((tI - A)/\beta) = \mathcal{P}_t(A/\beta).
\end{aligned}$$

The second to last equality follows from the quotient formula for Schur complements. This completes the proof.   □

Using this quotient formula for extended Perron complements and Lemma 10.4.2, we have the next result ([FN01]).

Theorem 10.4.5 *Let A be an n-by-n irreducible TN matrix, and let $\phi \neq \beta \subset N$ such that $\alpha = N \setminus \beta$ is contiguous. Then for any $t \in [\rho(A), \infty)$, the matrix*

$$\mathcal{P}_t(A/A[\beta]) = A[\alpha] + A[\alpha, \beta](tI - A[\beta])^{-1}A[\beta, \alpha]$$

is TN. In particular, the Perron complement $\mathcal{P}(A/A[\beta])$ is TN whenever $N \setminus \beta$ is contiguous.

The next result is a consequence of the previous theorem; its proof can be found in [FN01].

Corollary 10.4.6 *Let A be an n-by-n irreducible tridiagonal TN matrix. Then for any $\beta \subset N$, the matrix*

$$\mathcal{P}_t(A/A[\beta]) = A[\alpha] + A[\alpha, \beta](tI - A[\beta])^{-1}A[\beta, \alpha],$$

where $t \geq \rho(A)$, is irreducible tridiagonal and totally nonnegative.

The next fact involves an ordering between the compounds of extended Perron complements and Schur complements of TN matrices discussed earlier (also referred to as the compound ordering). Recall from Lemma 10.4.1 that if $\alpha = \{1, 2, \ldots, k\}$ or $\alpha = \{k, k+1, \ldots, n\}$, then

$$A[\alpha] \overset{(t)}{\geq} (A/A[\beta]) \overset{(t)}{\geq} 0,$$

where $\beta = N \setminus \alpha$ and provided that $A[\beta]$ is invertible. In the same spirit we have the following result, as in [FN01], which is included here with proof.

Theorem 10.4.7 *Let A be an n-by-n irreducible TN matrix, and let $\phi \neq \beta \subset N$ such that $\alpha = N \setminus \beta$ is a contiguous set. Then for any $t \in [\rho(A), \infty)$,*

$$\mathcal{P}_t(A/A[\beta]) \overset{(t)}{\geq} A[\alpha] \overset{(t)}{\geq} (A/A[\beta]) \overset{(t)}{\geq} 0.$$

Proof. Suppose $\alpha = \{1, \ldots, k\}$ and $\beta = \{k+1, \ldots, n\}$. It is enough to verify the inequality $\mathcal{P}_t(A/A[\beta]) \overset{(t)}{\geq} A[\alpha]$, as the remaining two inequalities are contained in Lemma 10.4.1. We begin with $\beta = \{n\}$. Recall from Lemma 10.4.2 that if

$$A = \begin{bmatrix} B & c \\ d^T & e \end{bmatrix} \quad \text{and} \quad X = \begin{bmatrix} B & -c \\ d^T & t-e \end{bmatrix},$$

then for any $\gamma, \delta \subset \{1, 2, \ldots, n-1\}$

$$\det \mathcal{P}_t[\gamma, \delta] = \frac{\det X[\gamma \cup \{n\}, \delta \cup \{n\}]}{t - e}$$

$$\geq \frac{t \det B[\gamma, \delta] - e \det B[\gamma, \delta]}{t - e}, \quad \text{by Fischer's inequality}$$

$$= \det B[\gamma, \delta] \quad (\text{here } B = A[\alpha]).$$

Thus $\mathcal{P}_t(A/A[\beta]) \overset{(t)}{\geq} A[\alpha]$, as desired. We are now ready to proceed with the case when β is not a singleton, namely, $\beta = \{k+1, \ldots, n\}$, $k < n-1$. Let $K = \mathcal{P}_t(A/A[\{n\}])$ and let $\gamma = \{1, \ldots, n-2\}$. Then K is irreducible and, by Lemma 10.4.2, we also know that K is a TN matrix. But then applying the initial part of the proof to K we see that

$$\mathcal{P}_t\left(K/[\{n-1\}]\right)[\gamma] \overset{(t)}{\geq} K[\gamma] \overset{(t)}{\geq} A[\gamma].$$

The last inequality follows since $\overset{(t)}{\geq}$ is inherited by submatrices. The claim of the theorem now follows by repeating this argument as many times as necessary and making use of the quotient formula in Theorem 10.4.4.

Thus far we have shown that if $\beta = \{1, \ldots, k\}$ or $\beta = \{k+1, \ldots, n\}$ and $\alpha = N \setminus \beta$, then

$$\mathcal{P}_t(A/A[\beta]) \overset{(t)}{\geq} A[\alpha] \overset{(t)}{\geq} (A/A[\beta]) \overset{(t)}{\geq} 0. \qquad (10.5)$$

More generally, suppose $\beta \subset N$ such that $\alpha = N \setminus \beta$ is a contiguous set. Then $\alpha = \{i, i+1, \ldots, i+k\}$, and hence $\beta = \{1, \ldots, i-1, i+k+1, \ldots, n\}$. Thus, by Theorem 10.4.4,

$$\mathcal{P}_t(A/\beta) = \mathcal{P}_t(\mathcal{P}_t(A/\{1, \ldots, i-1\})/\{i+k+1, \ldots, n\}).$$

Applying (10.5) twice we have

$$\mathcal{P}_t(A/\beta) = \mathcal{P}_t(\mathcal{P}_t(A/\{1, \ldots, i-1\})/\{i+k+1, \ldots, n\})$$
$$\overset{(t)}{\geq} \mathcal{P}_t(A/\{1, \ldots, i-1\})[\{i+k+1, \ldots, n\}]$$
$$\overset{(t)}{\geq} A[\alpha],$$

as desired. The remaining inequalities, namely $A[\alpha] \overset{(t)}{\geq} (A/A[\beta]) \overset{(t)}{\geq} 0$, follow from the remarks preceding Theorem 10.4.7. This completes the proof. $\square$

Bibliography

[ACDR04] Pedro Alonso, Raquel Cortina, Irene Díaz, and José Ranilla. Neville elimination: a study of the efficiency using checkerboard partitioning. *Linear Algebra Appl.*, 393:3–14, 2004.

[ACHR01] Pedro Alonso, Raquel Cortina, Vicente Hernández, and José Ranilla. A study of the performance of Neville elimination using two kinds of partitioning techniques. In *Proceedings of the Eighth Conference of the International Linear Algebra Society (Barcelona, 1999)*, volume 332/334, pages 111–117, 2001.

[AESW51] Michael Aissen, Albert Edrei, Isaac J. Schoenberg, and Anne Whitney. On the generating functions of totally positive sequences. *Proc. Nat. Acad. Sci. USA*, 37:303–307, 1951.

[AGP97] Pedro Alonso, Mariano Gasca, and Juan M. Peña. Backward error analysis of Neville elimination. *Appl. Numer. Math.*, 23(2):193–204, 1997.

[AKSM04] Víctor Ayala, Wolfgang Kliemann, and Luiz A. B. San Martin. Control sets and total positivity. *Semigroup Forum*, 69(1):113–140, 2004.

[And87] Tsuyoshi Ando. Totally positive matrices. *Linear Algebra Appl.*, 90:165–219, 1987.

[AP99] Pedro Alonso and Juan Manuel Peña. Development of block and partitioned Neville elimination. *C. R. Acad. Sci. Paris Sér. I Math.*, 329(12):1091–1096, 1999.

[Asn70] Bernard A. Asner, Jr. On the total nonnegativity of the Hurwitz matrix. *SIAM J. Appl. Math.*, 18:407–414, 1970.

[ASW52] Michael Aissen, Isaac J. Schoenberg, and Anne M. Whitney. On the generating functions of totally positive sequences. I. *J. Analyse Math.*, 2:93–103, 1952.

[AT01] Lidia Aceto and Donato Trigiante. The matrices of Pascal and other greats. *Amer. Math. Monthly*, 108(3):232–245, 2001.

[BCR98]   Jacek Bochnak, Michel Coste, and Marie-Françoise Roy. *Real Algebraic Geometry*, volume 36 of *Ergebnisse der Mathematik und ihrer Grenzgebiete (3) [Results in Mathematics and Related Areas (3)]*. Springer-Verlag, Berlin, 1998. Translated from the 1987 French original, revised by the authors.

[BF08]   Adam Boocher and Bradley Froehle. On generators of bounded ratios of minors for totally positive matrices. *Linear Algebra Appl.*, 428(7):1664–1684, 2008.

[BFZ96]   Arkady Berenstein, Sergey Fomin, and Andrei Zelevinsky. Parametrizations of canonical bases and totally positive matrices. *Adv. Math.*, 122(1):49–149, 1996.

[BG84]   Stanislaw Białas and Jürgen Garloff. Intervals of P-matrices and related matrices. *Linear Algebra Appl.*, 58:33–41, 1984.

[BHJ85]   Abraham Berman, Daniel Hershkowitz, and Charles R. Johnson. Linear transformations that preserve certain positivity classes of matrices. *Linear Algebra Appl.*, 68:9–29, 1985.

[BJ84]   Wayne W. Barrett and Charles R. Johnson. Possible spectra of totally positive matrices. *Linear Algebra Appl.*, 62:231–233, 1984.

[BJM81]   Lawrence D. Brown, Iain M. Johnstone, and K. Brenda MacGibbon. Variation diminishing transformations: a direct approach to total positivity and its statistical applications. *J. Amer. Statist. Assoc.*, 76(376):824–832, 1981.

[BP75]   Richard E. Barlow and Frank Proschan. *Statistical Theory of Reliability and Life Testing*. Holt, Rinehart and Winston, New York, 1975. Probability models, International Series in Decision Processes, Series in Quantitative Methods for Decision Making.

[BR05]   Arkady Berenstein and Vladimir Retakh. Noncommutative double Bruhat cells and their factorizations. *Int. Math. Res. Not.*, (8):477–516, 2005.

[Bre95]   Francesco Brenti. Combinatorics and total positivity. *J. Combin. Theory Ser. A*, 71(2):175–218, 1995.

[Bre00]   Claude Brezinski, editor. *Numerical Analysis 2000. Vol. II: Interpolation and extrapolation*, volume 122. Elsevier Science B.V., Amsterdam, 2000.

[BZ97]   Arkady Berenstein and Andrei Zelevinsky. Total positivity in Schubert varieties. *Comment. Math. Helv.*, 72(1):128–166, 1997.

[Car67]   David Carlson. Weakly sign-symmetric matrices and some determinantal inequalities. *Colloq. Math.*, 17:123–129, 1967.

[CC98]   Thomas Craven and George Csordas. A sufficient condition for strict total positivity of a matrix. *Linear and Multilinear Algebra*, 45(1):19–34, 1998.

[CFJ01]   Alissa S. Crans, Shaun M. Fallat, and Charles R. Johnson. The Hadamard core of the totally nonnegative matrices. *Linear Algebra Appl.*, 328(1-3):203–222, 2001.

[CFL02]   Wai-Shun Cheung, Shaun Fallat, and Chi-Kwong Li. Multiplicative preservers on semigroups of matrices. *Linear Algebra Appl.*, 355:173–186, 2002.

[CGP95]   Jésus M. Carnicer, Tim N. T. Goodman, and Juan M. Peña. A generalization of the variation diminishing property. *Adv. Comput. Math.*, 3(4):375–394, 1995.

[CGP99]   Jésus M. Carnicer, Tim N. T. Goodman, and Juan M. Peña. Linear conditions for positive determinants. *Linear Algebra Appl.*, 292(1-3):39–59, 1999.

[Cho05]   Inheung Chon. Strictly infinitesimally generated totally positive matrices. *Commun. Korean Math. Soc.*, 20(3):443–456, 2005.

[CP94a]   Jésus M. Carnicer and Juan M. Peña. Spaces with almost strictly totally positive bases. *Math. Nachr.*, 169:69–79, 1994.

[CP94b]   Jésus M. Carnicer and Juan M. Peña. Totally positive bases for shape preserving curve design and optimality of B-splines. *Comput. Aided Geom. Design*, 11(6):633–654, 1994.

[CP97]   Jésus M. Carnicer and Juan M. Peña. Bidiagonalization of oscillatory matrices. *Linear and Multilinear Algebra*, 42(4):365–376, 1997.

[CP08]   Vanesa Cortés and Juan M. Peña. A stable test for strict sign regularity. *Math. Comp.*, 77(264):2155–2171, 2008.

[CPZ98]   Jésus M. Carnicer, Juan M. Peña, and Richard A. Zalik. Strictly totally positive systems. *J. Approx. Theory*, 92(3):411–441, 1998.

[Cry73]   Colin W. Cryer. The LU-factorization of totally positive matrices. *Linear Algebra Appl.*, 7:83–92, 1973.

[Cry76]   Colin W. Cryer. Some properties of totally positive matrices. *Linear Algebra Appl.*, 15(1):1–25, 1976.

[dB76]   Carl de Boor. Total positivity of the spline collocation matrix. *Indiana Univ. Math. J.*, 25(6):541–551, 1976.

[dB82]   Carl de Boor. The inverse of a totally positive bi-infinite band matrix. *Trans. Amer. Math. Soc.*, 274(1):45–58, 1982.

[dBD85]   Carl de Boor and Ronald DeVore. A geometric proof of total positivity for spline interpolation. *Math. Comp.*, 45(172):497–504, 1985.

[dBJP82]   Carl de Boor, Rong Qing Jia, and Allan Pinkus. Structure of invertible (bi)infinite totally positive matrices. *Linear Algebra Appl.*, 47:41–55, 1982.

[dBP82]   Carl de Boor and Allan Pinkus. The approximation of a totally positive band matrix by a strictly banded totally positive one. *Linear Algebra Appl.*, 42:81–98, 1982.

[Dem82]   Stephen Demko. Surjectivity and invertibility properties of totally positive matrices. *Linear Algebra Appl.*, 45:13–20, 1982.

[DJ97]   Emily B. Dryden and Charles R. Johnson. Totally nonnegative completions. An unpublished paper from a National Science Foundation Research Experiences for Undergraduates program held at the College of William and Mary in the summer of 1997, 1997.

[DJK08]   Emily B. Dryden, Charles R. Johnson, and Brenda K. Korschel. Adjacent edge conditions for the totally nonnegative completion problem. *Linear Multilinear Algebra*, 56(3):261–277, 2008.

[DK01]   James Demmel and Plamen Koev. Necessary and sufficient conditions for accurate and efficient rational function evaluation and factorizations of rational matrices. In *Structured matrices in mathematics, computer science, and engineering, II (Boulder, CO, 1999)*, volume 281 of *Contemp. Math.*, pages 117–143. Amer. Math. Soc., Providence, RI, 2001.

[DK05]   James Demmel and Plamen Koev. The accurate and efficient solution of a totally positive generalized Vandermonde linear system. *SIAM J. Matrix Anal. Appl.*, 27(1):142–152 (electronic), 2005.

[DK08]   Froilán M. Dopico and Plamen Koev. Bidiagonal decompositions of oscillating systems of vectors. *Linear Algebra Appl.*, 428(11-12):2536–2548, 2008.

[DM88]   Nira Dyn and Charles A. Micchelli. Piecewise polynomial spaces and geometric continuity of curves. *Numer. Math.*, 54(3):319–337, 1988.

[DMS86]   Wolfgang Dahmen, Charles A. Micchelli, and Philip W. Smith. On factorization of bi-infinite totally positive block Toeplitz matrices. *Rocky Mountain J. Math.*, 16(2):335–364, 1986.

[DP05]   Dimitar K. Dimitrov and Juan Manuel Peña. Almost strict total positivity and a class of Hurwitz polynomials. *J. Approx. Theory*, 132(2):212–223, 2005.

[Edr52]   Albert Edrei. On the generating functions of totally positive sequences. II. *J. Analyse Math.*, 2:104–109, 1952.

[Edr53a] Albert Edrei. On the generation function of a doubly infinite, totally positive sequence. *Trans. Amer. Math. Soc.*, 74:367–383, 1953.

[Edr53b] Albert Edrei. Proof of a conjecture of Schoenberg on the generating function of a totally positive sequence. *Canadian J. Math.*, 5:86–94, 1953.

[eGJT08] Ramadán el Ghamry, Cristina Jordán, and Juan R. Torregrosa. Double-path in the totally nonnegative completion problem. *Int. Math. Forum*, 3(33-36):1683–1692, 2008.

[EP02]   Uri Elias and Allan Pinkus. Nonlinear eigenvalue-eigenvector problems for STP matrices. *Proc. Roy. Soc. Edinburgh Sect. A*, 132(6):1307–1331, 2002.

[Eve96]   Simon P. Eveson. The eigenvalue distribution of oscillatory and strictly sign-regular matrices. *Linear Algebra Appl.*, 246:17–21, 1996.

[Fal99]   Shaun M. Fallat. *Totally nonnegative matrices.* PhD thesis, College of William and Mary, 1999.

[Fal01]   Shaun M. Fallat. Bidiagonal factorizations of totally nonnegative matrices. *Amer. Math. Monthly*, 108(8):697–712, 2001.

[Fek13]   Michael Fekete. Uber ein problem von Laguerre. *Rend. Conti. Palermo*, 34:110–120, 1913.

[FFM03]   Shaun M. Fallat, Miroslav Fiedler, and Thomas L. Markham. Generalized oscillatory matrices. *Linear Algebra Appl.*, 359:79–90, 2003.

[FG05]   Shaun M. Fallat and Michael I. Gekhtman. Jordan structures of totally nonnegative matrices. *Canad. J. Math.*, 57(1):82–98, 2005.

[FGG98]   Shaun M. Fallat, Michael I. Gekhtman, and Carolyn Gold-
           beck. Ratios of principal minors of totally nonnegative matri-
           ces. An unpublished paper from a National Science Foundation
           Research Experiences for Undergraduates program held at the
           College of William and Mary in the summer of 1998, 1998.

[FGJ00]   Shaun M. Fallat, Michael I. Gekhtman, and Charles R. Johnson.
           Spectral structures of irreducible totally nonnegative matrices.
           SIAM J. Matrix Anal. Appl., 22(2):627–645 (electronic), 2000.

[FGJ03]   Shaun M. Fallat, Michael I. Gekhtman, and Charles R. Johnson.
           Multiplicative principal-minor inequalities for totally nonnega-
           tive matrices. *Adv. in Appl. Math.*, 30(3):442–470, 2003.

[FHGJ06]   Shaun M. Fallat, Allen Herman, Michael I. Gekhtman, and
           Charles R. Johnson. Compressions of totally positive matrices.
           SIAM J. Matrix Anal. Appl., 28(1):68–80 (electronic), 2006.

[FHJ98]   Shaun M. Fallat, H. Tracy Hall, and Charles R. Johnson. Char-
           acterization of product inequalities for principal minors of M-
           matrices and inverse M-matrices. *Quart. J. Math. Oxford Ser.
           (2)*, 49(196):451–458, 1998.

[FJ99]   Shaun M. Fallat and Charles R. Johnson. Sub-direct sums
           and positivity classes of matrices. *Linear Algebra Appl.*, 288(1–
           3):149–173, 1999.

[FJ00]   Shaun M. Fallat and Charles R. Johnson. Determinantal in-
           equalities: ancient history and recent advances. In *Algebra and
           its Applications (Athens, OH, 1999)*, volume 259 of *Contemp.
           Math.*, pages 199–212. Amer. Math. Soc., Providence, RI, 2000.

[FJ01]   Shaun M. Fallat and Charles R. Johnson. Multiplicative
           principal-minor inequalities for tridiagonal sign-symmetric P-
           matrices. *Taiwanese J. Math.*, 5(3):655–665, 2001.

[FJ07a]   Shaun M. Fallat and Charles R. Johnson. Hadamard duals,
           retractability and Oppenheim's inequality. *Oper. Matrices*,
           1(3):369–383, 2007.

[FJ07b]   Shaun M. Fallat and Charles R. Johnson. Hadamard powers and
           totally positive matrices. *Linear Algebra Appl.*, 423(2-3):420–
           427, 2007.

[FJM00]   Shaun M. Fallat, Charles R. Johnson, and Thomas L. Markham.
           Eigenvalues of products of matrices and submatrices in certain
           positivity classes. *Linear and Multilinear Algebra*, 47(3):235–
           248, 2000.

[FJS00] Shaun M. Fallat, Charles R. Johnson, and Ronald L. Smith. The general totally positive matrix completion problem with few unspecified entries. *Electron. J. Linear Algebra*, 7:1–20 (electronic), 2000.

[FK75] Shmual Friedland and Samuel Karlin. Some inequalities for the spectral radius of non-negative matrices and applications. *Duke Math. J.*, 42(3):459–490, 1975.

[FK00] Shaun M. Fallat and Nathan Krislock. General determinantal inequalities for totally positive matrices. An unpublished paper from an NSERC Undergraduate Summer Research Award held at the University of Regina in the summer of 2000, 2000.

[FL07a] Shaun Fallat and Xiao Ping Liu. A class of oscillatory matrices with exponent $n - 1$. *Linear Algebra Appl.*, 424(2-3):466–479, 2007.

[FL07b] Shaun Fallat and Xiao Ping Liu. A new type of factorization of oscillatory matrices. *Int. J. Pure Appl. Math.*, 37(2):271–296, 2007.

[Flo99] Michael S. Floater. Total positivity and convexity preservation. *J. Approx. Theory*, 96(1):46–66, 1999.

[FM97] Miroslav Fiedler and Thomas L. Markham. Consecutive-column and -row properties of matrices and the Loewner-Neville factorization. *Linear Algebra Appl.*, 266:243–259, 1997.

[FM00a] Miroslav Fiedler and Thomas L. Markham. A factorization of totally nonsingular matrices over a ring with identity. *Linear Algebra Appl.*, 304(1-3):161–171, 2000.

[FM00b] Miroslav Fiedler and Thomas L. Markham. Generalized totally positive matrices. *Linear Algebra Appl.*, 306(1-3):87–102, 2000.

[FM02] Miroslav Fiedler and Thomas L. Markham. Generalized totally nonnegative matrices. *Linear Algebra Appl.*, 345:9–28, 2002.

[FM04] Miroslav Fiedler and Thomas L. Markham. Two results on basic oscillatory matrices. *Linear Algebra Appl.*, 389:175–181, 2004.

[FN01] Shaun M. Fallat and Michael Neumann. On Perron complements of totally nonnegative matrices. *Linear Algebra Appl.*, 327(1-3):85–94, 2001.

[Fom01] Sergey Fomin. Loop-erased walks and total positivity. *Trans. Amer. Math. Soc.*, 353(9):3563–3583 (electronic), 2001.

[Fri85] Shmuel Friedland. Weak interlacing properties of totally positive matrices. *Linear Algebra Appl.*, 71:95–100, 1985.

[FT02] Shaun M. Fallat and Michael J. Tsatsomeros. On the Cayley transform of positivity classes of matrices. *Electron. J. Linear Algebra*, 9:190–196 (electronic), 2002.

[FvdD00] Shaun M. Fallat and Pauline van den Driessche. On matrices with all minors negative. *Electron. J. Linear Algebra*, 7:92–99 (electronic), 2000.

[FW07] Shaun M. Fallat and Hugo J. Woerdeman. Refinements on the interlacing of eigenvalues of certain totally nonnegative matrices. *Oper. Matrices*, 1(2):271–281, 2007.

[FZ99] Sergey Fomin and Andrei Zelevinsky. Double Bruhat cells and total positivity. *J. Amer. Math. Soc.*, 12(2):335–380, 1999.

[FZ00a] Sergey Fomin and Andrei Zelevinsky. Total positivity: tests and parametrizations. *Math. Intelligencer*, 22(1):23–33, 2000.

[FZ00b] Sergey Fomin and Andrei Zelevinsky. Totally nonnegative and oscillatory elements in semisimple groups. *Proc. Amer. Math. Soc.*, 128(12):3749–3759, 2000.

[Gar82a] Jürgen Garloff. Criteria for sign regularity of sets of matrices. *Linear Algebra Appl.*, 44:153–160, 1982.

[Gar82b] Jürgen Garloff. Majorization between the diagonal elements and the eigenvalues of an oscillating matrix. *Linear Algebra Appl.*, 47:181–184, 1982.

[Gar85] Jürgen Garloff. An inverse eigenvalue problem for totally nonnegative matrices. *Linear and Multilinear Algebra*, 17(1):19–23, 1985.

[Gar96] Jürgen Garloff. Vertex implications for totally nonnegative matrices. In *Total Positivity and its Applications (Jaca, 1994)*, volume 359 of *Math. Appl.*, pages 103–107. Kluwer, Dordrecht, 1996.

[Gar02] Jürgen Garloff. Intervals of totally nonnegative and related matrices. *PAMM*, 1(1):496–497, 2002.

[Gar03] Jürgen Garloff. Intervals of almost totally positive matrices. *Linear Algebra Appl.*, 363:103–108, 2003.

[Gas96] Mariano Gasca. Spline functions and total positivity. *Rev. Mat. Univ. Complut. Madrid*, 9(Special Issue, suppl.):125–139, 1996. Meeting on Mathematical Analysis (Spanish) (Avila, 1995).

[GG04] Graham M. L. Gladwell and Kazem Ghanbari. The total positivity interval. *Linear Algebra Appl.*, 393:197–202, 2004.

[GH87]   Jürgen Garloff and Volker Hattenbach. The spectra of matrices
             having sums of principal minors with alternating sign. *SIAM J.
             Algebraic Discrete Methods*, 8(1):106–107, 1987.

[Gha06]   Kazem Ghanbari. Pentadiagonal oscillatory matrices with two
             spectrum in common. *Positivity*, 10(4):721–729, 2006.

[GJ04]   Michael Gekhtman and Charles R. Johnson. The linear inter-
             polation problem for totally positive matrices. *Linear Algebra
             Appl.*, 393:175–178, 2004.

[GJdSW84]   Robert Grone, Charles R. Johnson, Eduardo M. de Sá, and
             Henry Wolkowicz. Positive definite completions of partial Her-
             mitian matrices. *Linear Algebra Appl.*, 58:109–124, 1984.

[GK37]   Feliks R. Gantmacher and Mark G Krein. Sur les matrices
             complètement non–negatives et oscillatoires. *Comp. Math.*,
             4:445–476, 1937.

[GK60]   Feliks R. Gantmacher and Mark G. Krein. *Oszillationsmatrizen,
             Oszillationskerne und kleine Schwingungen mechanischer Sys-
             teme*. Wissenschaftliche Bearbeitung der deutschen Ausgabe:
             Alfred Stöhr. Mathematische Lehrbücher und Monographien, I.
             Abteilung, Bd. V. Akademie-Verlag, Berlin, 1960.

[GK02]   Feliks P. Gantmacher and Mark G. Krein. *Oscillation matrices
             and kernels and small vibrations of mechanical systems*. AMS
             Chelsea, Providence, RI, 2002. Translation based on the 1941
             Russian original, Edited and with a preface by Alex Eremenko.

[Gla98]   Graham M. L. Gladwell. Total positivity and the QR algorithm.
             Linear Algebra Appl., 271:257–272, 1998.

[Gla02]   Graham M. L. Gladwell. Total positivity and Toda flow. *Linear
             Algebra Appl.*, 350:279–284, 2002.

[Gla04]   Graham M. L. Gladwell. Inner totally positive matrices. *Linear
             Algebra Appl.*, 393:179–195, 2004.

[GM87]   Mariano Gasca and Gunter Mühlbach. Generalized Schur-
             complements and a test for total positivity. *Appl. Numer. Math.*,
             3(3):215–232, 1987.

[GM96]   Mariano Gasca and Charles A. Micchelli, editors. *Total Posi-
             tivity and its Applications*, volume 359 of *Math. and its Appl.
             (Jaca 1994)*, Dordrecht, 1996. Kluwer Academic.

[GM00]   Mariano Gasca and Gunter Mühlbach. Elimination techniques:
             from extrapolation to totally positive matrices and CAGD. *J.
             Comput. Appl. Math.*, 122(1-2):37–50, 2000.

[GMP92]   Mariano Gasca, Charles A. Micchelli, and Juan M. Peña. Al-
          most strictly totally positive matrices. *Numer. Algorithms*,
          2(2):225–236, 1992.

[Goo86]   Gerald S. Goodman. A probabilistic representation of totally
          positive matrices. *Adv. in Appl. Math.*, 7(2):236–252, 1986.

[Goo96]   Tim N. T. Goodman. Total positivity and the shape of curves.
          In *Total positivity and its applications (Jaca, 1994)*, volume 359
          of *Math. Appl.*, pages 157–186. Kluwer Acad. Publ., Dordrecht,
          1996.

[GP92a]   Mariano Gasca and Juan M. Peña. On the characterization of
          totally positive matrices. In *Approximation theory, spline func-
          tions and applications (Maratea, 1991)*, volume 356 of *NATO
          Adv. Sci. Inst. Ser. C Math. Phys. Sci.*, pages 357–364. Kluwer
          Acad. Publ., Dordrecht, 1992.

[GP92b]   Mariano Gasca and Juan M. Peña. Total positivity and Neville
          elimination. *Linear Algebra Appl.*, 165:25–44, 1992.

[GP93a]   Mariano Gasca and Juan M. Peña. Scaled pivoting in Gauss and
          Neville elimination for totally positive systems. *Appl. Numer.
          Math.*, 13(5):345–355, 1993.

[GP93b]   Mariano Gasca and Juan M. Peña. Total positivity, QR factor-
          ization, and Neville elimination. *SIAM J. Matrix Anal. Appl.*,
          14(4):1132–1140, 1993.

[GP93c]   Mariano Gasca and Juan Manuel Peña. Sign-regular and to-
          tally positive matrices: an algorithmic approach. In *Multivari-
          ate approximation: from CAGD to wavelets (Santiago, 1992)*,
          volume 3 of *Ser. Approx. Decompos.*, pages 131–146. World Sci.
          Publishing, River Edge, NJ, 1993.

[GP94a]   Mariano Gasca and Juan M. Peña. Corner cutting algorithms
          and totally positive matrices. In *Curves and surfaces in geomet-
          ric design (Chamonix-Mont-Blanc, 1993)*, pages 177–184. A K
          Peters, Wellesley, MA, 1994.

[GP94b]   Mariano Gasca and Juan M. Peña. A matricial description of
          Neville elimination with applications to total positivity. *Linear
          Algebra Appl.*, 202:33–53, 1994.

[GP94c]   Mariano Gasca and Juan M. Peña. A test for strict sign-
          regularity. *Linear Algebra Appl.*, 197/198:133–142, 1994. Sec-
          ond Conference of the International Linear Algebra Society
          (ILAS) (Lisbon, 1992).

[GP95]   Mariano Gasca and Juan M. Peña. On the characterization of
          almost strictly totally positive matrices. *Adv. Comput. Math.*,
          3(3):239–250, 1995.

[GP96]   Mariano Gasca and Juan M. Peña. On factorizations of totally
          positive matrices. In *Total positivity and its applications (Jaca,
          1994)*, volume 359 of *Math. Appl.*, pages 109–130. Kluwer Acad.
          Publ., Dordrecht, 1996.

[GP02]   Laura Gori and Francesca Pitolli. On some applications of a
          class of totally positive bases. In *Wavelet analysis and appli-
          cations (Guangzhou, 1999)*, volume 25 of *AMS/IP Stud. Adv.
          Math.*, pages 109–118. Amer. Math. Soc., Providence, RI, 2002.

[GP06]   Mariano Gasca and Juan M. Peña. Characterizations and de-
          compositions of almost strictly positive matrices. *SIAM J. Ma-
          trix Anal. Appl.*, 28(1):1–8 (electronic), 2006.

[GS93]   Tim N. T. Goodman and Ambikeshaw Sharma. Factorization
          of totally positive, symmetric, periodic, banded matrices with
          applications. *Linear Algebra Appl.*, 178:85–107, 1993.

[GS97]   Michael I. Gekhtman and Michael Z. Shapiro. Completeness
          of real Toda flows and totally positive matrices. *Math. Z.*,
          226(1):51–66, 1997.

[GS04]   Tim N. T. Goodman and Qiyu Sun. Total positivity and refin-
          able functions with general dilation. *Appl. Comput. Harmon.
          Anal.*, 16(2):69–89, 2004.

[GT98]   Jiří Gregor and Jaroslav Tišer. On Hadamard powers of poly-
          nomials. *Math. Control Signals Systems*, 11(4):372–378, 1998.

[GT02]   Maite Gassó and Juan R. Torregrosa. A *PLU*-factorization of
          rectangular matrices by the Neville elimination. *Linear Algebra
          Appl.*, 357:163–171, 2002.

[GT04]   Maite Gassó and Juan R. Torregrosa. A totally positive fac-
          torization of rectangular matrices by the Neville elimination.
          SIAM J. Matrix Anal. Appl., 25(4):986–994 (electronic), 2004.

[GT06a]   Maite Gassó and Juan R. Torregrosa. Bidiagonal factorization
          of totally nonnegative rectangular matrices. In *Positive systems*,
          volume 341 of *Lecture Notes in Control and Inform. Sci.*, pages
          33–40. Springer, Berlin, 2006.

[GT06b]   Maite Gassó and Juan R. Torregrosa. Bidiagonal factorization
          of totally nonnegative rectangular matrices. 341:33–40, 2006.

[GT07]   Maite Gassó and Juan R. Torregrosa. A class of totally positive
            P-matrices whose inverses are M-matrices. *Appl. Math. Lett.*,
            20(1):23–27, 2007.

[GT08]   Maria T. Gassó and Juan R. Torregrosa. Bidiagonal factoriza-
            tions and quasi-oscillatory rectangular matrices. *Linear Algebra
            Appl.*, 429(8-9):1886–1893, 2008.

[GW76]   Leonard J. Gray and David G. Wilson. Construction of a Jacobi
            matrix from spectral data. *Linear Algebra and Appl.*, 14(2):131–
            134, 1976.

[GW96a]   Jürgen Garloff and David G. Wagner. Hadamard products of
            stable polynomials are stable. *J. Math. Anal. Appl.*, 202(3):797–
            809, 1996.

[GW96b]   Jürgen Garloff and David G. Wagner. Preservation of total
            nonnegativity under the Hadamard product and related topics.
            In *Total positivity and its applications (Jaca, 1994)*, volume 359
            of *Math. Appl.*, pages 97–102. Kluwer Acad. Publ., Dordrecht,
            1996.

[Hei94]   Berthold Heiligers. Totally nonnegative moment matrices. *Lin-
            ear Algebra Appl.*, 199:213–227, 1994.

[Hei01]   Berthold Heiligers. Totally positive regression: E-optimal de-
            signs. *Metrika*, 54(3):191–213 (electronic) (2002), 2001.

[HJ85]   Roger A. Horn and Charles R. Johnson. *Matrix Analysis*. Cam-
            bridge University Press, Cambridge, 1985.

[HJ91]   Roger A. Horn and Charles R. Johnson. *Topics in Matrix Anal-
            ysis*. Cambridge University Press, Cambridge, 1991.

[Hog07]   Leslie Hogben, editor. *Handbook of linear algebra*. *Discrete
            Mathematics and its Applications (Boca Raton)*. Chapman &
            Hall/CRC, Boca Raton, FL, 2007. Associate editors: Richard
            Brualdi, Anne Greenbaum, and Roy Mathias.

[Hol03]   Olga Holtz. Hermite-Biehler, Routh-Hurwitz, and total positiv-
            ity. *Linear Algebra Appl.*, 372:105–110, 2003.

[JB93]   Charles R. Johnson and Wayne W. Barrett. Determinantal in-
            equalities for positive definite matrices. *Discrete Math.*, 119(1-
            3):97–106, 1993. ARIDAM IV and V (New Brunswick, NJ,
            1988/1989).

[Jew89]   Ian Jewitt. Choosing between risky prospects: the characteri-
            zation of comparative statics results, and location independent
            risk. *Management Sci.*, 35(1):60–70, 1989.

[Jia83]   Rong Qing Jia. Total positivity of the discrete spline collocation matrix. *J. Approx. Theory*, 39(1):11–23, 1983.

[JK10]   Charles R. Johnson and Brenda K. Kroschel. Conditions for a totally positive completion in the case of a symmetrically placed cycle. *Electron. J. Linear Algebra*, To appear, 2010.

[JKL98]   Charles R. Johnson, Brenda K. Kroschel, and Michael Lundquist. The totally nonnegative completion problem. In *Topics in semidefinite and interior-point methods (Toronto, ON, 1996)*, volume 18 of *Fields Inst. Commun.*, pages 97–107. Amer. Math. Soc., Providence, RI, 1998.

[JN09]   Charles R. Johnson and Cris Negron. Totally positive completions for monotonically labeled block clique graphs. *Electron. J. Linear Algebra*, 18:146–161, 2009.

[JO04]   Charles R. Johnson and Dale D. Olesky. Sums of totally positive matrices. *Linear Algebra Appl.*, 392:1–9, 2004.

[JO05]   Charles R. Johnson and Dale D. Olesky. Rectangular submatrices of inverse M-matrices and the decomposition of a positive matrix as a sum. *Linear Algebra Appl.*, 409:87–99, 2005.

[Joh87]   Charles R. Johnson. Closure properties of certain positivity classes of matrices under various algebraic operations. *Linear Algebra Appl.*, 97:243–247, 1987.

[Joh98]   Charles R. Johnson. Olga, matrix theory and the Taussky unification problem. *Linear Algebra Appl.*, 280(1):39–49, 1998. With the assistance of Shaun Fallat, Special issue in memory of Olga Taussky Todd.

[JOvdD99]   Charles R. Johnson, Dale D. Olesky, and Pauline van den Driessche. Elementary bidiagonal factorizations. *Linear Algebra Appl.*, 292(1-3):233–244, 1999.

[JOvdD01]   Charles R. Johnson, Dale D. Olesky, and Pauline van den Driessche. Successively ordered elementary bidiagonal factorization. *SIAM J. Matrix Anal. Appl.*, 22(4):1079–1088 (electronic), 2001.

[JS00]   Charles R. Johnson and Ronald L. Smith. Line insertions in totally positive matrices. *J. Approx. Theory*, 105(2):305–312, 2000.

[JT03]   Cristina Jordán and Juan R. Torregrosa. Paths and cycles in the totally positive completion problem. In *Positive systems (Rome, 2003)*, volume 294 of *Lecture Notes in Control and Inform. Sci.*, pages 217–224. Springer, Berlin, 2003.

[JT04]   Cristina Jordán and Juan R. Torregrosa. The totally positive completion problem. *Linear Algebra Appl.*, 393:259–274, 2004.

[JTeG09]   Cristina Jordán, Juan R. Torregrosa, and Ramadán el Ghamry. The completable digraphs for the totally nonnegative completion problem. *Linear Algebra Appl.*, 430(5-6):1675–1690, 2009.

[JX93]   Charles R. Johnson and Christos Xenophontos. Irreducibility and primitivity of Perron complements: application of the compressed directed graph. In *Graph Theory and Sparse Matrix Computation*, volume 56 of *IMA Vol. Math. Appl.*, pages 101–106. Springer, New York, 1993.

[Kar64]   Samuel Karlin. Total positivity, absorption probabilities and applications. *Trans. Amer. Math. Soc.*, 111:33–107, 1964.

[Kar65]   Samuel Karlin. Oscillation properties of eigenvectors of strictly totally positive matrices. *J. Analyse Math.*, 14:247–266, 1965.

[Kar68]   Samuel Karlin. *Total Positivity. Vol. I.* Stanford University Press, Stanford, Calif, 1968.

[Kar71]   Samuel Karlin. Total positivity, interpolation by splines, and Green's functions of differential operators. *J. Approximation Theory*, 4:91–112, 1971.

[KD02]   Plamen Koev and James Demmel. Accurate solutions of totally positive linear systems application to generalized vandermonde systems. In *Householder Symposium XV (Peebles, Scotland, 2002)*, Peebles, Scotland, 2002.

[KL70]   Samuel Karlin and John Walter Lee. Periodic boundary-value problems with cyclic totally positive Green's functions with applications to periodic spline theory. *J. Differential Equations*, 8:374–396, 1970.

[KLM01]   Grzegorz Kubicki, Jenő Lehel, and Michał Morayne. Totally positive matrices and totally positive hypergraphs. *Linear Algebra Appl.*, 331(1-3):193–202, 2001.

[KM59]   Samuel Karlin and James McGregor. Coincidence probabilities. *Pacific J. Math.*, 9:1141–1164, 1959.

[Koe05]   Plamen Koev. Accurate eigenvalues and SVDs of totally nonnegative matrices. *SIAM J. Matrix Anal. Appl.*, 27(1):1–23 (electronic), 2005.

[Koe07]   Plamen Koev. Accurate computations with totally nonnegative matrices. *SIAM J. Matrix Anal. Appl.*, 29(3):731–751 (electronic), 2007.

[Kot50]   D. M. Kotelyanskiĭ. On the theory of nonnegative and oscillat-
            ing matrices. *Ukrain. Mat. Žurnal*, 2(2):94–101, 1950.

[Kot53]   D. M. Kotelyanskiĭ. On a property of sign-symmetric matrices.
            Uspehi Matem. Nauk (N.S.), 8(4(56)):163–167, 1953.

[Kot63a]   D. M. Koteljanskiĭ. A property of sign-symmetric matrices.
            Amer. Math. Soc. Transl. (2), 27:19–23, 1963.

[Kot63b]   D. M. Koteljanskiĭ. The theory of nonnegative and oscillating
            matrices. *Amer. Math. Soc. Transl. (2)*, 27:1–8, 1963.

[KP74]   Samuel Karlin and Allan Pinkus. Oscillation properties of gen-
            eralized characteristic polynomials for totally positive and pos-
            itive definite matrices. *Linear Algebra and Appl.*, 8:281–312,
            1974.

[KR80]   Samuel Karlin and Yosef Rinott. Classes of orderings of mea-
            sures and related correlation inequalities. I. Multivariate totally
            positive distributions. *J. Multivariate Anal.*, 10(4):467–498,
            1980.

[KR81]   Samuel Karlin and Yosef Rinott. Total positivity properties of
            absolute value multinormal variables with applications to con-
            fidence interval estimates and related probabilistic inequalities.
            Ann. Statist., 9(5):1035–1049, 1981.

[KR88]   Samuel Karlin and Yosef Rinott. A generalized Cauchy-Binet
            formula and applications to total positivity and majorization.
            J. Multivariate Anal., 27(1):284–299, 1988.

[KV06]   Olga M. Katkova and Anna M. Vishnyakova. On sufficient con-
            ditions for the total positivity and for the multiple positivity of
            matrices. *Linear Algebra Appl.*, 416(2-3):1083–1097, 2006.

[Lew80]   Mordechai Lewin. Totally nonnegative, *M*-, and Jacobi matri-
            ces. *SIAM J. Algebraic Discrete Methods*, 1(4):419–421, 1980.

[LF08]   Xiao Ping Liu and Shaun M. Fallat. Multiplicative principal-
            minor inequalities for a class of oscillatory matrices. *JIPAM. J.
            Inequal. Pure Appl. Math.*, 9(4):Article 92, 18, 2008.

[Lig89]   Thomas M. Liggett. Total positivity and renewal theory. In
            Probability, Statistics, and Mathematics, pages 141–162. Aca-
            demic Press, Boston, 1989.

[Liu08]   XiaoPing Liu. *Determinantal inequalities and factorizations of
            totally nonnegative matrices*. PhD thesis, University of Regina,
            2008.

[LL97]   Shiowjen Lee and J. Lynch. Total positivity of Markov chains and the failure rate character of some first passage times. *Adv. in Appl. Probab.*, 29(3):713–732, 1997.

[LM79]   Jens Lorenz and Wolfgang Mackens. Toeplitz matrices with totally nonnegative inverses. *Linear Algebra Appl.*, 24:133–141, 1979.

[LM02]   Chi-Kwong Li and Roy Mathias. Interlacing inequalities for totally nonnegative matrices. *Linear Algebra Appl.*, 341:35–44, 2002. Special issue dedicated to Professor T. Ando.

[Loe55]   Charles Loewner. On totally positive matrices. *Math. Z.*, 63:338–340, 1955.

[LS02]   Chi-Kwong Li and Hans Schneider. Applications of Perron-Frobenius theory to population dynamics. *J. Math. Biol.*, 44(5):450–462, 2002.

[Lus98]   George Lusztig. Introduction to total positivity. In *Positivity in Lie Theory: Open Problems*, volume 26 of *de Gruyter Exp. Math.*, pages 133–145. de Gruyter, Berlin, 1998.

[Lus08]   George Lusztig. A survey of total positivity. *Milan J. Math.*, 76:125–134, 2008.

[LY04]   Hong Bin Lü and Zhong Peng Yang. The Schur-Oppenheim strict inequality for tridiagonal totally nonnegative matrices. *J. Math. Study*, 37(2):193–199, 2004.

[Mar70a]   Thomas L. Markham. On oscillatory matrices. *Linear Algebra and Appl.*, 3:143–156, 1970.

[Mar70b]   Thomas L. Markham. A semigroup of totally nonnegative matrices. *Linear Algebra Appl.*, 3:157–164, 1970.

[Mel96]   Avraham A. Melkman. Another proof of the total positivity of the discrete spline collocation matrix. *J. Approx. Theory*, 84(3):265–273, 1996.

[Met73]   Kurt Metelmann. Ein Kriterium für den Nachweis der Totalnichtnegativität von Bandmatrizen. *Linear Algebra Appl.*, 7:163–171, 1973.

[Mey89]   Carl D. Meyer. Uncoupling the Perron eigenvector problem. *Linear Algebra Appl.*, 114/115:69–94, 1989.

[MG85]   Gunter Mühlbach and Mariano Gasca. A generalization of Sylvester's identity on determinants and some applications. *Linear Algebra Appl.*, 66:221–234, 1985.

[MG87]   Gunter Mühlbach and Mariano Gasca. A test for strict total positivity via Neville elimination. In *Current Trends in Matrix Theory (Auburn, Ala., 1986)*, pages 225–232. North-Holland, New York, 1987.

[Mør96]   Knut M. Mørken. On total positivity of the discrete spline collocation matrix. *J. Approx. Theory*, 84(3):247–264, 1996.

[MP77]   Charles A. Micchelli and Allan Pinkus. Total positivity and the exact n-width of certain sets in L^1. *Pacific J. Math.*, 71(2):499–515, 1977.

[MP99]   Esmeralda Mainar and Juan M. Peña. Corner cutting algorithms associated with optimal shape preserving representations. *Comput. Aided Geom. Design*, 16(9):883–906, 1999.

[MP00]   Esmeralda Mainar and Juan M. Peña. Knot insertion and totally positive systems. *J. Approx. Theory*, 104(1):45–76, 2000.

[MP07]   Esmeralda Mainar and Juan M. Peña. A general class of Bernstein-like bases. *Comput. Math. Appl.*, 53(11):1686–1703, 2007.

[Peñ95a]   Juan M. Peña. M-matrices whose inverses are totally positive. *Linear Algebra Appl.*, 221:189–193, 1995.

[Peñ95b]   Juan M. Peña. Matrices with sign consistency of a given order. *SIAM J. Matrix Anal. Appl.*, 16(4):1100–1106, 1995.

[Peñ97]   Juan M. Peña. On the Schur and singular value decompositions of oscillatory matrices. *Linear Algebra Appl.*, 261:307–315, 1997.

[Peñ98]   Juan M. Peña. On the relationship between graphs and totally positive matrices. *SIAM J. Matrix Anal. Appl.*, 19(2):369–377 (electronic), 1998.

[Peñ01]   Juan M. Peña. Determinantal criteria for total positivity. In *Proceedings of the Eighth Conference of the International Linear Algebra Society (Barcelona, 1999)*, volume 332/334, pages 131–137, 2001.

[Peñ02]   Juan M. Peña. Sign regular matrices of order two. *Linear Multilinear Algebra*, 50(1):91–97, 2002.

[Peñ03]   Juan M. Peña. On nonsingular sign regular matrices. *Linear Algebra Appl.*, 359:91–100, 2003.

[Peñ04]   Juan M. Peña. Characterizations and stable tests for the Routh-Hurwitz conditions and for total positivity. *Linear Algebra Appl.*, 393:319–332, 2004.

[Pin85]   Allan Pinkus. Some extremal problems for strictly totally pos-
            itive matrices. *Linear Algebra Appl.*, 64:141–156, 1985.

[Pin98]   Allan Pinkus. An interlacing property of eigenvalues of strictly
            totally positive matrices. *Linear Algebra Appl.*, 279(1-3):201–
            206, 1998.

[Pin04]   Allan Pinkus. Interpolation by matrices. *Electron. J. Linear
            Algebra*, 11:281–291 (electronic), 2004.

[Pin08]   Allan Pinkus. Zero minors of totally positive matrices. *Electron.
            J. Linear Algebra*, 17:532–542, 2008.

[Pin10]   Allan Pinkus. *Totally positive matrices*, volume 181 of *Cam-
            bridge Tracts in Mathematics*. Cambridge University Press,
            Cambridge, 2010.

[Pri68]   Harvey S. Prince. Monotone and oscillation matrices applied
            to finite difference approximations. *Math. Comp.*, 22:489–516,
            1968.

[Rad68]   Charles E. Radke. Classes of matrices with distinct, real char-
            acteristic values. *SIAM J. Appl. Math.*, 16:1192–1207, 1968.

[RH72]   John W. Rainey and George J. Habetler. Tridiagonalization
            of completely nonnegative matrices. *Math. Comp.*, 26:121–128,
            1972.

[Rie01]   Konstanze Rietsch. Quantum cohomology rings of Grassmanni-
            ans and total positivity. *Duke Math. J.*, 110(3):523–553, 2001.

[Rie03a]   Konstanze Rietsch. Total positivity, flag varieties and quantum
            cohomology. In *European women in mathematics (Malta, 2001)*,
            pages 149–167. World Sci. Publishing, River Edge, NJ, 2003.

[Rie03b]   Konstanze Rietsch. Totally positive Toeplitz matrices and
            quantum cohomology of partial flag varieties. *J. Amer. Math.
            Soc.*, 16(2):363–392 (electronic), 2003.

[Rie08]   Konstanze Rietsch. Errata to: "Totally positive Toeplitz matri-
            ces and quantum cohomology of partial flag varieties" [J. Amer.
            Math. Soc. **16** (2003), no. 2, 363–392; mr1949164]. *J. Amer.
            Math. Soc.*, 21(2):611–614, 2008.

[RS06]   Brendon Rhoades and Mark Skandera. Kazhdan-Lusztig im-
            manants and products of matrix minors. *J. Algebra*, 304(2):793–
            811, 2006.

[Sch30]   Isaac J. Schoenberg. Uber variationsvermindernde lineare trans-
            formationen. *Math. Z.*, 32:321–328, 1930.

[Sch47]   Isaac J. Schoenberg. On totally positive functions, Laplace in-
            tegrals and entire functions of the Laguerre-Polya-Schur type.
            Proc. Nat. Acad. Sci. U. S. A., 33:11–17, 1947.

[Sch86]   E. J. P. Georg Schmidt. On the total—and strict total—
            positivity of the kernels associated with parabolic initial-
            boundary value problems. *J. Differential Equations*, 62(2):275–
            298, 1986.

[Ska04]   Mark Skandera. Inequalities in products of minors of totally
            nonnegative matrices. *J. Algebraic Combin.*, 20(2):195–211,
            2004.

[Sob75]   Aleksander V. Sobolev. Totally positive operators. *Siberian
            Math. J.*, 16:636–641, 1975.

[SR03]   Mark Skandera and Brian Reed. Total nonnegativity and (3+1)-
            free posets. *J. Combin. Theory Ser. A*, 103(2):237–256, 2003.

[SS95]   Boris Z. Shapiro and Michael Z. Shapiro. On the boundary of
            totally positive upper triangular matrices. *Linear Algebra Appl.*,
            231:105–109, 1995.

[SS06]   Ernesto Salinelli and Carlo Sgarra. Correlation matrices of
            yields and total positivity. *Linear Algebra Appl.*, 418(2-3):682–
            692, 2006.

[Sta00]   Richard P. Stanley. Positivity problems and conjectures in alge-
            braic combinatorics. In *Mathematics: frontiers and perspectives*,
            pages 295–319. Amer. Math. Soc., Providence, RI, 2000.

[Ste90]   John R. Stembridge. Nonintersecting paths, Pfaffians, and plane
            partitions. *Adv. Math.*, 83(1):96–131, 1990.

[Ste91]   John R. Stembridge. Immanants of totally positive matrices are
            nonnegative. *Bull. London Math. Soc.*, 23(5):422–428, 1991.

[Ste92]   John R. Stembridge. Some conjectures for immanants. *Canad.
            J. Math.*, 44(5):1079–1099, 1992.

[Stu88]   Bernd Sturmfels. Totally positive matrices and cyclic polytopes.
            In *Proceedings of the Victoria Conference on Combinatorial Ma-
            trix Analysis (Victoria, BC, 1987)*, volume 107, pages 275–281,
            1988.

[SW49]   Isaac J. Schoenberg and Anne Whitney. Sur la positivité des
            déterminants de translation des fonctions de fréquence de Pólya,
            avec une application à un problème d'interpolation. *C. R. Acad.
            Sci. Paris*, 228:1996–1998, 1949.

[SW51]   Isaac J. Schoenberg and Anne Whitney. A theorem on polygons
           in n dimensions with applications to variation-diminishing and
           cyclic variation-diminishing linear transformations. *Compositio
           Math.*, 9:141–160, 1951.

[SW53]   Isaac J. Schoenberg and Anne Whitney. On Polya frequency
           function. III. The positivity of translation determinants with
           an application to the interpolation problem by spline curves.
           Amer. Math. Soc., 74(2):246–259, 1953.

[Vol05]   Yu. S. Volkov. Totally positive matrices in the methods of
           constructing interpolation splines of odd degree [Translation
           of Mat. Tr. **7** (2004), no. 2, 3–34; mr2124538]. *Siberian Adv.
           Math.*, 15(4):96–125, 2005.

[Wag92]   David G. Wagner. Total positivity of Hadamard products. *J.
           Math. Anal. Appl.*, 163(2):459–483, 1992.

[Whi52]   Anne M. Whitney. A reduction theorem for totally positive
           matrices. *J. Analyse Math.*, 2:88–92, 1952.

[YF07]   Zhongpeng Yang and Xiaoxia Feng. New lower bound of the
           determinant for Hadamard product on some totally nonnegative
           matrices. *J. Appl. Math. Comput.*, 25(1-2):169–181, 2007.

[ZY93]   Xiao Dong Zhang and Shang Jun Yang. An improvement of
           Hadamard's inequality for totally nonnegative matrices. *SIAM
           J. Matrix Anal. Appl.*, 14(3):705–711, 1993.

List of Symbols

E_{ij}:   standard basis matrix, page 3

I_n: identity matrix, page 3

$J_{m,n}$:   matrix of all ones, page 3

$K_{[p]}(\overline{x}, \overline{y})$: compound kernel of order p, page 18

LU: LU factorization, page 47

$L_i(s)$:   lower elementary bidiagonal matrix, page 45

$M_{m,n}$: m-by-n real matrices, page 1

$M_{m,n}(\mathbb{F})$: m-by-n matrices over $\mathbb{F}$, page 1

$M_n(\mathbb{F})$: n-by-n matrices over $\mathbb{F}$, page 1

N: $\{1, 2, \ldots, n\}$, page 2

P: set of P matrices, page 5

PF_n: Pólya frequency function of order n, page 17

PF_k: Pólya frequency function of order k, page 19

P_0:   set of P_0 matrices, page 5

P_n: Pascal matrix, page 13

RH: Routh-Hurwitz matrix, page 13

$S_n = \operatorname{diag}(1, -1, \ldots, \pm 1)$: alternating sign signature matrix, page 33

$U_j(s)$:   upper elementary bidiagonal matrix, page 45

$V(x_1, x_2, \ldots, x_n)$: Vandermonde matrix, page 12

$[a_{ij}]$:   entries of a matrix, page 1

$[p_1, p_2, \ldots, p_n]$: Plücker coordinate, page 31

ΔTN: triangular TN matrices, page 3

ΔTP: triangular TP matrices, page 3

$\alpha \leq \beta$: principal minor inequality, page 131

$\alpha(A)$:   principal minor product, page 131

α^c: complement of α, page 2

$\mathcal{C}^{(R)}$:   retractable subset of $\mathcal{C}$, page 199

$\mathcal{C}_R$:   set of all retractions of $\mathcal{C}$, page 199

$\mathcal{F}_0$:   set of Fischer matrices, page 154

$\det A$:   determinant of A, page 2

$\lambda_{\max}(A)$: largest eigenvalue of A, page 122

$\lambda_{\min}(A)$: smallest eigenvalue of A, page 122

$\leq$:   entrywise ordering, page 80

$\mathcal{D}_n$:   positive diagonal matrices, page 3

$\mathcal{J}$:   Jacobi matrix, page 6

$\mathcal{K}$:   Koteljanskiĭ matrices, page 130

$\mathcal{P}(A/A[\beta])$: Perron complement of $A[\beta]$ in A, page 213

$\mathcal{P}_t(A/A[\beta])$: extended Perron complement at t, page 213

$\oplus$:   direct sum, page 5

$\overline{A[\alpha,\beta]}$: convex hull of $A[\alpha,\beta]$, page 159

$\rho(A)$:   spectral radius of A, page 33

$\sigma(A)$:   eigenvalues of A, page 33

$\overset{*}{\leq}$:   checkerboard partial order, page 81

$\succ$:   majorization, page 121

$\tilde{C}_k(A)$: compression of A, page 149

$\tilde{T}$:   backward identity permutation matrix, page 34

$*$:   Hermitian adjoint, page 33

$d(\alpha)$:   dispersion of α, page 35

e: vector of all ones, page 3

e_i: ith standard basis vectors, page 3

$f_\alpha(L)$: set count for collection α, page 144

$f_\alpha(i)$: count of index i in α, page 135

m_{ij}:   multipliers for an elementary bidiagonal factorization, page 76

$v(x)$:   total sign variation of x, page 87

$v_M(x)$: maximum sign variation of x, page 87

$v_m(x)$: minimum sign variation of x, page 87

$x[\alpha]$: subvector of x, page 2

x_M: maximum sign variation realization of x, page 88

x_m: minimum sign variation realization of x, page 88

$\mathrm{Col}(A)$: column space of A, page 154

$\mathrm{GR}(n, 2n)$: Grassmannian, page 31

$\mathrm{Row}(A)$: row space of A, page 154

$\mathrm{TN}^{(C)}$: Hadamard core, page 170

$\mathrm{rank}(A)$: rank of A, page 107

$\mathrm{sgn}(a)$: sign of a, page 94

$\mathrm{p\text{-}rank}(A)$: principal rank of A, page 107

CRP: contiguous rank property, page 165

$\det A[\alpha, \beta]$: minor of A, page 2

$\det A[\alpha]$: principal minor of A, page 2

$\mathrm{diag}(x_i)$: diagonal matrix, page 2

EB: set of elementary bidiagonal matrices, page 45

$\mathrm{gap}(S, T)$: gap of S and T, page 133

GEB: set of generalized bidiagonal matrices, page 45

IITN_k: irreducible invertible TN_k matrices, page 2

IITN: irreducible invertible TN matrices, page 2

InTN_k: invertible TN_k matrices, page 2

InTN: invertible TN matrices, page 2

IrTN_k: irreducible TN_k matrices, page 2

IrTN: irreducible TN matrices, page 2

MLBC: monotonically labeled block-clique, page 189

OSC: oscillatory matrices, page 2

PRI: partitioned rank intersection property, page 162

PRP: principal rank property, page 155

PSD: positive semidefinite matrices, page 199

SRP: section rank property, page 163

STO: set-theoretic zero condition, page 136

TN′:   TN matrices with no zero lines, page 2

TN_k: all minors up to order k nonnegative, page 2

TN'_k: TN'_k matrices with no zero lines, page 2

TN_+:   TN sign pattern, page 37

TN_2^+: TN_2 matrices with positive contiguous prin. minors, page 181

TN: totally nonnegative matrices, page 2

TP_k: all minors up to order k positive, page 2

TP: totally positive matrices, page 2

Index